A History of Mathematics

A History of Mathematics
From Mesopotamia to Modernity

Luke Hodgkin

OXFORD
UNIVERSITY PRESS

Great Clarendon Street, Oxford, OX2 6DP,
United Kingdom

Oxford University Press is a department of the University of Oxford.
It furthers the University's objective of excellence in research, scholarship,
and education by publishing worldwide. Oxford is a registered trade mark of
Oxford University Press in the UK and in certain other countries

© Oxford University Press 2005

The moral rights of the authors have been asserted

First Edition published in 2005
First published in paperback 2013
Reprinted 2014

All rights reserved. No part of this publication may be reproduced, stored in
a retrieval system, or transmitted, in any form or by any means, without the
prior permission in writing of Oxford University Press, or as expressly permitted
by law, by licence or under terms agreed with the appropriate reprographics
rights organization. Enquiries concerning reproduction outside the scope of the
above should be sent to the Rights Department, Oxford University Press, at the
address above

You must not circulate this work in any other form
and you must impose this same condition on any acquirer

Published in the United States of America by Oxford University Press
198 Madison Avenue, New York, NY 10016, United States of America

British Library Cataloguing in Publication Data

Data available

ISBN 978–0–19–967676–7

Preface

This book has its origin in notes which I compiled for a course on the history of mathematics at King's College London, taught for many years before we parted company. My major change in outlook (which is responsible for its form) dates back to a day ten years ago at the University of Warwick, when I was comparing notes on teaching with the late David Fowler. He explained his own history of mathematics course to me; as one might expect, it was detailed, scholarly, and encouraged students to do research of their own, particularly on the Greeks. I told him that I gave what I hoped was a critical account of the whole history of mathematics in a series of lectures, trying to go beyond what they would find in a textbook. David was scornful. 'What', he said, 'do you mean that you stand up in front of those students and *tell stories?*' I had to acknowledge that I did.

David's approach meant that students should be taught from the start not to accept any story at face value, and to be interested in questions rather than narrative. It's certainly desirable as regards the Greeks, and it's a good approach in general, even if it may sometimes seem too difficult and too purist. I hope he would not be too hard on my attempts at a compromise. The aims of the book in this, its ultimate form, are set out in the introduction; briefly, I hope to introduce students to the history, or histories of mathematics as constructions which we make to explain the texts which we have, and to relate them to our own ideas. Such constructions are often controversial, and always provisional; but that is the nature of history.

The original impulse to write came from David Robinson, my collaborator on the course at King's, who suggested (unsuccessfully) that I should turn my course notes into a book; and providentially from Alison Jones of the Oxford University Press, who turned up at King's when I was at a loose end and asked if I had a book to publish. I produced a proposal; she persuaded the press to accept it and kept me writing. Without her constant feedback and involvement it would never have been completed.

I am grateful to a number of friends for advice and encouragement. Jeremy Gray read an early draft and promoted the project as a referee; the reader is indebted to him for the presence of exercises. Geoffrey Lloyd gave expert advice on the Greeks; I am grateful for all of it, even if I only paid attention to some. John Cairns, Felix Pirani and Gervase Franklin read parts of the manuscript and made helpful comments; various friends and relations, most particularly Jack Goody, John Hope, Jessica Hines and Sam and Joe Gold Hodgkin expressed a wish to see the finished product.

Finally, I'm deeply grateful to my wife Jean who has supported the project patiently through writing and revision. To her, and to my father Thomas who I hope would have approved, this book is dedicated.

Note on the paperback edition

This book was written seven years ago. It's frankly not a research text, but in putting it together I used the best sources available at the time. However, research moves on, and of course a number of scholars have published important new texts in the field. Rather than rewrite the bibliography, I'll just mention two recent general works which are (in my view) essential:

Robson, Eleanor, and Stedall, Jacqueline (2009). *The Oxford Handbook of the History of Mathematics.* Oxford: OUP.

Katz, V. J. (ed.) (2007). *The Mathematics of Egypt, Mesopotamia, China, India, and Islam: A Sourcebook.* Princeton: Princeton University Press.

These two books will in turn provide you with bibliographies leading you to further reading.

I enjoyed writing this book, and I have been grateful for feedback from the students, colleagues and friends who have read it. I hope that the issue of a paperback makes it more accessible to the many.

Contents

List of figures	xi
Picture Credits	xiv
Introduction	1
Why this book?	1
On texts, and on history	2
Examples	5
Historicism and 'presentism'	6
Revolutions, paradigms, and all that	8
External versus internal	10
Eurocentrism	12
1. Babylonian mathematics	14
1. On beginnings	14
2. Sources and selections	17
3. Discussion of the example	20
4. The importance of number-writing	21
5. Abstraction and uselessness	24
6. What went before	27
7. Some conclusions	30
Appendix A. Solution of the quadratic problem	30
Solutions to exercises	31
2. Greeks and 'origins'	33
1. Plato and the *Meno*	33
2. Literature	35
3. An example	36
4. The problem of material	39
5. The Greek miracle	42
6. Two revolutions?	44
7. Drowning in the sea of Non-identity	45
8. On modernization and reconstruction	47
9. On ratios	49
Appendix A. From the *Meno*	51
Appendix B. On pentagons, golden sections, and irrationals	52
Solutions to exercises	54

3. Greeks, practical and theoretical — 57
1. Introduction, and an example — 57
2. Archimedes — 60
3. Heron or Hero — 63
4. Astronomy, and Ptolemy in particular — 66
5. On the uncultured Romans — 69
6. Hypatia — 71
   Appendix A. From Heron's *Metrics* — 73
   Appendix B. From Ptolemy's *Almagest* — 75
   Solutions to exercises — 76

4. Chinese mathematics — 78
1. Introduction — 78
2. Sources — 80
3. An instant history of early China — 80
4. *The Nine Chapters* — 82
5. Counting rods—who needs them? — 85
6. Matrices — 88
7. The Song dynasty and Qin Jiushao — 90
8. On 'transfers'—when, and how? — 95
9. The later period — 98
   Solutions to exercises — 99

5. Islam, neglect and discovery — 101
1. Introduction — 101
2. On access to the literature — 103
3. Two texts — 106
4. The golden age — 108
5. Algebra—the origins — 110
6. Algebra—the next steps — 115
7. Al-Samaw'al and al-Kāshī — 117
8. The uses of religion — 123
   Appendix A. From al-Khwārizmī's algebra — 125
   Appendix B. Thābit ibn Qurra — 127
   Appendix C. From al-Kāshī, *The Calculator's Key*, book 4, chapter 7 — 128
   Solutions to exercises — 130

6. Understanding the 'scientific revolution' — 133
1. Introduction — 133
2. Literature — 134
3. Scholastics and scholasticism — 135
4. Oresme and series — 138
5. The calculating tradition — 140
6. Tartaglia and his friends — 143
7. On authority — 146

8. Descartes	149
9. Infinities	151
10. Galileo	153
Appendix A	155
Appendix B	156
Appendix C	157
Appendix D	158
Solutions to exercises	159

7. The calculus — 161

1. Introduction	161
2. Literature	163
3. The priority dispute	165
4. The Kerala connection	167
5. Newton, an unknown work	169
6. Leibniz, a confusing publication	172
7. The *Principia* and its problems	176
8. The arrival of the calculus	178
9. The calculus in practice	180
10. Afterword	182
Appendix A. Newton	183
Appendix B. Leibniz	185
Appendix C. From the *Principia*	186
Solutions to exercises	187

8. Geometries and space — 189

1. Introduction	189
2. First problem: the postulate	194
3. Space and infinity	197
4. Spherical geometry	199
5. The new geometries	201
6. The 'time-lag' question	203
7. What revolution?	205
Appendix A. Euclid's proposition I.16	207
Appendix B. The formulae of spherical and hyperbolic trigonometry	209
Appendix C. From Helmholtz's 1876 paper	210
Solutions to exercises	210

9. Modernity and its anxieties — 213

1. Introduction	213
2. Literature	214
3. New objects in mathematics	214
4. Crisis—what crisis?	217
5. Hilbert	221
6. Topology	223

7. Outsiders	228
Appendix A. The cut definition	231
Appendix B. Intuitionism	231
Appendix C. Hilbert's programme	232
Solutions to exercises	232
10. A chaotic end?	**235**
1. Introduction	235
2. Literature	236
3. The Second World War	238
4. Abstraction and 'Bourbaki'	240
5. The computer	243
6. Chaos: the less you know, the more you get	246
7. From topology to categories	249
8. Physics	251
9. Fermat's Last Theorem	254
Appendix A. From Bourbaki, 'Algebra', Introduction	256
Appendix B. Turing on computable numbers	256
Solutions to exercises	258
Conclusion	260
Bibliography	263
Index	271

List of figures

Introduction 1
    1. Euclid's proposition II.1   5

Chapter 1. Babylonian mathematics 14
    1. A mathematical tablet   15
    2. Tally of pigs   16
    3. The 'stone-weighing' tablet YBC4652   18
    4. Cuneiform numbers from 1 to 60   23
    5. How larger cuneiform numbers are formed   23
    6. The 'square root of 2' tablet   25
    7. Ur III tablet (harvests from Lagash)   28

Chapter 2. Greeks and 'origins' 33
    1. The *Meno* argument   34
    2. Diagram for Euclid I.35   37
    3. The five regular solids   46
    4. Construction of a regular pentagon   53
    5. The 'extreme and mean section' construction   53
    6. How to prove 'Thales' theorem'   55

Chapter 3. Greeks, practical and theoretical 57
    1. Menaechmus' duplication construction   58
    2. Eratosthenes' 'mesolabe'   59
    3. Circumscribed hexagon   63
    4. Angle bisection for polygons   63
    5. Heron's slot machine   64
    6. The geocentric model   67
    7. The chord of an angle   68
    8. The epicycle model   69
    9. Figure for 'Heron's theorem'   74
    10. Diagram for Ptolemy's calculation   75
    11. The diagram for Exercise 5   77

Chapter 4. Chinese mathematics 78
    1. Simple rod numbers   86
    2. 60390 as a rod-number   86
    3. Calculating a product by rod-numbers   87
    4. Li Zhi's 'round town' diagram   91
    5. Diagram for Li Zhi's problem   92

6. Watchtower from the *Shushu jiuzhang*	93
7. Equation as set out by Qin	94
8. The 'pointed field' from Qin's problem	96
9. Chinese version of 'Pascal's triangle'	97

Chapter 5. Islam, neglect and discovery — 101

1. MS of al-Kāshī	105
2. Abū-l-Wafā's construction of the pentagon	108
3. Al-Khwārizmī's first picture for the quadratic equation	112
4. Diagram for Euclid's proposition II.6	113
5. Table from al-Samaw'al (powers)	118
6. Table from al-Samaw'al (division of polynomials)	119
7. Al-Khwārizmī's second picture	127
8. The figure for Thābit ibn Qurra's proof	127
9. Al-Kāshī's seven regular solids	128
10. Al-Kāshī's table of solids	129
11. The method of finding the qibla	131

Chapter 6. Understanding the 'scientific revolution' — 133

1. Arithmetic book from Holbein's *The Ambassadors*	142
2. Graph of a cubic curve	151
3. Kepler's diagram from *Astronomia Nova*	153
4. Descartes' curve-drawing machine	156
5. Kepler's infinitesimal diagram for the circle	158
6. Archimedes' proof for the area of a circle	159

Chapter 7. The calculus — 161

1. Indian calculation of the arc	167
2. Tangent at a point on a curve	169
3. Infinitely close points, infinite polygons, and tangents	171
4. The exponential/logarithmic curve of Leibniz	175
5. Newton's diagram for *Principia* I, proposition 1	178
6. The catenary, and the problem it solves	181
7. Cardioid and an element of area	183
8. Newton's picture of the tangent	184
9. Newton's 'cissoid'	184
10. Leibniz's illustration for his 1684 paper	185

Chapter 8. Geometries and space — 189

1. The figure for Euclid's postulate 5	190
2. Saccheri's three 'hypotheses'	191
3. 'Circle Limit III' by Escher	192
4. Geometry on a sphere	195
5. Ibn al-Haytham's idea of proof for postulate 5	196
6. Descriptive geometry	198
7. Perspective and projective geometry	199
8. Lambert's quadrilateral	201

9. Lobachevsky's diagram	202
10. The parallax of a star	205
11. The diagram for Euclid I.16	208
12. A 'large' triangle on a sphere, showing how proposition I.16 fails	208
13. The elements for solving a spherical triangle	209
14. Proof of the 'angles of a triangle' theorem	211
15. Figure for Exercise 1(b)	211
16. Figure for Exercise 2	211
17. Figure for Exercise 7	212

Chapter 9. Modernity and its anxieties — 213

1. Dedekind cut	215
2. The Brouwer fixed point theorem	220
3. Circle, torus and sphere	224
4. Torus and knotted torus	224
5. The 'dodecahedral space'	225
6. A true lover's knot	225
7. Elementary equivalence of projections	226
8. The three Reidemeister moves	227
9. Two equivalent knots—why?	228
10. Graph of a hyperbola	230

Chapter 10. A chaotic end? — 235

1. A 'half-line angle'	242
2. Trigonometric functions from Bourbaki	242
3. The 'butterfly effect' (Lorenz)	247
4. 'Douady's rabbit'	248
5. The Smale horseshoe map	250
6. A string worldsheet, or morphism	251
7. The classical helium atom	253
8. Elliptic curve (real version)	256
9. Torus, or complex points on a projective elliptic curve	256

Picture Credits

The author thanks the following for permission to reproduce figures and illustrations in this text:

The Schøyen Collection, Oslo and London, for tablet MS1844 (fig. 1.1), bpk/Staatliche Museen zu Berlin - Vorderasiatisches Museum, for tablet VAT16773 (fig. 1.2), the Yale Babylonian Collection for tablets YBC 4652 and YBC7289 (figs. 1.3 and 1.6), Duncan Melville for the tables of cuneiform numerals (figs. 1.4 and 1.5), the Musé du Louvre for tablet AO03448 (fig. 1.7); the Department of History and Philosophy of Science, Cambridge for fig. 3.6, Springer Publications, New York for fig. 3.10; World Scientific Publishing for fig. 4.3; MIT Press for fig. 4.6; Roshdi Rashed for figs. 5.5 and 5.6; the Trustees of the National Gallery, London, for fig. 6.1; C. H. Beck'sche Verlagsbuchhandlung, Munich for figs. 6.3 and 6.5; Dover Publications, New York for fig. 6.4; the Regents of the University of California for fig. 7.5, and Cambridge University Press for figs. 7.8 and 7.9; the M. C. Escher Company, the Netherlands for fig. 8.3; Donu Arapura for fig. 9.4; Mladen Bestvina for figs. 9.6 and 9.8 (created with Knotplot); James Gleick for fig. 10.3; Robert Devaney for fig. 10.4.

Every effort has been made to contact and acknowledge the copyright owners of all figures and illustrations presented in this text, any omissions will be gladly rectified.

Introduction

Why this book?

[M. de Montmort] was working for some time on the *History of Geometry*. Every Science, every Art, should have its own. It gives great pleasure, which is also instructive, to see the path which the human spirit has taken, and (to speak geometrically) this kind of progression, whose intervals are at first extremely long, and afterwards naturally proceed by becoming always shorter. (Fontenelle 1969, p. 77)

With so many histories of mathematics already on the shelves, to undertake to write another calls for some justification. Montmort, the first modern mathematician to think of such a project (even if he never succeeded in writing it) had a clear Enlightenment aim: to display the accelerating progress of the human spirit through its discoveries. This idea—that history is the record of a progress through successive less enlightened ages up to the present—is usually called 'Whig history' in Anglo-Saxon countries, and is not well thought of. Nevertheless, in the eighteenth century, even if one despaired of human progress in general, the sciences seemed to present a good case for such a history, and the tradition has survived longer there than elsewhere. The first true historian of mathematics, Jean Étienne Montucla, underlined the point by contrasting the history of mathematical discovery with that which we more usually read:

Our libraries are overloaded with lengthy narratives of sieges, of battles, of revolutions. How many of our heroes are only famous for the bloodstains which they have left in their path!... How few are those who have thought of presenting the picture of the progress of invention, or to follow the human spirit in its progress and development. Would such a picture be less interesting than one devoted to the bloody scenes which are endlessly produced by the ambition and the wickedness of men?...

It is these motives, and a taste for mathematics and learning combined, which have inspired me many years ago in my retreat... to the enterprise which I have now carried out. (Montucla 1758, p. i–ii)

Montucla was writing for an audience of scholars—a small one, since they had to understand the mathematics, and not many did. However, the book on which he worked so hard was justly admired. The period covered may have been long, but there was a storyline: to simplify, the difficulties which we find in the work of the Greeks have been eased by the happy genius of Descartes, and this is why progress is now so much more rapid. Later authors were more cautious if no less ambitious, the major work being the massive four-volume history of Moritz Cantor (late nineteenth century, reprinted as (1965)). Since then, the audience has changed in an important way. A key document in marking the change is a letter from Simone Weil (sister of a noted number theorist, among much else) written in 1932. She was then an inexperienced philosophy teacher with extreme-left sympathies, and she allowed them to influence the way in which she taught.

Dear Comrade,

As a reply to the Inquiry you have undertaken concerning the historical method of teaching science, I can only tell you about an experiment I made this year with my class. My pupils, like most other pupils, regarded the various sciences as compilations of cut-and-dried knowledge, arranged in the manner indicated by the textbooks. They had *no idea* either of the connection between the sciences, or of the methods by which they were created...

I explained to them that the sciences were not ready-made knowledge set forth in textbooks for the use of the ignorant, but knowledge acquired in the course of the ages by men who employed methods entirely different from those used to expound them in textbooks... I gave them a rapid sketch of the development of mathematics, taking as central theme the duality: continuous–discontinuous, and describing it as the attempt to deal with the continuous by means of the discontinuous, measurement itself being the first step. (Weil 1986, p. 13)

In the short term, the experiment was a failure; most of her pupils failed their baccalaureate and she was sacked. In the long term, her point—that science students gain from seeing their study not in terms of textbook recipes, but in its historical context—has been freed of its Marxist associations and has become an academic commonplace. Although Weil would certainly not welcome it, the general agreement that the addition of a historical component to the course will produce a less limited (and so more marketable) science graduate owes something to her original perception.

It is some such agreement which has led to the proliferation of university courses in the history of science, and of the history of mathematics in particular. Their audience will rarely be students of history; although they are no longer confined to battles and sieges, the origins of the calculus are still too hard for them. Students of mathematics, by contrast, may find that a little history will serve them as light relief from the rigours of algebra. They may gain extra credit for showing such humanist inclinations, or they may even be required to do so. A rapid search of the Internet will show a considerable number of such courses, often taught by active researchers in the field. While one is still ideally writing for the general reader (are you out there?), it is in the first place to students who find themselves on such courses, whether from choice or necessity, that this book is addressed.

On texts, and on history

Insofar as it stands in the service of life, history stands in the service of an unhistorical power, and, thus subordinate, it can and should never become a pure science such as, for instance, mathematics is...

History pertains to the living man in three respects; it pertains to him as a being who acts and strives, as a being who preserves and reveres, as a being who suffers and seeks deliverance. (Nietzsche 1983, p. 67)

American history practical math
Studyin hard and tryin to pass. (Berry 1957)

Chuck Berry's words seem to apply more to today's student of history, mathematics, or indeed the history of mathematics, than Nietzsche's; history pertains to her or him as a being who goes to lectures and takes exams. And naturally where there is a course, the publisher (who also has a living to make) appears on the scene to see if a textbook can be produced and marketed. Probably, the first history designed for use in teaching, and in many ways the best, was Dirk Struik's admirably short text (1986) (288pp., paperback); it is probably no accident that Struik the pioneer held to a more mainstream version of Simone Weil's far-left politics. This was followed by John Fauvel and Jeremy Gray's sourcebook (1987), produced together with a series of short texts from the Open University. This performed the most important function, stressed in the British National Curriculum for history, of foregrounding primary material and enabling students to see for themselves just how 'different' the mathematics of others might appear.[1] Since then, broadly, the textbooks have become longer, heavier, and more expensive. They certainly sell well, they have been produced by professional historians of mathematics, and they are exhaustive in their coverage.[2] What then is lacking? To explain this requires some thought about what 'History' is, and what we would like to learn from it. From this, hopefully, the aims which set this book off from its competitors will emerge.

E. H. Carr devoted a short classic to the subject (2001), which is strongly recommended as a preliminary to thinking about the history of mathematics, or of anything else. In this, he begins by making a measured but nonetheless decisive critique of the idea that history is simply the amassing of something called 'facts' in the appropriate order. Telling the story of the brilliant Lord Acton, who never wrote any history, he comments:

What had gone wrong was the belief in this untiring and unending accumulation of hard facts as the foundation of history, the belief that facts speak for themselves and that we cannot have too many facts, a belief at that time so unquestioning that few historians then thought it necessary—and some still think it unnecessary today—to ask themselves the question 'What is history?' (Carr 2001, p. 10)

If we accept for the moment Carr's dichotomy between historians who ask the question and those who consider that the accumulation of facts is sufficient, then my contention would be that most specialist or local histories of mathematics do ask the question; and that the long, general and all-encompassing texts which the student is more likely to see do not. The works of Fowler (1999) and Knorr (1975) on the Greeks, of Youschkevitch (1976), Rashed (1994), and Berggren (1986) on Islam, the collections of essays by Jens Høyrup (1994) and Henk Bos (1991) and many others in different ways are concerned with raising questions and arguing cases. The case of the Greeks is particularly interesting, since there are so few 'hard' facts to go on. As a result, a number of handy speculations have acquired the status of facts; and this in itself may serve as a warning. For example, it is usually stated that Eudoxus of Cnidus invented the theory of proportions in Euclid's book V. There is evidence for this, but it is rather slender. Fowler is suspicious, and Knorr more accepting, but both, as specialists, necessarily argue about its status. In all *general* histories, it has acquired the status of a fact, because (in Carr's terms) if history is about facts, you must have a clear line which separates them from non-facts, and speculations, reconstructions, and arguments disrupt the smoothness of the narrative.

As a result, the student is not, I would contend, being offered *history* in Carr's sense; the distinguished authors of these 750-page texts are writing (whether from choice or the demands of the market) in the Acton mode, even though in their own researches their approach is quite different. Indeed, in this millennium, they can no longer write like Montucla of an uninterrupted progress from beginning to present day perfection, and they are aware of the need to be fair to other civilizations. However, the price of this academic good manners is the loss of any argument at all. One is reminded of Nietzsche's point that it is necessary, for action, to forget—in this case, to forget some of the detail. And there are two grounds for attempting a different approach, which

---

1. There are a number of other useful sourcebooks, for example, by Struik (1969) but Fauvel and Gray is justly the most used and will be constantly referred to here.
2. Ivor Grattan-Guinness's recent work (1997) escapes the above categorization by being relatively light, cheap, and very strongly centred on the neglected nineteenth century. Although appearing to be a history of everything, it is nearer to a specialist study.

have driven me to write this book:

1. The supposed 'humanization' of mathematical studies by including history has failed in its aim if the teaching lacks the critical elements which should go with the study of history.
2. As the above example shows, the live field of doubt and debate which is *research* in the history of mathematics finds itself translated into a dead landscape of certainties. The most interesting aspect of history of mathematics as it is practised is omitted.

At this point you may reasonably ask what better option this book has to offer. The example of the 'Eudoxus fact' above is meant to (partly) pre-empt such a question by way of illustration. We have not, unfortunately, resisted the temptation to cover too wide a sweep, from Babylon in 2000 BCE to Princeton 10 years ago. We have, however, selected, leaving out (for example) Egypt, the Indian contribution aside from Kerala, and most of the European eighteenth and nineteenth centuries. Sometimes a chapter focuses on a culture, sometimes on a historical period, sometimes (the calculus) on a specific event or turning-point. At each stage our concern will be to raise questions, to consider how the various authorities address them, perhaps to give an opinion of our own, and certainly to prompt you for one.

Accordingly, the emphasis falls sometimes on history itself, and sometimes on *historiography*: the study of what the historians are doing. Has the Islamic contribution to mathematics been undervalued, and if so, why? And how should it be described? Was there a 'revolution' in mathematics in the seventeenth century—or at any other time, for that matter; by what criteria would one decide that one has taken place? Such questions are asked in this book, and the answers of some writers with opinions on the subjects are reported. Your own answers are up to you.

Notice that we are not offering an alternative to those works of scholarship which we recommend. Unlike the texts cited above (or, in more conventional history, the writings of Braudel, Aries, Hill, or Hobsbawm) this book does not set out to argue a case. The intention is to send you in search of those who have presented the arguments. Often lack of time or the limitations of university libraries will make this difficult, if not impossible (as in the case of Youschkevitch's book (Chapter 5), in French and long out of print); in any case the reference and, hopefully, a fair summary of the argument will be found here.

This approach is reflected in the structure of the chapters. In each, an opening section sets the scene and raises the main issues which seem to be important. In most, the following section, called 'Literature', discusses the sources (primary and secondary) for the period, with some remarks on how easy they may be to locate. Given the poverty of many libraries it would be good to recommend the Internet. However, you will rarely find anything substantial, apart from Euclid's *Elements* (which it is certainly worth having); and you will, as always with Internet sources, have to wade through a great mass of unsupported assertions before arriving at reliable information. The St Andrews archive (www-gap.dcs.st-and.ac.uk/ history/index.html) does have almost all the biographies you might want, with references to further reading. If your library has any money to spare, you should encourage it to invest in the main books and journals; but if you could do that,[3] this book might even become redundant.

---

3. And if key texts like Qin Jiushao's Jiuzhang Xushu (Chapter 4) and al-Kāshī's Calculator's Key (Chapter 5) were translated into English.

Examples

> For a long time I had a strong desire in studying and research in sciences to distinguish some from others, particularly the book [Euclid's] *Elements of Geometry* which is the origin of all mathematics, and discusses point, line, surface, angle, etc. (Khayyam in Fauvel and Gray 6.C.2, p. 236)

> At the age of eleven, I began Euclid, with my brother as my tutor. This was one of the great events of my life, as dazzling as first love. I had not imagined there was anything so delicious in the world. From that moment until I was thirtyeight, mathematics was my chief interest and my chief source of happiness. (Bertrand Russell 1967, p. 36)

Perhaps the central problem of the history of mathematics is that the texts we confront are at once strange and (with a little work) familiar. If we read Aristotle on how stones move, or on how one should treat slaves, it is clear that he belongs to a different time and place. If we read Euclid on rectangles, we may be less certain. Indeed, one could fill a whole chapter with examples taken from the *Elements*, the most famous textbook we have and one of the most enigmatic. Because our history likes to centre itself on discoveries, it is common to analyse the ingenious but hypothetical discoveries which underlie this text, rather than the text itself. And yet the student can learn a great deal simply by considering the unusual nature of the document and asking some questions. Take proposition II.1:

> If there are two straight lines, and one of them is cut into any number of segments whatever, then the rectangle contained by the two straight lines equals the sum of the rectangles contained by the uncut straight line and each of the segments.
> Let A and BC be two straight lines, and let BC be cut at random at the points D and E.
> I say that the rectangle A by BC equals the sum of the rectangle A by BD, the rectangle A by DE, and the rectangle A by EC.

If we draw the picture (Fig. 1), we see that Euclid is saying *in our terms* that $a(x+y+z) = ax+ay+az$; what in algebra is called the distributive law. Some commentators would say (impatiently) that that is, essentially, what he is saying; others would say that it is important that he is using a geometric language, not a language of number; such differences were expressed in a major controversy of the 1970s, which you will find in Fauvel and Gray section 3.G. Whichever point of view we take, we can ask why the proposition is expressed in these terms, and how it might have been understood (a) by a Greek of Euclid's time, thought to be about 300 BCE and (b) by one of his readers at any time between then and the present. Euclid's own views on the subject are unavailable, and are therefore open to argument. And (it will be argued in Chapter 2), the question of what statements like proposition II.1 might mean is given a particular weight by:

1. the poverty of source material—almost no writings from before Euclid's time survive;
2. the central place which Greek geometry holds in the Islamic/Western tradition.

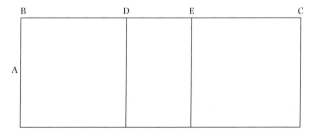

Fig. 1 The figure for Euclid's proposition II.1.

A second well-known example, equally interesting, confronted the Greeks in the nineteenth century. A classical problem dealt with by the Greeks from the fifth century onwards was the 'doubling of the cube': given a cube C, to construct a cube D of double the volume. Clearly this amounts to multiplying the side of C by $\sqrt[3]{2}$. A number of constructions for doing this were developed, even perhaps for practical reasons (see Chapter 3). As we shall discuss later, while Greek writers seemed to distinguish solutions which they thought better or worse for particular reasons, they never seem to have thought the problem insoluble—it was simply a question of which means you chose.

A much later understanding of the Greek tradition led to the imposition of a rule that the construction should be done with ruler and compasses only. This excluded all the previous solutions; and in the nineteenth century following Galois's work on equations, it was shown that the ruler-and-compass solution was impossible. We can therefore see three stages:

1. a Greek tradition in which a variety of methods are allowed, and solutions are found;
2. an 'interpreted' Greek tradition in which the question is framed as a ruler-and-compass problem, and there is a fruitless search for a solution in these restricted terms;
3. an 'algebraic' stage in which attention focuses on proving the impossibility of the interpreted problem.

All three stages are concerned with the same problem, one might say, but at each stage the game changes. Are we doing the same mathematics or a different mathematics? In studying the history, should we study all three stages together, or relate each to its own mathematical culture? Different historians will give different answers to these questions, depending on what one might call their philosophy; to think about these answers and the views which inform them is as important as the plain telling of the story.

Historicism and 'presentism'

Littlewood said to me once, [the Greeks] are not clever schoolboys or 'scholarship candidates', but 'Fellows of another college'. (Hardy 1940, p. 21)

There is not, and cannot be, number as such. We find an Indian, an Arabian, a Classical, a Western type of mathematical thought and, corresponding with each, a type of number—each type fundamentally peculiar and unique, an expression of a specific world-feeling, a symbol having a specific validity which is even capable of scientific definition, a principle of ordering the Become which reflects the central essence of one and only one soul, viz., the soul of that particular Culture. (Spengler 1934, p. 59)

In the rest of this introduction we raise some of the general problems and controversies which concern those who write about the history of science, and mathematics in particular. Following on from the last section in which we considered how far the mathematics of the past could be 'updated', it is natural to consider two approaches to this question; historicism and what is called 'presentism'. They are not exactly opposites; a glance at (say) the reviews in *Isis* will show that while historicism is sometimes considered good, presentism, like 'Whig history', is almost always bad. It is hard to be precise in definition, since both terms are widely applied; briefly, historicism asserts that the works of the past can only be interpreted in the context of a past culture, while presentism tries to relate it to our own. We see presentism in Hardy and Littlewood's belief that the ancient Greeks were Cambridge men at heart (although earlier Hardy has denied that status to the 'Orientals'). By contrast, Spengler, today a deeply unfashionable thinker, shows a radical historicism in going so far as to claim that different cultures (on which he was unusually well-informed) have different concepts of number. It is unfair, as we shall see, to use him as representative—almost no one would make such sweeping claims as he did.

The origins of the history of mathematics, as outlined above (p. 1), imply that it was at its outset presentist. An Enlightenment viewpoint such as that of Montucla saw Archimedes (for example) as engaged on the same problems as the moderns—he was simply held back in his efforts by not having the language of Newton and Descartes. 'Classical' historicism of the nineteenth-century German school arose in reaction to such a viewpoint, often stressing 'hermeneutics', the interpretation of texts in relation to what we know of their time of production (and indeed to how we evaluate our own input). Because it was generally applied (by Schleiermacher and Dilthey) to religious or literary texts, it was not seen as leading to the radical relativism which Spengler briefly made popular in the 1920s; to assert that a text must be studied in relation to its time and culture is not necessarily to say that its 'soul' is completely different from our own—indeed, if it were, it is hard to see how we could hope to understand it. Schleiermacher in the early nineteenth century set out the project in ambitious terms:

The vocabulary and the history of an author's age together form a whole from which his writings must be understood as a part. (Schleiermacher 1978, p. 113)

And we shall find such attempts to understand the part from the whole, for example, in Netz's study (1999, chapters 2 and 3) of Greek mathematical practice, or Martzloff's attempt (1995, chapter 4) to understand the ancient Chinese texts. The particular problem for mathematics, already sketched in the last section, is its apparent timelessness, the possibility of translating any writing from the past into our own terms. This makes it *apparently* legitimate to be unashamedly presentist and consider past writing with no reference to its context, as if it were written by a contemporary; a procedure which does not really work in literature, or even in other sciences.

To take an example: a Babylonian tablet of about 1800 BCE may tell us that the side of a square and its area add to 45; by which (see Chapter 1) it means $\frac{45}{60} = \frac{3}{4}$. There may follow a recipe for solving the problem and arriving at the answer 30 (or $\frac{30}{60} = \frac{1}{2}$) for the side of the square. Clearly we can interpret this by saying that the scribe is solving the quadratic equation $x^2 + x = \frac{3}{4}$. In a sense this would be absurd. Of equations, quadratic or other, the Babylonians knew nothing. They operated in a framework where one solved particular types of problems according to certain rules of procedure. The tablet says in these terms: Here is your problem. Do this, and you arrive at the answer. A historicist approach sees Babylonian mathematics as (so far as we can tell) framed in these terms. You can find it in Høyrup (1994) or Ritter (1995).[4]

However, the simple dismissal of the translation as unhistorical is complicated by two points. The first is straightforward: that it can be done and makes sense, and that it may even help our understanding to do so. The second is that (although we have no hard evidence) it seems that there could be a transmission line across the millennia which connects the Babylonian practice to the algebra of (for example) al-Khwārizmī in the ninth century CE. In the latter case we seem to be much more justified in talking about equations. What has changed, and when? A presentist might

---

4. Høyrup is even dubious about the terms 'add' and 'square' in the standard translation of such texts, claiming that neither is a correct interpretation of how the Babylonians saw their procedures.

argue that, since Babylonian mathematics has become absorbed into our own (and this too is open to argument), it makes sense to understand it in our own terms.

The problem with this idea of translation, however, is that it is a dictionary which works one way only. We can translate Archimedes' results on volumes of spheres and cylinders into our usual formulae, granted. However, could we then imagine explaining the arguments, using calculus, by which we now prove them to Archimedes? (And if we could, what would he make of non-Euclidean geometry or Gödel's theorem?) At some point the idea that he is a fellow of a different college does seem to come up against a difference between what mathematics meant for the Greeks and what it means for us.

As with the other issues raised in this introduction, the intention here is not to come down on one side of the dispute, but to clarify the issues. You can then observe the arguments played out between historians (explicitly or implicitly), and make up your own mind.

Revolutions, paradigms, and all that

Though most historians and philosophers of science (including the later Kuhn!) would disagree with some of the details of Kuhn's 1962 analysis, it is, I think, fair to say that Kuhn's overall picture of the growth of science as consisting of non-revolutionary periods interrupted by the occasional revolution has become generally accepted. (Gillies 1992, p. 1)

From Kuhn's sociological point of view, astrology would then be socially recognised as a science. This would in my opinion be only a minor disaster; the major disaster would be the replacement of a rational criterion of science by a sociological one. (Popper 1974, p. 1146f)

If we grant that the subject of mathematics does change, how does it change, and why? This brings us to Thomas Kuhn's short book *The Structure of Scientific Revolutions*, a text which has been fortunate, even if its author has not. Quite unexpectedly it seems to have appealed to the *Zeitgeist*, presenting a new and challenging image of what happens in the history of science, in a way which is simple to remember, persuasively argued, and very readable. Like Newton's Laws of Motion, its theses are few enough and clear enough to be learned by the most simple-minded student; briefly, they reduce to four ideas:

Normal science. Most scientific research is of this kind, which Kuhn calls 'puzzle-solving'; it is carried out by a community of scholars who are in agreement with the framework of research.

Paradigm. This is the collection of allowable questions and rules for arriving at answers within the activity of normal science. What force might move the planets was not an allowable question in Aristotelian physics (since they were in a domain which was not subject to the laws of force); it became one with Galileo and Kepler.

Revolutions. From time to time—in Kuhn's preferred examples, when there is a crisis which the paradigm is unable to deal with by common agreement—the paradigm changes; a new community of scholars not only change their views about their science, but change the kinds of questions and answers they allow. This change of the paradigm is a scientific revolution. Examples include physics in the sixteenth/seventeenth century, chemistry around 1800, relativity and quantum theory in the early twentieth century.

Incommensurability. After a revolution, the practitioners of the new science are again practising normal science, solving puzzles in the new paradigm. They are unable to communicate with their pre-revolutionary colleagues, since they are talking about different objects.

Consider... the men who called Copernicus mad because he proclaimed that the earth moved. They were not either just wrong or quite wrong. Part of what they meant by 'earth' was fixed position. Their earth, at least, could not be moved. (Kuhn 1970a, p. 149)

Setting aside for the moment the key question of whether any of this might apply to mathematics, its conclusions have aroused strong reactions. Popper, as the quote above indicates, was prepared to use the words 'major disaster', and many of the so-called 'Science Warriors' of the 1990s[5] saw Kuhn's use of incommensurability in particular as opening the floodgates to so-called 'relativism'. For if, as Kuhn argued in detail, there could be no agreement across the divide marked by a revolution, then was one science right and the other wrong, or—and this was the major charge—was one indifferent about which was right? Relativism is still a very dangerous charge, and the idea that he might have been responsible for encouraging it made Kuhn deeply unhappy. Consequently, he spent much of his subsequent career trying to retreat from what some had taken to be evident consequences of his book:

I believe it would be easy to design a set of criteria—including maximum accuracy of predictions, degree of specialization, number (but not scope) of concrete problem solutions—which would enable any observer involved with neither theory to tell which was the older, which the descendant. For me, therefore, scientific development is, like biological development, unidirectional and irreversible. One scientific theory is not as good as another for doing what scientists normally do. In that sense I am not a relativist. (Kuhn 1970b, p. 264)

It is often said that writers have no control over the use to which readers put their books, and this seems to have been very much the case with Kuhn. The simplicity of his theses and the arguments with which he backed them up, supported by detailed historical examples, have continued to win readers. It may be that the key terms 'normal science' and 'paradigm' under the critical microscope are not as clear as they appear at first reading, and many readers subscribe to some of the main theses while holding reservations about others. Nonetheless, as Gillies proclaimed in our opening quote, the broad outlines have almost become an orthodoxy, a successful 'grand narrative' in an age which supposedly dislikes them.

So what of mathematics? It is easy to perceive it as 'normal science', if one makes a sociological study of mathematical research communities present or past; but has it known crisis, revolution, incommensurability even? This is the question which Gillies' collection (1992) attempted to answer, starting from an emphatic denial by Michael Crowe. His interesting, if variable, 'ten theses' on approaching the history of mathematics conclude with number 10, the blunt assertion: 'Revolutions never occur in mathematics' (Gillies 1992, p. 19). The argument for this, as Mehrtens points out in his contribution to the volume, is not a strong one. Crowe aligns himself with a very traditional view, citing (for example) Hankel in 1869:

In most sciences, one generation tears down what another has built... In mathematics alone each generation builds a new storey to the old structure. (Cited in Moritz 1942, p. 14)

Other sciences may have to face the problems of paradigm change and incommensurability, but ours does not. It seems rather complacent as a standpoint, but there is some evidence. One test case appealed to by both Crowe and Mehrtens is that of the 'overthrow' of Euclidean geometry in the nineteenth century with the discovery of non-Euclidean geometries (see chapter 8). The point made by Crowe is that unlike Newtonian physics—which Kuhn persuasively argued could not be

---

5. This refers to a series of arguments, mainly in the United States, about the supposed attack on science by postmodernists, sociologists, feminists, and others. See (Ashman and Barringer 2000)

seen as 'true' in the same sense after Einstein—Euclidean geometry is still valid, even if its status is now that of one acceptable geometry among many.

This point, of course, links to those raised in the previous sections. How far is Euclid's geometry the same as our own? An interesting related variant on the 'revolution' theme, which concerns the same question, is the status of geometry as a subject. Again in Chapter 8, we shall see that geometry in the time of Euclid was (apparently) an abstract study, which was marked off from the study of 'the world' in that geometric lines were unbounded (for example), while space was finite. By the time of Newton, space had become infinite, and geometry was much more closely linked to what the world was like. Hence, the stakes were higher, in that there could clearly only be one world and one geometry of it. The status of Euclidean geometry as one among many, to which Crowe refers, is the outcome of yet another change in mathematics, *later* than the invention of the non-Euclidean geometries: the rise of the axiomatic viewpoint at the end of the nineteenth century and the idea that mathematics studied not the world, but axiom-systems and their consequences.

It may be that neither of these radical changes in the role of geometry altered the 'truth-claims' of the Euclidean model. Nonetheless, there is a case for claiming that they had a serious effect on what geometry was about, and so could be treated as paradigm shifts. Indeed, we shall see early nineteenth-century writers treating geometry as an applied science; in which case, one imagines, the Kuhnian model would be applicable.

As can be seen, to some extent the debate relates to questions raised earlier, in particular how far one adheres to a progressive or accumulative view of the past of mathematics. There have been subsequent contributions to the debate in the years since Gillies' book, but there is not yet a consensus even at the level that exists for Kuhn's thesis.

External versus internal

[In Descartes' time] mathematics, under the tremendous pressure of social forces, increased not only in volume and profundity, but also rose rapidly to a position of honor. (Struik 1936, p. 85)

I would give a chocolate mint to whoever could explain to me why the social background of the small German courts of the 18th century, where Gauss lived, should inevitably lead him to deal with the construction of the 17-sided regular polygon. (Dieudonné 1987)

An old, and perhaps unnecessary dispute has opposed those who in history of science consider that the development of science can be considered as a logical deduction in isolation from the demands of society ('internal'), and those who claim that the development is at some level shaped by its social background ('external'). Until about 30 years ago, Marxism and various derivatives were the main proponents of the external viewpoint, and the young Dirk Struik, writing in the 1930s, gives a strong defence of this position. Already at that point Struik is too good a historian not to be nuanced about the relations between the class struggle and mathematical renewal under Descartes:

In [the] interaction between theory and practice, between the social necessity to get results and the love of science for science's sake, between work on paper and work on ships and in fields, we see an example of the dialectics of reality, a simple illustration of the unity of opposites, and the interpenetration of polar forms...The history and the structure of mathematics provide example after example for the study of materialist dialectics. (Struik 1936, p. 84)

The extreme disfavour under which Marxism has fallen since the 1930s has led those who believe in some influence of society to abandon classes and draw on more acceptable concepts such as milieus, groups, and actors; and Dieudonné has died without conceding that anyone had earned his chocolate mint. Yet in a sense the struggle has sharpened, under the influence of what has been called the 'Edinburgh School' or the 'strong program in the sociology of knowledge' (SPSK), originally propounded in the 1980s by Barry Barnes and David Bloor. For Marxists believed that scientific knowledge (including Marxism) was objective, and hence the rising classes would be inspired to find out true facts (as Struik's examples of logarithms and Cartesian geometry illustrate); as Mao famously said:

Where do correct ideas come from? Do they fall from the sky? No. Are they innate in the mind? No. They come from social practice, and from it alone. They come from three kinds of social practice, the struggle for production, the class struggle and scientific experiment. (Mao Zedong 1963, p. 1)

Notice that Mao too allows for 'internal' factors; the use of scientific experiment to arrive at correct ideas. The Edinburgh school has led the way in an increased scepticism, even relativism on the issue of scientific truth, and in seeing, in the limit, *all* knowledge as socially determined. In one way such a view might be easier for mathematicians to accept than for physicists (say), since the latter consider it important for their justification that electrons, quarks, and so on should be objects 'out there' rather than social constructions. Mathematicians, one would think, are less likely to feel the same way about (say) the square root of minus one, however useful it may be in electrical engineering. In this respect, Leopold Kronecker's famous saying that 'God made the natural numbers; all else is the work of man' places him as a social constructivist before his time.

A deliberately hard test case in a recent text by some of the school goes to work on the deduction of '$2 + 2 = 4$', on proof in general, underlying assumptions, logical steps in proof, and so on.

So-called 'self-evidence' is historically variable... Rather than endorsing one of the claims to self-evidence and rejecting the other, the historian can take seriously the unprovability of the claims that are made at this level, and search out the immediate *causes* of the credibility that is attached or withheld from them. Self-evidence should be treated as an 'actors' category'... (Barnes et al. 1996, p. 190)

Because they are sociologists rather than historians, the Edinburgh school tend not to have an underlying theory of historical change; hence they are stronger on identifying difference across cultures or periods than on identifying the basis on which change takes place. While influenced by Kuhn, and so seeing some sort of a crisis or breakdown in the consensus as motivating, they feel that the actors and their social norms must have something to do with it. However, the society in crisis may be simply the mathematical research community, in which case we are still in a modified 'internal' model similar to that of Kuhn (cf. the disputes about the axiom of choice cited in Barnes et al. 1996, pp. 191–2); or it may be influenced by the wider community, as in the case of Joan Richards' study of the relation between Euclidean geometry and the Victorian established church (see chapter 8). As Paul Forman, responsible for one of the best studies of the interaction of science and society (1971), has pointed out recently (1995), the accusation of relativism seems to have driven many advocates of the strong programme into a partial retreat from a position which was never very historically explicit.

And yet, the hard-line internalist position is still considered inadequate by many historians, even if they are not sure what mixture of determinants they should put in its place. Often in the last two centuries, internal determinants seem paramount,[6] though in operational research, computing

---

6. One could, for example, point out that knot theory, while first developed in the 1870s by an electrical engineer (Tait) to deal with a physical problem, has proceeded according to an apparent internal logic of its own since then. See chapter 9.

and even chaos theory one could see outside forces at work. In earlier history, when we have the evidence (and we often do not) it often seems the other way round. In his commentary on the 'Rectangular Arrays' (matrices) section of the *Nine Chapters* (see chapter 4), Liu Hui analyses a problem on different grades of paddy. He says, 'It is difficult to comprehend in mere words, so we simply use paddy to clarify'. Does he mean that the authors of the classical text first hit on the idea of using matrix algebra and then applied it to grades of paddy for ease of exposition? We have no evidence, but it seems easier to believe that the discovery went the other way round, from problems about paddy (or something) to matrices.

It is easy to say that among most responsible historians now the tendency is to take both internal and external determinants seriously in any given situation and to give them their appropriate weight. The problem is that with the eclipse of Marxism and with doubts about Kuhn's relevance to mathematics, there is no very well organized version of either available to the historian. We shall continue to appeal to Marxism (and indeed to Kuhn) where we find either of them relevant in what follows.

Eurocentrism

> I propose to show ... that the standard treatment of the history of non-European mathematics exhibits a deep-rooted historiographical bias in the selection and interpretation of facts, and that mathematical activity outside Europe has in consequence been ignored, devalued or distorted. (Joseph 1992, p. 3)

> His willingness to concoct historically insupportable myths that are pleasing to his political sensibilities is obvious on every page. His eagerness to insinuate himself into the good graces of the supposed educators who incessantly preach the virtues of 'multiculturalism' and the vices of 'eurocentrism' is palpable and pervasive. (Review on mathbook.com)

It would appear that the argument set out by Joseph has not been won yet. I have no way of judging the book under review (it is not Joseph's) in the second quote, but there is an underlying suggestion that the reviewer has heard more than enough about eurocentrism and is pleased to find a book which is both anti-eurocentrist and intellectually shoddy, thereby supporting his or her suspicions. This is the 'fashionable nonsense'[7] school of reviewing, and it is not going to go away; in fact, the current anti-Islamic trend in the West, and specifically in the United States, may lend it more support.

What is eurocentrism (for those who have not heard yet)? In general terms, it is the privileging of (white) European/American discourse over others, most often African or Asian; in history, it might mean privileging the European account of the Crusades, or of the Opium Wars, or any imperialist episode over the 'other side'. For what it might mean in mathematics, we should go back to Joseph who, at the time he began his project (in the 1980s), had a strong, passionate, and undeniable point. If we count as the 'European' tradition one which consists *solely* of the ancient Greeks and the modern Europeans—and we shall soon see how problematic that is—a glance through many major texts in the history of mathematics showed either ignorance or undervaluing of the achievements of those outside that tradition. We shall discuss this in more detail later (Chapter 5), but his book was important; it is the only book in the history of mathematics written from a strong personal conviction, and it is valuable for that reason alone. It also stands as the single most influential work in changing attitudes to non-European mathematics. The sources, such as Neugebauer on the

---

7. The title of a book (Sokal and Bricmont 1998) which is devoted to attacking what it sees as sloppy thinking about science by postmodernists, feminists, post-colonialists, and many others.

Egyptians and Babylonians, or Youschkevitch on the Islamic tradition, may have been available for some time before, but Joseph drew their findings into a forceful argument which since (like Kuhn's work) its main thrust is easy to follow has made many converts. After sketching the views which he intends to counter, Joseph characterizes three historical models which can be used to describe the transmission of mathematical knowledge.

First, the 'classical Eurocentric trajectory' already referred to: mathematics passed directly from the ancient Greeks to the Renaissance Europeans;

Second, the 'modified Eurocentric trajectory': Greece drew to some extent on the mathematics of Egypt and Babylonia; while after Greek learning had come to an end, it was preserved in the Islamic world to be reintroduced at the Renaissance;

Third, Joseph's own 'alternative trajectory'. This—with a great many arrows in the transmission diagram—stresses the central role of the Islamic world in the Middle Ages as a cultural centre in touch with the learning of India, China, and Europe and acting both as transmitter and receiver of knowledge. The more we know, particularly of the Islamic world, the more this appears to be a reasonably accurate picture, and while Joseph's tone can be polemical and some of his detailed points have been questioned, his arguments are rarely overstated. We are learning more of the mathematics of India, China, and Islam, as of the Greeks' predecessors, and scholars are becoming better able to read their texts and understand their way of thinking about mathematics.

The body of the book is given over to a detailed account of the various non-European cultures and their contributions. Interestingly, his account is now to be found substantially unchanged (if with more detail) in most of the standard textbooks. The culture warriors may rage against fashionable anti-Eurocentrism, but as far as mainstream teaching of the history of mathematics is concerned, it seems to have been absorbed successfully. Again, we shall return to this point later.

The specific reasons for Eurocentrism in the history of mathematics (setting aside traditional racism and other prejudices) have been two-fold. The first is the very high value accorded to the work of the ancient Greeks specifically, the second the emphasis on discovery and proof of results. These are indeed linked: much of the Greek work was organized in the form of result + proof. All the same, there is an important point to be made here; namely, that after the Greeks it was the Arabs who continued the tradition, with propositions and proofs in the Euclidean mode. (Khayyam's geometric work on the cubic equations is a model of the form.) If we contrast Islamic mathematics of around 1200 with that of western Europe, we would have no doubt that the former was, in our terms, 'Western', and the latter a primitive outsider. However, this has not, until recently, helped the integration of the great Islamic mathematicians into the Western tradition; and if it did, it would still leave the Indians and Chinese, with very different practices, outside it.

Indeed, the problem of Eurocentrism could be seen in Kuhnian terms as one of paradigms. The Greek paradigm, or a version of it, is one which has in some form persisted into modern Western mathematics[8] and hence traditional histories have constructed themselves around that paradigm, either leaving out or subordinating ways of doing mathematics which did not fit. It is only more recently that a more culturally aware (historicist?) history has been able to ask how other cultures thought of the practice of mathematics, and to escape the trap of evaluating it against a supposed Greek or Western ideal.

---

8. Not at all times; Descartes, Newton in his early work, and Leibniz initiated a tradition in which the Euclidean mode was at least temporarily abandoned. See chapters 6 and 7.

1 Babylonian mathematics

1 On beginnings

Obviously the pioneers and masters of hydraulic society were singularly well equipped to lay the foundations for two major and interrelated sciences: astronomy and mathematics. (Wittfogel, *Oriental Despotism*, p. 29, cited Høyrup 1994, p. 47)

Based on intensive cereal agriculture and large-scale breeding of small livestock, all in the hands of a centralized power, [this civilization] was quickly caught up in a widespread economy which made necessary the meticulous control of infinite movements, infinitely complicated, of the goods produced and circulated. It was to accomplish this task that writing developed; indeed for several centuries, this was virtually its only use. (M. Bottéro, cited in Goody 1986, p. 49)

When did mathematics begin? Naive questions like this have their place in history; the answer is usually a counter-question, in this case, what do you mean by 'mathematics'? A now rather outdated view restricts it to the logical-deductive tradition inherited from the Greeks, whose beginnings are discussed in the next chapter. The problem then is that much interesting work which we would commonly call 'mathematics' is excluded, from the Leibnizian calculus (strong on calculation but short on proofs) to the kind of exploratory work with computers and fractals which is now popular in studying complex systems and chaotic behaviour. Many cultures before and since the Greeks have used mathematical operations from simple counting and measuring onwards, and solved problems of differing degrees of difficulty; the question is how one draws the line to demarcate when mathematics proper started, or if indeed it is worth drawing.[1] As we shall see, the early history of Greek mathematics is hard to reconstruct with certainty. In contrast, the history of the much more ancient civilizations of Iraq (Sumer, Akkad, Babylon) in the years from 2500 to 1500 BCE provides a quite detailed, if still patchy record of different stages along a route which leads to mathematics of a kind. Without retracing the whole history in detail, in this chapter we can look at some of these stages as illustrations of the problem raised by our initial question/questions. Mathematics of what kind, and what for? And what are the conditions which seem to have favoured its development?

Before attempting to answer any of these questions, we need some minimal historical background. Various civilizations, with different names, followed each other in the region which is now Iraq, from about 4000 to 300 BCE (the approximate date of the Greek conquest). Our evidence about them is entirely archaeological—the artefacts and records which they left, and which have been excavated and studied by scholars. From a very early date, for whatever reason, they had, as the quotation from Bottéro describes, developed a high degree of hierarchy, slave or semi-slave labour, and obsessive bureaucracy, in the service of a combination of kings, gods, and

---

1. This relates to the questions raised recently in the field of 'ethnomathematics'; mathematical practices used, often without explicit description or justification, in a variety of societies for differing practical ends from divination to design. For these see, for example, Ascher (1991); because the subject is mainly concerned with contemporary societies, it will not be discussed in this book.

their priests. Writing of the most basic kind was developed around 3 300 BCE, and continued using a more developed form of the original 'cuneiform' (wedge-shaped) script for 3000 years, in different languages. The documents have been unusually well preserved because the texts were produced by making impressions on clay tablets, which hardened quickly and were preserved even when thrown away or used as rubble to fill walls (see Fig. 1). A relatively short period in the long history has provided the main mathematical documents, as far as our present knowledge goes. As usual, we should be careful; our knowledge and estimation of the field has changed over the past 30 years and we have no way of knowing (a) what future excavation or decipherment will turn up and (b) what texts, currently ignored, will be found important by future researchers. In this period—from 2500 to 1750 BCE—the Sumerians, founders of a south Iraqi civilization based on Uruk, and inventors of writing among other things—were overthrown by a Semitic-speaking people, the Akkadians, who as invaders often do, adopted the Sumerian model of the state and used Sumerian (which is not related to any known language, and which gradually became extinct) as the language of culture. A rough guide will show the periods from which our main information on mathematics derives:

2500 BCE *'Fara period'*. The earliest (Sumerian) school texts, from Fara near Uruk; beginning of phonetic writing.

2340 BCE *'Akkadian dynasty'*. Unification of all Mesopotamia under Sargon (an Akkadian). Cuneiform is adapted to write in Akkadian; number system further developed.

2100 BCE *'Ur III'*. Re-establishment of Ur, an ancient Sumerian city, as capital. Population now mixed, with Akkadians in the majority. High point of bureaucracy under King Šulgi.

1800 BCE *'Old Babylonian', or OB*. Supremacy of the northern city of Babylon under (Akkadian) Hammurapi and his dynasty. The most sophisticated mathematical texts.

MS 1844

Fig. 1 A mathematical tablet (Powers of 70 multiplied by 2. Sumer, C. 2050 BC).

Fig. 2 Tablet VAT16773 (c. 2500 BCE).; numerical tally of different types of pigs.

Each dynasty lasted roughly a hundred years and was overthrown by outsiders, following a common pattern; so you should think of less-centralized intervals coming between the periods listed above. However, there was a basic continuity to life in southern Iraq, with agriculture and its bureaucratic-priestly control probably continuing without much change throughout the period.

In the quotation set at the beginning of the chapter, the renegade Marxist Karl Wittfogel advanced the thesis that mathematics was born out of the need of the ancient Oriental states of Egypt and Iraq to control their irrigation. In Wittfogel's version this 'hydraulic' project was indeed responsible for the whole of culture from the formation of the state to the invention of writing. The thesis has been attacked over a long period, and now does not stand much scrutiny in detail (see, for example, the critique by Høyrup 1994, p. 47); but a residue which bears examining (and which predates Wittfogel) is that the ancient states of Egypt and Iraq had a broadly similar priestly bureaucratic structure, and evolved both writing and mathematics very early to serve (among other things) bureaucratic ends. Indeed, as far as our evidence goes, 'mathematics' precedes writing, in that the earliest documents are inventories of goods. The development of counting-symbols seems to take place at a time when the things counted (e.g. different types of pigs in Fig. 2) are described by pictures rather than any phonetic system of writing. The bureaucracy needed accountancy before it needed literature—which is not necessarily a reason for mathematicians to feel superior.[2]

On this basis, there could be a case for considering the questions raised above with reference to ancient Egypt as well—the organization of Egyptian society and its use of basic mathematical procedures for social control were similar, if slightly later. However, the sources are much

---

2. There were certainly early poems celebrating heroic actions, the *Gilgamesh* being particularly famous. But in many societies, such poems are not committed to writing, and this seems to have been the case with the *Gilgamesh* for a long time—before it too was pressed into service by the bureaucracy to be learned by heart in schools.

poorer, largely because papyrus, the Egyptian writing-material, lasts so badly; there are two major mathematical papyri and a handful of minor ones from ancient Egypt. It is also traditional to consider Babylonian mathematics more 'serious' than Egyptian, in that its number-system was more sophisticated, and the problems solved more difficult. This controversy will be set aside in what follows; fortunately, the re-evaluations of the Babylonian work which we shall discuss below make it outdated. The Iraqi tradition is the earliest, it is increasingly well-known, discussed, and argued about; and on this basis we can (with some regret) restrict attention to it.

2 Sources and selections

Even with great experience a text cannot be correctly copied without an understanding of its contents ... It requires years of work before a small group of a few hundred tablets is adequately published. And no publication is 'final'. (Neugebauer 1952, p. 65)

We need to establish the economic and technical basis which determined the development of Sumerian and Babylonian applied mathematics. This mathematics, as we can see today, was more one of 'book-keepers' and 'traders' than one of 'technicians' and 'engineers'. Above all, we need to research not simply the mathematical texts, but also the mathematical content of economic sources systematically. (Vaiman 1960, p. 2, cited Robson 1999, p. 3)

The quotations above illustrate how the study of ancient mathematics has developed. In the first place, crucially, there would not be such a study at all if a dedicated group of scholars, of whom Neugebauer was the best-known and most articulate, had not devoted themselves to discovering mathematical writings (generally in well-known collections but ignored by mainstream orientalists); to deciphering their peculiar language, their codes, and conventions; and to trying to form a coherent picture of the whole activity of mathematics as illustrated by their material—overwhelmingly, exercises and tables used by scribes in OB schools. These pioneers played a major role in undermining a central tenet of Eurocentrism, the belief that serious mathematics began with the Greeks. They pictured a relatively unified activity, practised over a short period, with some interesting often difficult problems. However, it is the fate of pioneers that the next generation discovers something which they had neglected; and Vaiman as a Soviet Marxist was in a particularly good position to realize that the neglected mathematics of book-keepers and traders was needed to complete the rather restricted picture derived from the scribal schools. For various reasons—its simplicity, based on a small body of evidence, and its supposed greater mathematical interest—the older (Neugebauer) picture is easy to explain and to teach; and you will find that most accounts of ancient Iraqi mathematics (and, for example, the extracts in Fauvel and Gray) concentrate on the work of the OB school tradition. In this chapter, trying to do justice to the older work and the new, we shall begin by presenting what is known of the classical (OB) period of mathematics; and then consider how the picture changes with the new information which we have on it and on its more practical predecessors.

At the outset—and this is implicit in what Neugebauer says—we have to face the problem of 'reading texts'. The ideal of a history in the critical liberal tradition, such as this aims to be, is that on any question the reader should be pointed towards the main primary sources; the main interpretations and their points of disagreement; and perhaps a personal evaluation. The reader is then encouraged to think about the questions raised, form an opinion, and justify it with reference to the source material. Was it possible to be an atheist in the sixteenth century; when was non-Euclidean geometry discovered, and by whom? There is plenty of material to support

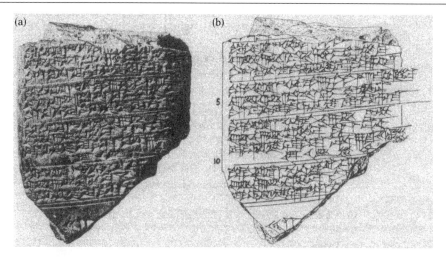

Fig. 3 The 'stone-weighing' tablet YBC4652; (a) photograph and (b) line drawing.

arguments on such questions, and there are writers who have used the material to develop a case. When we approach Babylonian mathematics, we find that this model does not work. There are, it is true, a large number of documents. They are partly preserved, sometimes reconstructed clay tablets, written in a dead language—Sumerian or Akkadian or a mixture—using the cuneiform script. It should also be noted that their survival is a matter of chance, and that we have few ways of knowing whether the selection which we have is representative. There seem to be gaps in the record, and most of our studies naturally are directed at the periods from which most evidence has survived.

Unless we want to spend years acquiring specialist knowledge, we must necessarily depend on experts to tell us how (a) to read the tablets, (b) to decipher the script, and (c) to translate the language.

It is useful to begin with an example. The tablet pictured (Fig. 3) is called YBC4652 (YBC for Yale Babylonian Catalogue). Here is the text of lines 4–6, which is cited in Fauvel and Gray as 1.E.1(20). The language is Akkadian, the date about 1800 BCE.

na$_4$ ì-pà ki-lá nu-na-tag 8-bi ì-lá 3 gín bí-daḫ-ma

igi-3-gál igi-13-gál a-rá 21 e-tab bi-daḫ-ma

ì-lá 1 ma-na sag na$_4$ en-nam sag na$_4$ $4\frac{1}{2}$ gín

Note that the figures in this quotation correspond to Babylonian numerals, of which more will follow later[3]; that is, where in the translation below the phrase 'one-thirteenth' appears, a more accurate translation would be '13-fraction', which shows that the *word* thirteen is not used. There is a special sign for $\frac{1}{2}$. The translation reads as follows (words in brackets have been supplied by the translator):

I found a stone, (but) did not weigh it; (after) I weighed (out) 8 times its weight, added 3 gín

one-third of one-thirteenth I multiplied by 21, added (it), and then

I weighed (it): 1 ma-na. What was the origin(al weight) of the stone? The origin(al weight) of the stone was $4\frac{1}{2}$ gín.

---

3. Except for the '4' in 'na$_4$', which seems to be a reference to the meaning of 'na' we are dealing with.

As you can see, from tablet to drawing to written Akkadian text to translation we have stages over which you and I have no control. We must make the best of it.

There are subsidiary problems; for example, we need to accept a dating on which there is general agreement, but whose basis is complicated. If a source gives the dates of King Ur-Nammu of the Third Dynasty as 'about 2111–2095 BCE', where do these figures come from, and what is the force of 'about'? Most scholars are ready to give details of all stages, but we are in no position to check. The restricted range of the earlier work perhaps made a consensus easier. In the last 30 years, divergent views have appeared. Even the traditional interpretation of the OB mathematical language has been questioned. An excellent account of this history is given by Høyrup (1996). In general the present-day historians of mathematics in ancient Iraq are models of what a secondary source should be for the student; they discuss their methods, argue, and reflect on them. But given the problems of script and language we have referred to, when experts do pronounce, by interpreting a document as a 'theoretical calculation of cattle yields', for example, rather than an actual count (see Nissen et al. 1993, pp. 97–102), the reader can hardly disagree, however odd the idea of doing such a calculation in ancient Ur may seem.

On a core of OB mathematics there is a consensus, which dates back to the pioneering work of Neugebauer and Thureau-Dangin in the first half of the twentieth century. There may be an argument about whether it is appropriate to use the word 'add' in a translation, but in the last instance there is agreement that things are being added. This is helpful, because it does give us a coherent and reliable picture of a practice of mathematics in a society about which a good deal is known. However, it is necessarily restricted in scope, and the sources which are usually available do not always make that fact clear. For example, most texts which you will see commented and explained come from the famous collection *Mathematical Cuneiform Texts* (Neugebauer and Sachs 1946). This is a selection, almost all from the OB period, and the selection was made according to a particular view of what was interesting. If you look at an account of Babylonian mathematics in almost any history book, what you see will have been filtered through the particular preoccupations of Neugebauer and his contemporaries, for whom OB mathematics was fascinating in part (as will be explained below) because it appeared both difficult and in some sense useless. The broader alternative views which have been mentioned do not often find their way into college histories.

It should be added that Neugebauer and Sachs's book is itself long out of print, and almost no library stocks it; your chances of seeing a copy are slim. Because the texts are so repetitive, the selections (from what is already a selection) given in textbooks, in particular Fauvel and Gray, give a pretty good picture of OB mathematics as it was known 50 years ago. All the same, they *are* selections from a large body of texts. Other useful reading—again not necessarily accessible in most libraries—is to be found in the works of Høyrup (1994), Nissen et al. (1993), and Robson (1999). There is a useful selection of Internet material (and general introduction) at http://it.stlawu.edu/~dmelvill/mesomath/; and in particular you can find various bibliographies, particularly the recent one by Robson (http://it.stlawu.edu/~dmelvill/mesomath/biblio/erbiblio.html).

Exercise 1. *(which we shall not answer). Consider the example given above; try to correlate the original text with (a) the pictures and (b) the translation. (Note that the line drawing is much clearer than the photograph; but, given that someone has made it, have we any reason to suspect its clarity?) Can you find out anything about either the script or the meaning of the words in the original as a result? How much editing seems to have been done, and how comprehensible is the end product?*

Exercise 2. *(which will be dealt with below). Clearly what we have here, in the translation, is a question and its answer. If I add the information that there are 60 gín in 1 ma-na, what do you think the question is, and how would you get at the answer?*

3 Discussion of the example

As is often observed, the problem above appears 'practical' (it is about weights of stones) until you look at it more closely. It was set, we are told, as an exercise in one of the schools of the Babylonian empire where the caste known as 'scribes' who formed the bureaucracy were trained in the skills they needed: literacy,[4] numeracy, and their application to administration. The usual answer to Exercise 2 is as follows. You have a stone of unknown weight (you did not weigh it); in our language, you would call the weight x gín. You then multiply the weight by 8 (how?) and add 3 gín, giving a weight of $8x + 3$. However, worse is yet to come. You now 'multiply one-third of one-thirteenth' by 21. What this means is that you take the fraction $\frac{1}{3} \times \frac{1}{13} \times 21 = \frac{21}{39}$ and multiply that by the $8x + 3$. You are not told that, but the tablets explain no more than they have to, and the problem does not come right without it, so we have to assume that the language which may seem ambiguous to us was not so to the scribes. Adding this, we have:

$$8x + 3 + \frac{21}{39}(8x + 3) = 60$$

Here we have turned the ma-na into 60 gín.

Clearly, as a way of weighing stones, this is preposterous; but perhaps it is not so very different from many equally artificial arithmetic problems which are set in schools, or were until recently. Effectively—and this is a point which we could deduce without much help from experts, although they concur in the view—such exercises were 'mental gymnastics' more than training for a future career in stone-weighing.

An advantage of beginning with the Babylonians is that their writing gives us a strong sense of historical *otherness*. Even if we can understand what the question is aiming at, the way in which it is put and the steps which are filled in or omitted give us the sense of a different culture, asking and answering questions in a different way, although the answer may be in some sense the same. In this respect, such writing differs from that of the Greeks, who we often feel are speaking a similar language even when they are not. You are asked a question; the type of question points you to a procedure, which you can locate in a 'procedure text'. To carry it out, you use calculations derived from 'table texts'; these tell you (to simplify) how to multiply numbers, to divide, and to square them. As James Ritter says:

the systematization of both procedure and table texts served as a means to the same end: that of providing a network or grille through which the mathematical world could be seized and understood, at least in an operational sense. (Ritter 1995, p. 42)

It is worth noting that part of Ritter's aim in the text from which the above passage is taken is to situate the mathematical texts in relation to other forms of procedure, from medicine to divination, in OB society: they all provide the practitioner with 'recipes' of form: if you are confronted with

---

4. This included not only their own language but a dead language, Sumerian, which carried higher status; as civil servants in England 100 years ago had to learn Latin.

problem A, then do procedure B. The 'point' of the sum, then, is not mysterious, and indeed we can recognize in it some of our own school methods. First, scribes are trained to follow rules; second, they are required to use them to do something difficult. As usual, such an ability marks them off as workers by brain rather than by hand, and fixes their relatively privileged place in the social order. We know something of the arduous training and the beatings that went with it; but not what happened to those trainees who failed to make the grade.

What is mysterious in this particular case is the way in which one is supposed to get to the answer from the question, since the tablet gives no clue. Here the term 'procedure text' is rather a misnomer, but other tablets are more explicit on harder problems. With our knowledge of algebra, we can say (as you will find in the books) that the equation above leads to:

$$(8x+3)\frac{39+21}{39} = 60$$

and so, $8x + 3 = 39$, and $x = 4\frac{1}{2}$. The fact that 39 and 21 add to 60, one would suppose, could not have escaped the setter of the problem; but language, such as I have just used would have been quite impossible. What method would have been available? The Egyptians (and their successors for millennia) solved simple linear equations, such as (as we would say) $4x + 3 = 87$ by 'false position': guessing a likely answer, finding it is wrong, and scaling to get the right one. This seems not to work easily in this case. To spend some time thinking about how the problem *could* have been solved is already an interesting introduction to the world of the OB mathematician.

Having looked at just one example, let us broaden out to the general field of OB mathematics. What were its methods and procedures, what was distinctive about it? And second, do the terms 'elementary' and 'advanced' make sense in the context of what the Babylonians were trying to do; and if so, which is appropriate?

4 The importance of number-writing

As we have already pointed out, Neugebauer and his generation were working on a restricted range of material. To some extent this was an advantage, in that it had some coherence; but even so, there were typical problems in determining provenance and date, because they were processing the badly stored finds of many earlier archaeologists who had taken no trouble to read what they had brought back. It is well worth reading the whole of Neugebauer's chapter on sources, which contains a long diatribe on the priorities and practices of museums, archaeological funds, and scholars:

Only minute fractions of the holdings of collections are catalogued. And several of the few existing rudimentary catalogues are carefully secluded from any outside use. I would be surprised if a tenth of all tablets in museums have ever been identified in any kind of catalogue. The task of excavating the source material in museums is of much greater urgency[5] than the accumulation of new uncounted thousands of texts on top of the never investigated previous thousands. I have no official records of expenditures for expeditions at my disposal, but figures mentioned in the press show that a preliminary excavation in one season costs about as much as the salary of an Assyriologist for 12 to 15 years. And the result of every such dig is frequently more tablets than can be handled by one scholar in 15 years. (Neugebauer 1952, pp. 62–3)

---

5. Partly because, as Neugebauer has said earlier, tablets deteriorate when excavated and removed from the climate of Iraq.

There is probably better conservation of tablets now than when the above was written, but the long delay in publishing is still a problem[6]; and there are grounds for new pessimism now that one hears that tablets are being removed from sites in Iraq and traded, presumably with no 'provenance' or indication of place and date, over the Internet. (For a discussion by Eleanor Robson of these and other problems which face historians in the aftermath of the Iraq war see http://www.dcs.warwick.ac.uk/bshm/Iraq/iraq-war.htm.)

The best-known of the OB tablets can be seen as rather special. What can be recognized in them are several features that subsequent scholars felt could be identified as truly 'mathematical':

1. The use of a sophisticated system for writing numbers;
2. The ability to deal with quadratic (and sometimes, if rather by luck, higher order) equations;
3. The 'uselessness' of problems, even if they were framed in an apparently useful language, like the one above.

None of these characteristics are present (so far as we know) in the mathematics of the immediately preceding period, which in itself is noteworthy. Let us consider them in more detail.

The number system

You will find this described, usually with admiration, in numerous textbooks. The essence was as follows. Today we write our numbers in a 'place-value' system, derived from India, using the symbols $0, 1, \ldots, 9$; so that the figure '3' appearing in a number means 3, 30, 300, etc. (i.e. $3 \times 10^0, 3 \times 10^1, 3 \times 10^2, \ldots$) depending on where it is placed. The Babylonians used a similar system, but the base was 60 instead of 10 ('sexagesimal' not 'decimal'), and they therefore based it on signs corresponding to the numbers $1, \ldots, 59$—without a 'zero' sign. The signs were made by combining symbols for 'ten' and 'one'—a relic of an earlier mixed system, but obviously practical, in that what was needed was some easily comprehensible system of 59 signs. (see Fig. 4) You might, as an exercise, think of how to design one. The place-value system was constructed, like ours, by setting these basic signs side by side; we usually transliterate them and add commas, so that they can be read as in Fig. 5. '1, 40' means, then, what we would call $1 \times 60 + 40 = 100$; '2, 30, 30' means $2 \times 60^2 + 30 \times 60 + 30 = 7200 + 1800 + 30 = 9030$. 60 plays the role which 10 plays in our system.

There are, though, important differences from our practice. First, it is not explicitly clear that '30' on its own, with no further numbers involved necessarily means what we should call 30. It may mean $30 \times 60 (= 1800)$ or $30 \times 60^2 (=108,000), \ldots$. In a problem, it will be 30 somethings—a measurement of some kind, which is stated explicitly, for example, length or area in appropriate units; and this will usually make clear which meaning it should have. This is not the case with 'table texts' (e.g. the '40 times table'), which often concern simple numbers. Furthermore—compare our decimals—'30' can also mean $30 \times \frac{1}{60} = \frac{1}{2}$, and often does.[7] Or $30 \times \frac{1}{60} \times \frac{1}{60}$ and so on. If the answer was written as 30, you should—and this is an idea which we can recognize from our own practice—be able to deduce what '30' meant from the context.

---

6. Robson (1999) cites an example of a collection of OB proverb texts which were published in the 1960s with no acknowledgement by the scholarly editor that they had calculations on the back.

7. Although there were also symbols for the commonest fractions like $\frac{1}{2}$—see the above example—and (it seems) rules about when you used them.

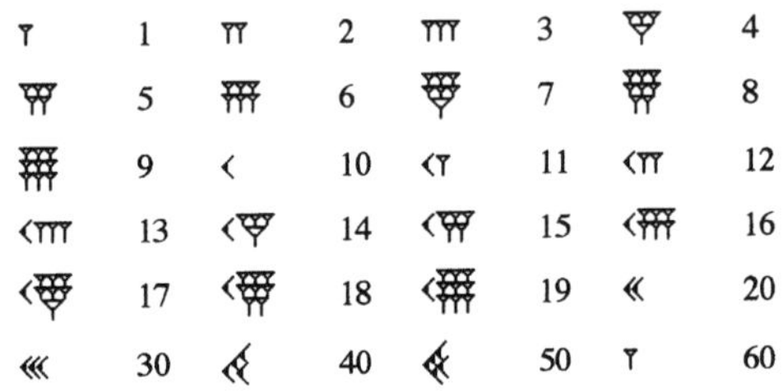

BABYLONIAN MATHEMATICS

Fig. 4 The basic cuneiform numbers from 1 to 60.

Cuneiform	Transliteration	Decimal value
	1,15	75
	1,40	100
	16,43	1003
	44,26,40	160000
	1,24,51,10	305470

Fig. 5 How larger cuneiform numbers are formed.

You can find the details of how the system works in various textbooks; in particular, there are plenty of examples in Fauvel and Gray. (Notice that the sum which I gave above was one in which it was not needed—why?) Again following a general convention, modern editors make things easier for readers by inserting a semi-colon where they deduce the 'decimal point' must have come, and inserting zeros as in '30, 0' or '0; 30'. So '1, 20' means 80, but '1; 20' means $1 + \frac{20}{60} = 1\frac{1}{3}$. There would be no distinction in a Babylonian text; both would appear as '1 20'.

To help themselves, the Babylonians, as we do, needed to learn their tables. They were, it would seem, in a worse situation than us, since there were in principle 59 tables to learn, but they probably used short cuts. A scribe 'on site' would quite possibly have carried tablets with the important multiplication tables on them, as an engineer or accountant today will carry a pocket calculator or palmtop; and in particular the vital table of 'reciprocals'. This lists, for 'nice' numbers x, the value of the reciprocal $\frac{1}{x}$, and starts:

2	30
3	20
4	15
5	12
6	10
7, 30	8
8	7, 30
9	6, 40

Using this table it is possible to divide simply by multiplying by the reciprocal; dividing by 4 is multiplying by 15 (and of course thinking about what the answer means in practical terms—what size of number one should expect).

This way of writing numbers is so advanced and sophisticated that it has impressed most commentators, particularly mathematicians. The absence of a decimal point, as I have said, is not a serious problem in practical calculations; but it could raise questions when one is asked, for example, to take the square root (we will see this was done too) of 15. If '15' means $\frac{1}{4}$, then it has square root $30 = \frac{1}{2}$, but if it means '15', of course, it does not have an exact square root. However, the scribe would find the square root by looking in a table, and only one answer would appear, for any number.

The more serious problem which is often pointed out is the absence of a sign for 'zero'. In principle, $60\frac{1}{2}$, which should in our terms be '1 0 30' (one sixty, no units, 30 sixtieths) would be written '1 30', which could also mean '90' (or '$1\frac{1}{2} = 90 \times \frac{1}{60}$'). It is hard to know how often this caused confusion. One case is given by Damerow and Englund (in Nissen et al. 1993, pp. 149–50) of a scribe who is finding the powers of '1, 40', or what we would call 100. At the sixth stage one of the figures should be a '0', and is omitted. Hence this calculation, and the subsequent ones (he continues to 100^{10}) are wrong. However, you can see (why?) that this mistake would occur less often than in our decimal system if we happened to 'forget' zeros, and so confused 105 and 15.

Exercise 3. *Explain (a) how the table of reciprocals works, (b) why it does not contain '7'.*

Exercise 4. *Work out $(1, 40)/(8)$ using the table, given that the reciprocal of 8 is 7, 30. (Check that this is indeed the reciprocal; and verify that you have the right answer, given that $1, 40 = 100$ in our terms.)*

Exercise 5. *(a) What is the square root of 15 if '15' means 15×60? (b) Show that, in Babylonian terms, there cannot be two different interpretations of a number which have different (exact) square roots.*

5 Abstraction and uselessness

The discovery of the sexagesimal system is sometimes described, by those who like the word, as a revolution. How it came about is unclear, but it does seem to have arisen quite suddenly out of a number of near- or pseudo-sexagesimal systems, around the beginning of the OB period. Damerow and Englund (Nissen et al. 1993, pp. 149–50) seem to consider it impractical, and claim it did not outlast the OB period—which is difficult to reconcile with their admission that it was used by the Greek astronomers. Here, indeed, we find our first example of the problem of connecting similar practices across time. Sexagesimals were used in Babylon in 1800 BCE, and again, mainly in astronomy, 1500 years later. (They were still being used—with multiplication tables—by Islamic writers in the fifteenth century CE (see Chapter 5) under the name 'astronomers' numbers'.) It seems almost certain that this was a direct line of descent from Babylon to Greece. More dubious claims are often made, though, in situations where the same result (e.g. 'Pythagoras' theorem') is known to two different societies—that there must have been either communication or a common ancestor. Such arguments are central (for example) to van der Waerden's fascinating but eccentric (1983); always controversial, they have to be evaluated on the basis of the evidence.

Equations

Here, if anywhere, the mathematicians can be allowed to judge what it is to be sophisticated. In examples like the one above, we see probably for the first time the idea of an unknown quantity—an unweighed stone, in this case. The Egyptians were using the same idea a little afterwards, and may have arrived at it independently; but they did not succeed in the next step, which was a general method for solving quadratic-type problems. It makes sense to use this term, rather than 'quadratic equations', since the problems are very varied in nature; the 'quadratic equation' as we know it, a combination of squares, things, and constants, begins its history properly in the Islamic period. Fauvel and Gray's 1.E.(f) problem 7 starts:

I have added up seven times the side of my square and eleven times the area: 6; 15

In other words, we have a square, and we are told that seven times the unknown side x ($7x$) added to eleven times the area ($11x^2$) gives 6; 15 or $6\frac{1}{4}$. This leads to a simple quadratic equation, which we would write $7x + 11x^2 = 6\frac{1}{4}$, with answer $x = 0; 30 = \frac{1}{2}$. For how it is solved, which in particular shows where square roots were used, see Appendix A.

In addition to the relatively common equation texts, we have some texts which seem to show extra mathematical sophistication, some of which is still subject to debate. One is the notorious 'Plimpton 322'; for the original decoding of this see Fauvel and Gray and for a recent counter-argument, Robson (2001); we shall not consider this here, although it is an interesting introduction to the disagreements of historians. A simpler case is the 'square root of 2' tablet, which seems straightforward in its interpretation (Fig. 6). The picture shows a square; its side is marked 30 (or $\frac{1}{2}$), and the diagonal has two sexagesimal numbers marked. One is a good approximation to $\sqrt{2}$ (1, 24, 51, 10), the other to the diagonal $\sqrt{2}/2$ (42, 25, 35). Nearly the same sexagesimal numbers will appear again when we deal with Islamic mathematicians over 3000 years later; for now it is worth raising the question of what these numbers were used for, and how they were arrived at. In the absence of any written procedures, we can at least admire the result.

'Uselessness'

Sometimes mathematicians need to be reminded that mathematics, to be worthwhile, does not *have* to be useless; and they have often had a two-faced attitude on the subject, pointing (e.g. when

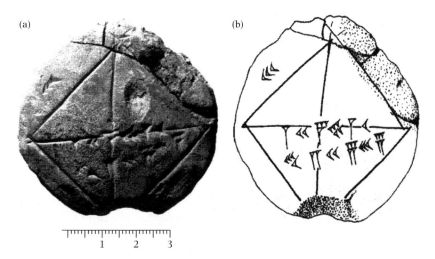

Fig. 6 The 'square root of 2' tablet YBC7289.

requesting a research grant) to results which were thought useless at the time and afterwards discovered to have an application. Well-known examples include Riemannian geometry and relativity, finite fields and the manufacture of CDs, etc. It has been a part of the case for the seriousness of Babylonian mathematics that their problems, while apparently practical, were clearly not designed for the real world. Rather, they were exercises in technique dressed up in practical language (because that was the only language available). The point is often made, and can hardly be contested. Our first example (stone-weighing) is a good illustration. So is the quadratic equation above—it would be hard to think of circumstances in which one would want to add lengths to areas, and the Greeks, with a more strict idea of geometry, did not have a language in which to do it. In another example often cited, the student is given the amount of earth required to fill a ramp, and asked to find its dimensions—exactly the opposite of the practical question. No Babylonian text theorizes this impracticality as such, or makes a virtue of it; while Plato, as we shall see, makes a distinction between real mathematics and that which is used by artisans, the Babylonian scribes to all appearances were trained for a career of useful tasks by solving problems with no application.

What was the point of this? To answer this question would require some thought about what the 'point' of any mathematical procedure is. At one level, we can imagine that the ability to deal with increasingly difficult problems, regardless of their meaning, could be used as an examination-type filtering mechanism within the scribal schools, marking off the bright students from the mediocre ones; or, outside the schools, it could be a form of competition between 'freelance' scribes (they existed too) who were trying to attract clients. This virtuosity is part of a whole package of skills which were important for self-definition and for status:

According to the 'Examination Text A', the accomplished scribe must know everything about bilingual [that is, Sumerian/Akkadian] texts; he must know occult writings, and occult meanings of signs in Akkadian as well as Sumerian; he must be familiar with the concepts of musical practice, and he must understand the distorted idiom of various crafts and trades. Into the bargain then comes mathematics... All that, as a totality, has a name (of course Sumerian): nam-lú-ulù, 'humanity'. (Høyrup 1994, p. 65)

This 'external' explanation does not, however, account for the particular choice of impractical quadratic equations for the display of accomplishment. Here we have, almost, an example of Kuhn's 'normal science'. A technique—the solution of linear and quadratic problems using sexagesimal numbers and tables—becomes available, for reasons which are unclear; and the scholars who make up the community are defined by their ability to solve puzzles using the technique. In addition, they may find the problems interesting or challenging, in a career dedicated to routine tasks (but here we are indeed speculating). In principle, hard puzzles can generate harder ones without limit; in terms of the historical record, it seems that either invasion or loss of interest or both put an end to the practice.

The idea of 'uselessness' is one which needs to be treated with some care, however. It is easy, considering some of the OB calculations, to deduce that their apparent practicality is a fake and that they are simply occasions for what Høyrup calls displays of 'scribal virtuosity'. This is the traditional view, and many of the texts support it. However, Robson's detailed new publication (1999), containing a wide variety of tablets, is the basis for arguing a more complex view. An example is a long tablet (BM96957 + VAT6598) containing a succession of problems about brick walls. These depend crucially for their solution on one of the basic scribes' numbers: the conversion factors from (volume of wall) to (number of bricks in the wall) and back—an eminently practical figure, and one which was certainly often used. The problems start with questions which give the measurements of the wall (length, width, and height), and ask how many bricks. Naturally, the OB scribes (like us) used different units for width (cubits, compare inches), and length and height (nindan, compare feet); the calculation was therefore not always a straightforward one leading to a certain number of cubic nindan and dividing by the number of bricks in a cubic nindan. Such a question seems both simple and practical, and just the kind of thing which a scribe in the brick-wall construction trade might be asked. However, question 5 on the same tablet is:

A wall. The height is $1\frac{1}{2}$ nindan, the bricks 45 sar$_b$ [brick measure]. The length exceeds the width of the wall by 2; 20 nindan. What are the length and the width of my wall? (Robson 1999, p. 232)

The details of brick-measure and height belong to everyday practice, but it seems very unlikely that one would ever need to answer a question of this type in a practical situation. Somewhere along the list of problems on the tablet a link to real-world wall-building has been broken.

Exercise 6. *If you are told that 72 sar$_b$ of bricks occupy a volume of 1 cubic nindan, (a) show that this is equivalent to a quadratic problem and (b) find the answer.*

6 What went before

The last example shows that there may still be more to learn about the OB period. In recent times, a much fuller picture has emerged of the earlier period of mathematics, and it is currently perhaps the most interesting area of research. What we have is still more a series of snapshots than a record of discovery; in archaeology it is almost unknown to find an innovation which can be accounted for, much less attributed to an 'author'; but it allows us to question the idea of Babylonian mathematics as the earliest serious practice, based on the criteria I have given.

In the first place, we know that the profession of scribe, and the scribal schools, existed for some time before (the usual estimate is around 2500 BCE for the beginning of the institution). Even in this very early period, when the number system, while quite clear and flexible, was much less advanced than the sexagesimal one, we find that the schools had discovered the idea of setting problems which were both difficult and useless, if in a different way—in fact, the mixed nature of the number system made questions which we might think easy harder. They require simple division of a very large (i.e. impractical) number by a number which makes problems. Specifically,

that the content of a silo containing 2400 'great gur', each of 480 sila, be distributed in rations of 7 sila per man (the correct result is found in no. 50: 164,571 men, and a remainder of 3 sila)... (Høyrup 1994, p. 76)

A sila being roughly (it is thought) a litre, we are dealing with over a million litres, and the proposed division by 7 (with remainder!) is an exercise in obsessive accuracy rather than a practical problem. In the words of Jöran Friberg

the obvious implication is that the 'current fashion' among mathematicians about four and a half millennia ago was to study non-trivial division problems involving large (decimal or sexagesimal) numbers and 'non-regular' divisors such as 7 and 33. (Cited in Høyrup 1994, p. 76)

Friberg uses the term 'mathematicians' to describe those scribes and teachers who discussed such problems; and such a usage not only sets the origin of mathematics as an independent practice much earlier, but makes it appear much more 'trivial' to us. If the Babylonians can be grudgingly admired for solving quadratic equations, can we extend a similar recognition to the scribes of Fara for doing rather long divisions? There has indeed been quite a controversy about what the Fara scribes were supposed to do in answering the question; see Powell in Fauvel and Gray 1.E.5, or for a more recent view, Melville, 'Ration Computations at Fara: Multiplication or Repeated Addition' in Steele and Imhausen (2002). Again, this question is perhaps best left unanswered, or as a point for discussion. Friberg would probably justify calling the scribes 'mathematicians' not in terms of their use of unrealistic examples, but in the formation of a community—again that Kuhnian word—with a common project, whose language was a language of numbers. Training for practical purposes seems, here too, to have generated a class of impractical exercises, if entirely different from those which followed 500 years later.

However, this impracticality, characteristic of the school-texts which have survived, disappears when we look at a different family of texts, the accounts from the harsh period known as Ur III, which were the work of *practising* scribes and administrators. (What kind of texts survive from which period is at least partly chance, depending on the kind of site excavated.) Dating from the twenty-first century BCE, these are in time between the Fara texts and the OB ones, and they are both utilitarian and highly 'mathematized'. The period, under King Šulgi, was one of increasingly rigid centralized control of production; the aim, for a variety of industries—seed production, cattle raising, fishing, milling, and so on—is to calculate the expected yield and the extent to which the farmers or managers fulfil their targets. Analogies with old Soviet planning or indeed modern Western management come to mind. Accounts were complicated by the fact that almost any quantity had a special system of units to measure it. However, the scribe is, on the whole, up to the calculation; as usual, there are tables of conversion factors to help. Here is an example which, according to Damerow and Englund (Nissen et al. 1993, pp. 141–2), represents 'the calculation of the harvest yield of the province of Lagash for the third year recorded in the text' (Fig. 7). We begin

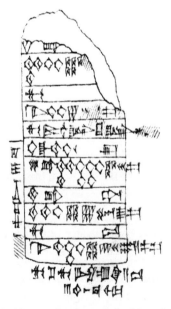

Fig. 7 The tablet recording harvests from Lagash, AO 3448.

by setting out the area:

  1 (šár-gal)   1 (šar'u)   1 (šár)   1 (bùr) field surface

Then follow the 'targets'; the amount which this area should produce:

  the barley involved: 3 (šar'u)   5 (šár)   3 (geš'u)   3 (u) gur

Finally, the actual amount produced, and the shortfall:

Therefrom
2 (šar'u) 1 (šár) 4 (geš'u) 7 (géš) 4 (u) 2 (gur) 1 (barig) 4 (bán) gur delivered.
Deficit: 1 (šar'u) 3 (šár) 4 (geš'u) 3 (géš) 2 (u) 7 (gur) 3 (barig) 2 (bán) gur

A first observation is that a quite unnecessary number of units of measurement seem to be involved (and there are yet more ...). They are of course exotic to us, but at 4000 years' distance we can expect that. The first row gives the area of the fields producing barley. According to Nissen et al. 1993, pp. 141–2, 1 bùr is about 6.3 hectares; and

1 šár = 60 bùr
1 šár'u = 10 šár
1 šár-gal = 6 šár'u.

The total area is therefore (work it out) 4261 bùr or 26,844 hectares. The calculation of 'the barley involved' in the second row is the 'target'; it assumes that an area of 1 bùr produces 30 gur (9000 litres) of grain. For the grain measure we have:

(1 bán = 10 litres)
1 barig = 6 bán
1 gur = 5 barig
1 u = 10 gur
1 géš = 6 u
1 geš'u = 10 géš
1 šár = 6 geš'u
1 šar'u = 10 šár

As you can see, the units do not proceed by uniform steps, and even multiplying the area by the factor of 30 gur and translating it into volume units to get the target volume is quite complicated. Hence the figures 1, 1, 1, 1 in the first row translate into 3, 5, 3, 3 in the second.

We now have to subtract the actual output from the target; and the actual figure involves a rather excessive eight units of measurement (all the ones listed above).[8]

This should be enough to convince you that, while Ur III accountants' arithmetic was 'elementary', it was far from simple, and considerable skill was required to get the deficit right. (Happily, there was, it seems, not always a deficit; apparently in the first of the three years listed on the tablet the harvest was more than expected. On the other hand—see Englund (1991)—the targets set for labourers in factories seem generally to have been unrealistically high and calculated

---

8. But before we condemn the Sumerians for their complexity, it should be noted that schools in England 50 years ago taught a system of 8 units of length—line, inch, foot, yard, rod (or pole, or perch), chain, furlong, mile—and that the factors relating them were more complicated than the Sumerian 5s, 6s, and 10s.

to ruin their overseers to the greater profit of the state.) Of course, we still sometimes face problems of this multi-unit kind, such as when we try to find the time lapse between 1.25 p.m. on January 28 and 11.15 a.m. on February 2 in days, hours, and minutes; but these are rare and the metric system is reducing them.

Exercise 7. *Trace the calculation of 'the barley involved' through and check it.*

Exercise 8. *Calculate the deficit, using the table of barley measures, and find the two places where the scribe has made a mistake.*

7 Some conclusions

The above example is worth some consideration, if only because you will not often find such work discussed. In a sense the mathematics is trivial, in another clearly not; it is highly organized, and it needs to be accurate (although mistakes were not uncommon). It is as much a product of the bureaucracy and the organization of scribes as are the more interesting and mathematically impressive OB examples with which we started, and which you will usually meet; and its basic tools—multiplication and subtraction, with 'conversion factors' to make it more difficult—have also had a long history, and are still with us. The rationality of the OB system is often mentioned to boost its credentials as the earliest real mathematics, as is the fact that it survives in our measurements of time (minutes and seconds) and angle (degrees, minutes, and seconds). However, even today we often in practice find we have to operate with mixed systems of measurement, and work out the relevant sums as best we can. We could call such a procedure irrational (but on what grounds?); it does not make the mathematics easier. Only those who have never made mistakes in such conversions (e.g. miles and yards to and from metric) can dismiss them as not mathematical.

Appendix A. Solution of the quadratic problem

The solution given (from Neugebauer, also in Fauvel and Gray I.E.(f), problem 7) is as follows. The intrusive semicolons have been omitted; you will have to work out where they should come. On the other hand, the procedure is translated into algebraic notation in brackets, so that it can be followed more easily.

You write down 7 and 11. You multiply 6,15 by 11: 1,8,45. (Multiply the constant term by the coefficient of x^2.)

You break off half of 7. You multiply 3,30 and 3,30. (Square half the x-coefficient.)

You add 12,15 to 1,8,45. Result 1,21. (12,15 is the result of the squaring, so the 1, 21 is what we would call $(b/2)^2 + ac$, if the equation is $ax^2 + bx = c$.)

This is the square of 9. You subtract 3,30, which you multiplied, from 9. Result 5,30. (This is $-(b/2) + \sqrt{(b/2)^2 + ac}$; in the usual formula, we now have to divide this by $a = 11$, which we proceed to do.)

The reciprocal of 11 cannot be found. By what must I multiply 11 to obtain 5,30? The side of the square is 30. ('Simple' division was multiplying by the reciprocal, for example, dividing by 4 is multiplying by 15, as we have seen. If there is no reciprocal, you have to work it out by intelligence or guesswork, as is being done here.)

Solutions to exercises

1. I shall not answer, while exercise 2 is answered in the text.
3. If the number in the left column is x, that in the right column is y where $x \cdot y = 1$. How does this work? For example, $4 \cdot 15 = 60$ (which is 1, or 1, 0 if you want to use the notation of modern translation), and $8 \times (7, 30) = 8 \times 7 + 8 \times (0, 30) = 56 + 4 = 60$ again. More generally, one could think of x and y as solving some equation $x \cdot y = 60^k$; the value of k is immaterial, since in Babylonian notation we cannot, for example, tell the different answers 15 and '0,15' ($= \frac{1}{4}$) apart.

    This process works if such a y can be found, that is, if x divides some power of 60 exactly. (More exactly, we choose an interpretation of x which is a whole number, not a fraction.) This will be true if (and only if) all the factors of x are 2s, 3s, and 5s. It will therefore not work for 7.

4. $(1, 40)/(8)$ would be calculated as $1, 40 \times 7, 30$ (times the reciprocal). Use the formulae: $7 \times 40 (= 280) = 4, 40$ and $30 \times 40 = 20, 0$; and take care of place value. You find the product is

$$7, 0, 0 + 4, 40, 0 + 30, 0 + 20, 0 = 12, 30, 0$$

If you were a Babylonian scribe, and knew that the '1, 40' meant 100, you would have no difficulty in interpreting this answer as $12\frac{1}{2}$.

5. 1. Of course, $15 \times 60 = 900$, which is a square, indeed the square of 30. So the statement 'square root of 15 is 30' is true also for this interpretation of 15.
   2. This is a standard fact about place-value systems (unless the 'base' is a square). The different interpretations of any number are, say a basic 'x' and $x \times 60^k$. If $x = y^2$, then (since 60 is not a square), $x \times 60^k$ is a square if and only if k is even, say $k = 2l$; when $x \times 60^k$ is the square of $y \times 60^l$. So, in Babylonian terms, the square root of x is always y.
6. The Babylonian answer is given in Robson (2000, p. 232); it is hard to follow, since the text switches between 'sar$_v$' (a volume unit), nindan, and cubits (at 12 cubits to a nindan). Above, we have given the conversion factor from bricks to volume in cubic nindan instead of sar$_v$ to reduce the number of measures. Here is a simplified version of the answer in our notation.

    45 sar$_b$ of bricks occupy $\frac{15}{24}$ cubic nindan, so that is the volume. The height is $1\frac{1}{2}$, so the area (length times width) is $\frac{5}{12}$ square nindan. If l is the length and w the width, $l = w + \frac{7}{3}$, and $lw = \frac{5}{12}$; so

$$w\left(w + \frac{7}{3}\right) = \frac{5}{12}; \qquad 12w^2 + 28w = 5$$

Clearly this is quadratic, and the solutions are $w = \frac{1}{6}$ and $w = -\frac{5}{2}$. Realistically, the wall has width $\frac{1}{6}$ nindan (2 cubits) and length $2\frac{1}{2}$ nindan.

7. We could, as above, reduce everything to the simplest units; but that is probably not what was done. To proceed 'properly', start from the right (this may not have been usual, but it is our habit). 1 bùr gives 30 gur or 3 u, from the table. 1 šár (60 bùr) therefore gives 60×3 u, or 3 geš'u (since there are 60 u in 1 geš'u). 1 šar'u gives 10 times this, which is 30 geš'u, or 5 šár. And finally, 1 šár-gal gives six times this, which is 30 šár, or 3 šar'u.

8. We now have to subtract. This time again, it is more correct to borrow along the line. However, since you can risk making mistakes quite easily given the number of 0s in the top row, I will reduce everything to bán. The amount due is 3,834,900 bán, the amount delivered is 2,353,870 bán; and the deficit is 1,481,030 bán; which is 1,080,000 (1 šar'u) + 3,240,000 = 3 × 108,000 (3 šár) + 72,000 = 4 × 18,000 (4 geš'u) + 3600 = 2 × 1800 (2 géš) + 1200 = 4 × 300 (4 u) + 210 = 7 × 30 (7 gur) + 18 = 3 × 6 (3 barig) + 2 bán. This agrees with the scribe's calculation apart from the figures for u and géš.

2 Greeks and 'origins'

> Socrates: Then as between the calculating and measurement employed in building or commerce and the geometry and calculation employed in philosophy—well, should we say there is one sort of each, or should we recognize two sorts?
> Protarchus: On the strength of what has been said, I should give my vote for there being two.
> (Plato, *Philebus*, tr. in Fauvel and Gray 2.E.4, p. 75)

1 Plato and the *Meno*

One feature of mathematics which has remained fairly constant from the earliest times to the present day is a general view that its aim is to use 'numbers' to solve problems which arise in the world. However, another idea has been widespread among mathematicians at least since the time of the ancient Greeks, and its statement dates back to Plato, whose different view is summarized above: there is a down-to-earth mathematics which you use for accounts and measuring, and there is a superior mathematics, which I use for some other purpose. What we know of this view, what its implications were, and its early history are the subject of this chapter. Plato—a philosopher, whose dates are usually given as roughly 427–348 BCE and who was mostly writing in the early fourth century—is one of the central figures in the history of Greek mathematics. There are a number of reasons for this. A simple one is that Plato dealt in some detail with mathematical questions in his works; and, while mathematics had supposedly been practised for 200 years before his time, his *Dialogues* are the earliest first-hand documents which we have. Almost equally important is that, as the quotation indicates, Plato defined a particular view of what mathematics was, or should be. A rough characterization is that real mathematics is more abstract—numbers are no longer numbers of 'things' or measurements of length, area, or time, but have an independent existence as objects which you reason with. As Socrates says earlier in the same dialogue:

> The ordinary arithmetician, surely, operates with unequal units; his 'two' may be two armies or two cows or two anythings from the smallest thing in the world to the biggest; while the philosopher will have nothing to do with him, unless he consents to make every single instance of his unit equal to every other of its infinite number of instances.
> (Plato, *Philebus*, tr. in Fauvel and Gray 2.E.4, p. 75)

And while not all of his successors agreed with this approach, those who did were those who had most influence. This is especially true of some very late writers (after 300 CE) who are the main authorities for what we know of the history.

One of Plato's longest and clearest mathematical discussions, often referred to, is in the dialogue called the *Meno*. (For the mathematical part, see Fauvel and Gray 2.E.1 (pp. 61–67); Fowler (1999) has text with variations of his own construction; and the whole dialogue is online for example,

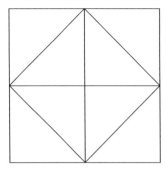

Fig. 1 The *Meno* argument. The large square has side 4 feet (area 16 square feet), the four small squares have side 2 feet (area 4 square feet). The four diagonals form a square of area 8 square feet.

at http://classics.mit.edu /Plato/meno.html.) What is done in this dialogue is a good introduction to Greek mathematics—or the kind which is considered 'typical', the classics if you like. The other kinds, referred to in our opening quotation, will be discussed in the next chapter. Although the problem and the solution would easily have been available to the Egyptians or Babylonians a thousand years earlier, what seems suddenly to be new is the appeal to argument and discussion. The philosophical point of the dialogue is an idea about 'knowing'. Socrates has a strange theory that the truths which we know have not been learned but were always present in our minds and we simply bring them to consciousness or 'remember' them. (He is referring to a particular kind of truth—knowledge about triangles or the Good, not mere facts like 'it's raining'.) With this aim, he calls over a supposedly ignorant slave-boy, and asks him how, if you are given a square of side 2 feet, you can construct a square twice the size. Since the original square has an area of 4 square feet, then the one you construct must have an area of 8. The slave-boy suggests squares of side 4 feet (wrong, because its area is 16 square feet) and 3 feet (again wrong, area 9 square feet). He then becomes perplexed, and admits to not knowing. Socrates then—we assume—draws the figure shown (Fig. 1), and continues to 'find out' from the boy that it contains the answer to the problem.

His arguments are given in Appendix A to this chapter, and are conceptually quite simple.

1. Each of the four squares in the diagram is 2×2 square feet, and so has an area 4 square feet.
2. So each of the eight triangles (half of a square) has an area of 2 square feet.
3. Now look at the square made by the four diagonals. It consists of four triangles, so its area must be 4×2 or 8 square feet.
4. It is therefore twice the area of the (2×2 square feet) square we started off with so it is the 'square of double the size' we were looking for.

It is easy to find fault with the way the dialogue is conducted: Socrates is in fact leading the witness inadmissibly, putting the answer which he himself knows into the boy's mouth, and then claiming that he has done nothing of the kind. However, more purely mathematical objections arise, and they relate to some key ideas about the nature of Greek mathematics. In particular, a question which Socrates does not deal with—which is interesting given the precise use of numbers like '2 feet', '3 feet', and so on—is what the length of the diagonal (the side of the 'eight-foot square') is. Today, we would say that, since the area is 8, the side must be $\sqrt{8} = 2.828\ldots$, which is not a whole number, or even a 'rational' number (a fraction p/q, where p, q are whole numbers). You can approximate it as closely as you like by fractions, but the result will never be exact.

As Michel Serres points out, in a detailed discussion of the dialogue:

Nobody asks the asker: how long? He questions the ignorant slave about a content about which nobody, however, bothers him. He found the side all right, but he did not measure it. Socrates is cheating: he knows that he will not find the exact length. (Serres 1995, p. 105)

In fact, one reason why this particular problem has been chosen for the dialogue is perhaps—but here, as usual, we have to start attributing motives to the Greeks—because it shows the limitations of numbers, and the superior power of geometrical methods. The boy will never arrive at the right answer by guessing different numbers; but Socrates can draw a picture which solves it. The philosophers' mathematics not only uses a more abstract idea of 'number', but when number becomes a problem, it can dispense with it.[1]

Greeks said that ratios such as diagonal-to-side were 'alogoi'—without a reason, irrational. More simply, they said that the side and the diagonal were 'incommensurable', that is, there is no shorter line l having the property that both side and diagonal are exact multiples of l (are 'measured' by l). We shall return to these terms, and the problems they pose, later.

Exercise 1. *Is it obvious that the figure which Socrates constructs in Fig. 1 is a square? Why?*

Exercise 2. *Why does 'diagonal is incommensurable with side' mean the same to us as '$\sqrt{2}$ is not a fraction'?*

Exercise 3. *Why is $\sqrt{2}$ not a fraction anyway?*

2 Literature

It is striking that near the end of the twentieth century there should appear two books arguing that much of the history of Greek mathematics written during that century is wrong. Reviel Netz argues that it is wrong because historians have not understood the crucial roles that language and diagrams played in shaping the deductive structure that is Greek mathematics' most striking characteristic. David Fowler argues that it is wrong because a key component of the mathematics that developed in and around Plato's academy was lost in Hellenistic times and was not rediscovered until the Renaissance. (Berggren 2003)

The literature on ancient Greek mathematics, as the above quotation reminds us, is large and in constant flux. We shall consider the specific problem of the ancient Greek texts themselves shortly. For the moment, let us concentrate on what material is available on the period, as primary and secondary sources. The standard reference text is certainly Heath's (1981), a reprint of a 1921 classic. Because it is so old, and so much a standard work, it is the basis for most later authors' arguments, disagreements, and conjectural reconstructions. The main *primary* sources for the period we are considering—up to and including Euclid, around 300 BCE—are the works of Plato, Aristotle, and Euclid himself. Fauvel and Gray give plentiful extracts from all of them, and all can be found easily on the Internet. It is very strongly recommended that you read some of these texts, which vary from

---

1. There is evidence that Plato did know that the side of the eight-foot square was not a rational number, from the *Theaetetus*—see extract in Fauvel and Gray 2.E.3, pp. 73–4. But the question is not raised in the *Meno*. Why not?

quite easy to spectacularly difficult. This will give some idea of the achievements of Greek mathematics, of its range, and of the limits within which it operated. In Fauvel and Gray you will also find some useful extracts from the very late (*c.*450 CE) commentator Proclus, for whom see later.

With regard to secondary literature, the situation is better, since the question of the earliest Greeks, their aims, and achievements, has seemed so important—we will discuss why later. There are sections on the Greeks in both van der Waerden (1961) and Neugebauer (1952). They are sometimes dated, and in van der Waerden's case, given to conjecture using unreliable ancient sources. The major modern works which cover the question can be dauntingly detailed, but if you find yourself developing, as one does, an interest in Greek mathematics, they will draw you in. The central works are probably Knorr's (1975) attempt to account for the form of the *Elements* and Fowler's (1999) more informal but very scholarly reconstruction of what geometry might have been like *before* Euclid. Netz (referred to by Berggren above) (1999) gives an intriguingly different approach, by a consideration of actual Greek proofs and how they work, followed by a 'sociology' of Greek mathematicians based on the very fragmentary evidence we have about them. We shall refer to other texts when they are useful; but these will do for the present. You have a reasonable chance, given the prestige of Greek mathematics, of finding some or all of them in your library.

3 An example

Being in a Gentleman's Library, Euclid's *Elements* lay open, and 'twas the *47 El. libri I* [Pythagoras' theorem]. He [Thomas Hobbes] read the Proposition. *By G—*, sayd he (he would now and then sweare an emphaticall Oath by way of emphasis) *this is impossible*! So he reads the Demonstration of it, which referred him back to such a Proposition; which proposition he read. That referred him back to another, which he also read. *Et sic deinceps* [and so on] that at last he was demonstratively convinced of that trueth. This made him in love with Geometry. (J. Aubrey, *Brief Lives*, quoted in Fauvel and Gray 3.F.2)

While the *Meno* is a very illuminating discussion on a mathematical subject, it is too informal to be a good illustration of the mainstream Greek mathematics which is our primary concern. Socrates' arguments make no attempt to go back to first principles, and the points he makes about areas of triangles are treated as obvious (which they are) rather than justified in painful detail. The mathematical argument of the *Meno*, if not its philosophical one, would have been easily accessible to an Egyptian.

To see how 'classical' Greek mathematics claims to work, it is best to start, at least, with Euclid's *Elements*. (For texts see the bibliography.) This is a strange and complex work—some would say a composite, or scissors-and-paste compilation of previous works; but it has been the most read and commented of all mathematical works in history, so it deserves a central position in any account. For that reason, we shall privilege it over the harder works of Archimedes and Apollonius, the other main classics. It is also a sensible idea in the first place to consider it in itself as a text rather than speculating on its origins; such speculation is natural in a history which focuses on 'discovery', but other histories are available. A whole book could be written about the *Elements*, and many have; the most recent and scholarly are Knorr (1975) (already mentioned) and Artmann (1999). The work, as its title suggests, is supposed to give the student the essentials of mathematics, carefully deduced from 'first principles', statements which are either in some sense obvious, or which the reader/student can reasonably allow to be true. ('All right angles are equal', for example.) Here, as

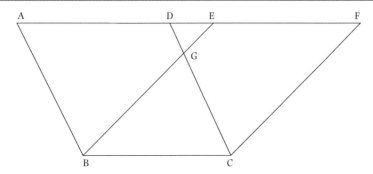

Fig. 2 The diagram for Euclid proposition I.35.

an example of the approach, is a single short proposition. The aim is a simple one: to show that two parallelograms with the same base and the same height have equal area. In Euclid's language, they are just called 'equal'.

Proposition I.35. *Parallelograms which are on the same base and in the same parallels equal one another.*
  Let ABCD and EBCF be parallelograms on the same base BC and in the same parallels AF and BC.
  I say that ABCD equals the parallelogram EBCF. [Here is Euclid's diagram (Fig. 2). The proof then continues.]
  Since ABCD is a parallelogram, therefore AD equals BC. (Proposition I.34.)
  For the same reason EF equals BC, so that AD also equals EF. And DE is common, therefore the whole AE equals the whole DF. (Common Notions 1, Common Notions 2.)
  But AB also equals DC. Therefore, the two sides EA and AB equal the two sides FD and DC, respectively, and the angle FDC equals the angle EAB, the exterior equals the interior. Therefore, the base EB equals the base FC, and the triangle EAB equals the triangle FDC. (Propositions I.34, I.29, I.4.) [So at this point, we have proved the triangles EAB and FDC are what is usually called 'congruent'—all sides and angles are the same. This implies that they have equal areas, of course.]
  Subtract DGE from each. Then the trapezium ABGD which remains equals the trapezium EGCF. (Common Notions 3.)
  Add the triangle GBC to each. Then the whole parallelogram ABCD equals the whole parallelogram EBCF. (Common Notions 2.)
  Therefore, parallelograms which are on the same base and in the same parallels equal one another. Q.E.D.

This result stands *instead of* the numerical formula which we, like the ancient Egyptians and Babylonians, would use to relate the area of a parallelogram (and, more commonly, a triangle), to the length of the base and to the height.
  This use of the term 'equal', to mean having the same area, is not in Euclid's list of definitions; you are supposed to know what it means. Here we see it as retrospectively defined by 'adding' and 'subtracting' the triangles DGE and GBC, starting from two triangles which themselves are equal in all respects—all sides and all angles correspond. So, 'equal' means having equal area; but paradoxically, 'area' is undefined. In a sense the process is a sleight of hand.
  Notice also that the point G is not defined, except by the diagram; and also, that the proof as stated only works if G *is* as shown, that is, if CD crosses BE between the parallels. You can adjust the proof for the case where it does not. Is this a 'mistake'? Such minor oversights do occur, and

have worried commentators, but they are not—perhaps you will agree—very important unless you suppose the *Elements* to be perfect.

As you see—and this is Euclid's most famous and influential characteristic—every statement is backed up by a reference back to a previous result, or in this case to a 'common notion'. (If you like, like Thomas Hobbes in our opening quote, you can read 'backwards' from an apparently unbelievable statement until you arrive at something you can believe.) For example, Common Notion 1, clearly important in the proof, states that 'if equals are added to equals, the results will be equal'. This deductive structure (statement of proposition → proof, justified by reference back → Q.E.D.) is still in use today. It has been usual to treat Greek mathematics as part of our Western tradition, and so as similar to our own, and indeed the style of exposition and the material are familiar; but the actual aim of the proposition is unusual from our viewpoint, as it would have been for the Egyptians; it tells us only that two parallelograms have equal areas, instead of saying what those areas are. Does this mean that Euclid did not think of areas as numbers? There is evidence that he did not. To do so, you have to have a fixed unit of measure (one square foot or one square metre, or whatever), and this the *Elements* is unwilling to do.[2]

In fact, the *Elements* is not a user-friendly text, to put it mildly; there are no examples, no explanations of aims and objectives, or of the uses to which any result can be put. All this the reader has to invent. And, because Euclid did not explain his aim, we have to use other sources to try and find out what he was trying to do and why. Here are two other examples of the thinking in the *Elements* to illustrate what I mean:

1. Today, we usually think of 'Pythagoras's theorem' as a statement about numbers; again, this theorem was known in much the same sense by the Egyptians and Babylonians. In these terms, the theorem says that if a right-angled triangle has short sides of lengths a and b, and long side (hypotenuse) c, then $a^2 + b^2 = c^2$. But the theorem as stated in Euclid I.47 (see Fauvel and Gray 3.B.5, p. 115–6) is about actual *squares*: if you draw the squares on the three sides, then you can cut up the squares on the two short sides and piece them together to make up the square on the hypotenuse.

2. Again, when we want to describe the area of a circle, we—like the Egyptians—describe it by the formula πr^2, where r is the radius and π is a number which we can approximate in various ways, depending on how accurate we want to be ($\frac{22}{7}$, for example, or 3.14159.; the Egyptian approximation was worse, but not too bad.) There is no such formula in Euclid; the only statement about areas of circles is (XII.2—see Fauvel and Gray 3.E.3, p. 136–7): 'Circles are to one another as the squares on their diameters'.

This rather enigmatic sentence means in plain—slightly simplified—terms that the area of a circle C is in a fixed ratio to the area of the square S on its diameter. (Which is four times the square on the radius.) We could recover the usual statement if we called the ratio '$\pi/4$', but that would not be in agreement with the whole use of the term 'ratio' in Euclid—a 'ratio', whatever it is, is not the same as a number, as we shall explain later.

Confused? It is not surprising; and the *Elements*, despite their prestige, have probably generated more confusion than any other Greek mathematical work, among readers in the Islamic, medieval, and modern European worlds. How the ancient Greeks understood them is rather unclear to us.

These examples should give some sense of the difference between Greek mathematics and our own; and this is a useful starting point, since our problems in understanding it help us to begin

---

2. However, for comparison, we have Socrates' use of numbers for areas in the *Meno*—'the four-foot square' and so on.

with a historicist sense of difference. The classical works were using methods which are alien to us to achieve ends, which in the main, we no longer have.[3]

With this in mind, let us consider what we know of Greek methods, and how far conjecture, hearsay, and the like can help us.

Exercise 4. *What steps would be needed to deduce a formula for the area of a parallelogram from proposition I.35 above?*

4 The problem of material

Acting as many inventors are known to have done in the case of their discoveries, they have perhaps feared that their method being so very easy and simple, would if made public, diminish, not increase public esteem. Instead they have chosen to propound, as being the fruits of their skill, a number of sterile truths, deductively demonstrated with great show of logical subtlety, with a view to winning an amazing admiration, thus dwelling indeed on the results obtained by way of their method, but without disclosing the method itself—a disclosure which would have completely undermined that amazement. (Descartes *Rules for the Direction of the Mind* (1968a), Rule 4)

Our Greek sources, with much more material at their disposal than we have, had little difficulty telling stories of how mathematics developed, however inventive we might find them. By the seventeenth century when Descartes wrote, it was a different matter. New translations of the ancient Greek writers clarified both their importance and the excessive difficulty of their work, in comparison with the algebraic methods which could be seen as an alternative. Never one to defer to any older authority, Descartes imagined the Greeks using easy methods of discovery, similar to his own prescriptions for solving problems, and then making the results look hard so as to mystify posterity. Drawing on hints in Greek authors, particularly Pappus, he opposed 'analysis' (a method of discovery, in which you consider the properties of the thing you want to find and deduce what it must be like) to the classical Euclidean 'synthesis' (a method of disclosure, in which you state what the result must be, without explanation, and prove you are right). His claim was that the Greeks had used analysis, as he would do in his own work, but destroyed the works in which it was used. By such arguments, he not only propagandized for his own work, but started the fertile genealogy of speculation about why the Greeks had chosen to do mathematics in such a strange way. The tradition of 'scholarly' history, which began in the late nineteenth and early twentieth century with Paul Tannéry (1887) and Sir Thomas Heath (1921), had more respect for the Greeks and less of an axe to grind, but little more material to go on.[4] In his pioneering work, Tannéry described the problem of sources:

These writings [Euclid, Archimedes, etc.] cannot teach us the history of science; they leave us ignorant about its origin, of its first developments, just as, since important works have been lost, they give us no way of gauging, without having recourse to conjectures, the direction of research in higher geometry and the level of understanding which was reached.

The history of Greek geometry must therefore appeal to other sources; it must subject those sources to a methodical critique such as one applies in other similar cases. This is the aim which I have set myself. (Tannéry 1887)

---

3. It is enough to recall the classical Greek problem of 'squaring the circle', that is, constructing a square whose area is equal to that of a given circle. A great deal of work was expended on this in ancient times (see Knorr 1986, for example). Grossly misunderstood by some medieval writers (see chapter 5), it is now—for reasons which are as much to do with our different perspective as with our greater knowledge—not a concern for mathematics.

4. The only exception was Heath's discovery of an unknown manuscript of Archimedes' 'The Method', published for the first time in 1906. This work, which will be discussed in the next chapter, in some ways confirms Descartes's suspicions.

Although he has been criticized since for the way in which he used his often unreliable sources, Tannéry was on the whole careful and many of his successors have been less so. Writing before the Babylonian and Egyptian contributions had come to light, he saw the Greeks as the earliest known mathematicians, and as such particularly important. But even today, the importance, and the difficulties of source material, remain much the same.

The works of Euclid and his successors have been 'classics' for three major civilizations—their own, the medieval Islamic, and the modern European. They have the advantage of being readable (in a language which is related to a modern European language), often accessible in libraries and now, at least in extracts, on the Internet, and clearly central to any history of mathematics; whatever reservations one may have about Eurocentrism, and the role of the Greeks as overvalued 'Europeans', their methods and discoveries have had a decisive influence. However, as Tannéry pointed out, their study needs supplementing to give us a 'history', and the construction of such a history raises serious questions of method and approach.

In the first place, Greek mathematics is supposed to have started in the early sixth century BCE At this time, the Greek world consisted of what is now Greece, western Turkey, and a few colonies, particularly Sicily. However, there are no surviving mathematical texts earlier than the discussions in Plato's dialogues (say about 380 BCE) and no substantial writings before those of Euclid, about 70 years later. By Euclid's time, the Greek world had changed completely as a result of Alexander the Great's conquests, and included what we now call the Middle East; and the centre of learning had shifted from Athens to Alexandria in Egypt. While there are a number of substantial mathematical texts from Euclid onwards (see table given below), there are still major gaps in the record before it comes to an end, about 600 CE Some works said to have been written are lost, and many well-known mathematicians have no surviving works, known or not. Here is a list, with dates, of the main 'important' Greek mathematicians and one rather arbitrarily chosen memorable work for each, where known. Euclid, Archimedes, Ptolemy, and others wrote a number of works, some lost, some not; the table is simply there as a rough guide. An entry in italics means that the work(s) have not survived.

Mathematician	Works	Rough Date
Thales	*None known*	480 BCE
Pythagoras	*None known*	500 BCE
Democritus	*None known*	450 BCE
Hippocrates	*Elements*	420 BCE
Plato	Dialogues	380 BCE
Eudoxus	*Phaenomena*	350 BCE
Euclid	Elements	310 BCE
Archimedes	Sphere and cylinder	240 BCE
Apollonius	Conics	210 BCE
Hipparchus	*Star catalogue*	150 BCE
Heron	Metrics	50 CE
Ptolemy	Syntaxis ('almagest')	150 CE
Pappus	Collection	200 CE

Second, as Tannéry recognized, our sources for most of what happened are of two kinds. The first, of course, is the works themselves; the second is descriptive material about the works of mathematicians scattered among other writers. While Plato and Aristotle date from the fourth-century BCE (and are in different ways very important sources), the rest are over 500 years later, and are commentators with particular philosophical allegiances. They have often been treated, as they should be, with suspicion, but for much of the early history they are all we have. A particularly important source is Proclus, who, around 450 CE, wrote a commentary on Euclid's book I, prefaced by a general history of mathematics up to Euclid's time and an exposition of what, in general, the aim of the *Elements* was (the most recent edition is Proclus 1970). For a number of statements, Proclus is the only source; but besides his late date, his allegiance to Plato's philosophy and to later embroideries of it make him a source to be treated with caution.

Third, all the works which exist are not—like the works of the Babylonians or Egyptians—'contemporary documents' (manuscripts of Archimedes from his time or near it, say). As in ancient Egypt, papyrus was used for writing, and it is extremely perishable. Manuscripts were copied and recopied over centuries, and the earliest surviving copy of Euclid's *Elements* is from the ninth-century CE, over 1000 years after the work was written. (Details on the earliest manuscripts are given in Fowler 1999, p. 218ff) Worse, mathematicians see no need to respect a text they are copying if they can explain it or express its meaning more simply, and for some works the earliest manuscripts have certainly been edited in this way. This problem is not as serious as the others, but it exists.

More interestingly, we know very little about what kind of people in the ancient Greek world did mathematics, and why. Archimedes wrote a few interesting letters to patrons and colleagues, but he gives no explanation of why mathematicians are engaged in their solitary activity; and, as Netz points out (1999, p. 284–5), the community he refers to is extremely small. And this, one would think, was at the high point of Greek mathematics. Most of the time, the mathematicians seem not to have left records describing their practice, their students, their aims, and thoughts about their work. In this respect, our situation is worse than for ancient Egypt, say, where mathematics, so far as we know, was done by a socially well-defined group of people for particular reasons. A useful start at 'profiling' the world of Greek mathematicians is done by Netz (1999, ch. 7), but it is necessarily speculative.[5]

In general, if we want a good basis for writing the history of any activity (witchcraft in the seventeenth century, say, or the Vietnam War), we would like to have documents which describe what the participants were doing; how contemporary observers saw it; and the general social setting in which the events took place. What we have for Greek mathematics is nothing like this. Rather, it consists of an impressive collection of major texts—less than 20 in number—with some more minor ones, and some late and unreliable stories. The material spans a period of 1000 years or more. If we want to think about it in a historical way, how do we do so? More contentiously, one could ask: why do we want to?

Without even attempting the second question, we should make a particular point about 'reconstruction', which is one of the few methods available for dealing with the first. Whenever the historical record is weak, one wants to fill in the gaps, supplementing slender materials with imagination so as to form an idea of what Boudicca's chariots were like, what was in the Holy Grail, or the identity of Jack the Ripper. In mathematics, we attempt a reconstruction of a particular form: briefly, to deduce an unknown proof from the fact that we are told that one existed. This can

---

5. In particular, Netz's conclusion that the number of Greek mathematicians over the whole recorded history was at most 1000 gives the classic works, and in particular the *Elements*, a tiny readership; unless one distinguishes a higher-level 'creative' mathematician from someone educated well-enough to read the basic works.

be illuminating; but, unless a document which confirms the reconstruction turns up, it must necessarily be provisional, leaving the field open for other competitors.

5 The Greek miracle

The king moreover (so they say) divided the country among all the Egyptians by giving each an equal square parcel of land... And any man who was robbed by the river of a part of his land would come to Sesostris and declare what had befallen him; then the king would send men to look into it and measure the space by which the land was diminished, so that thereafter it should pay in proportion to the tax originally imposed. From this, to my thinking, the Greeks learned the art of geometry... (Herodotus, cited in Fauvel and Gray 1.D.4 (a), p. 21)

What the Greeks discovered — the greatest discovery made by man — is the power of reason. It was the Greeks of the classical period, which was at its height during the years from 600 to 300 B.C., who recognized that man has an intellect, a mind which, aided occasionally by observation or experimentation, can discover truths. (Kline (1980), p. 9.)

The problem of why Greek mathematicians wrote their works as they did tends to be framed as a problem of origins: there must have been some event, some discovery perhaps, which led to what is called the 'Greek revolution' or 'Greek miracle' in mathematics. The quotation from Morris Kline is a typical example. There is a case for using revolutionary language, on the analogy of the sixteenth century 'scientific revolution'; and one might try to adopt a version of Kuhn's theory of scientific revolutions. An obvious difficulty is our ignorance of when or how it might have taken place. The Greek authors are not responsible for the idea that they were revolutionary, but that should not stop us from adopting it if it seems reasonable. The 'paradigm' (to use Kuhn's term) of doing mathematics for the known Greek authors is a different one from any which went before. The evidence that we have, suggests that their predecessors and indeed contemporaries in Egypt and Mesopotamia were better at calculation, but quite unconcerned with formal proof.[6] However, if we accept that there was a revolution, the study of its origin has been a very problematic one, and it does not seem well adapted to Kuhn's framework in which the idea of a 'scientific community' who pursue normal science is an essential component. We have a good record of what could be called a scientific community in the ancient Near East, even if it may not have been much like our own, but our knowledge of such a community at the dawn of Greek mathematics is almost non-existent.

Before the 1920s, and the publication of the Egyptian and Babylonian texts, it was universally believed that the Greek revolution actually founded mathematics as a science—and hence, the first truly scientific discourse. The belief lingers on among those who, either because they wish to define *scientific* mathematics in such a narrow way that it excludes the sophisticated procedures of the Babylonians, or because they have an ideological investment in a Western origin, cannot accept the evidence that Greek mathematicians, however revolutionary, were *transforming* an earlier Middle Eastern practice. To take one example, an important French 'structuralist' tradition adopts the hard science/prescience distinction made by Gaston Bachelard in the 1930s (the so-called 'epistemological break' which founds a science, see Bachelard 2003). In the 1960s, Louis Althusser

---

6. One should always here remember how very fragmentary our sources are. Both Herodotus (see the quote at the opening of this section) and Aristotle described mathematics as an Egyptian import (see Fauvel and Gray extracts 1.D.4), while Plato in his *Laws* (extract 2.E.6), probably for propagandist reasons, claimed that the teaching of mathematics was better in Egypt than in Athens. However, the state of mathematics in Egypt *at that time* (fifth–fourth century BCE) is unknown to us.

in *For Marx* gave a naive version ('Thales opened up the "continent" of mathematics for scientific knowledge', 1996, p. 14); it is fair to say that the history of science was not his primary concern. And the tradition is still present in a more sophisticated form in the 'archaeology' of Michel Foucault: having described the stages of formation of a science as 'discursive practice' from the lowest ('positivity') to the highest ('formalization'), he continues:

> mathematics [is] the only discursive practice to have crossed at one and the same time the thresholds of positivity, epistemologization, scientificity and formalization ... hence the fact that the beginning of mathematics is questioned not so much as a historical event as a principle of history: a geometry emerging suddenly, once and for all, from the trivial practices of land-measuring. (Foucault (2002), 188–9.)

As Paul Ernest points out (1998, p. 230) '[Foucault] has fallen victim to the popular myth about the origins of mathematics in Greece'—and rather late in the day, it might be added. It was perhaps one of Kuhn's major achievements to question the simplistic division implied here between what is scientific and what is not, to make clear that there is not a single scientific practice founded once for all, and to exhibit science as subject to its own 'breaks', which cannot necessarily be described as straightforward advances or retreats. The idea of a 'revolution' as founding Greek culture is by no means confined to mathematics. It has long been a central dogma in the history of Western culture that the Greeks—around the fifth-century BCE—were responsible for the invention of the scientific method, philosophy, rational argument, democracy, and much more. That view has been recently challenged, at least in part, in works such as Martin Bernal's *Black Athena* (1987, 1991), which argue that there was a close relation in culture and even in language between Greece and the Middle East, particularly Egypt.[7] However, even if the changes which may have brought abstract rational thought into being are not specifically Greek, they are significant; G. E. R. Lloyd has done the best recent work in trying to describe them (see in particular 1979), and to distinguish a hypothetical 'before' and 'after'. Attempts at non-Kuhnian (sociological/'external') explanations have included:

1. The introduction of alphabetic writing—adapted from the Phoenicians around the eighth century BCE, but probably only brought into general circulation some time later in the transition from an 'oral' to a 'written' culture (Goody 1986).
2. The invention of coined money, about the sixth century BCE. According to the Marxist thesis of Alfred Sohn-Rethel (1978), this led to the 'abstraction' of things from values, and hence to abstract thought in general—for a recent version see Seaford (2004).
3. The institution of the (more or less democratic) city–state with its tradition of public political argument. This is the central thrust of Lloyd's books on the subject, although he is careful to avoid single-thesis explanations.

All of these theses may have some force, and it may, as in the case of the revolution of the sixteenth century, be necessary to think in terms of a combination of factors. Any means for deciding the question is almost completely lost in the fog of historical conjecture, but this has not stopped this particularly fascinating historical problem from being the subject of wild speculation

---

7. For the controversy surrounding, Bernal's theories, which is far from resolved, see for example, the website www.blackathena.com. For several illuminating discussions of the Greeks, their innovations, and their possible indebtedness to others, see the articles by Bernal et al. 1992 collected in *Isis*, vol. 83, pp. 554–607.

or dogmatic assertion. To be more cautious, to insist as Lloyd does on how little we know, is to risk being found unexciting.

To some extent the status of the revolution in mathematics is linked to the others, particularly philosophy (given Plato's views on mathematics as a model for reasoning). From a different viewpoint, though, mathematics could gain by being considered separately. In the first place, the innovations in mathematics were of a very specific nature and have led to quite particular accounts of the revolution; and in the second, to see mathematics as the product of a *general* revolutionary process makes it too homogeneous.

6 Two revolutions?

[The Greeks] were certainly not the first to develop a complex mathematics—only the first to use, and then also to give a formal analysis of, a concept of rigorous mathematical demonstration. (Lloyd 1979, p. 232)

Greek mathematical deduction was shaped by two tools: the lettered diagram and the mathematical language. (Netz, 1999, p. 89)

So far, I have concentrated on the most obvious novelty in Greek mathematics—the use of an ordered sequence of deductions, which in Euclid appears to be from what is considered self-evident. This interpretation of Euclid is strengthened by Aristotle's prescriptions (for mathematics and deductive science in general), for which see Fauvel and Gray 2.H.1, p. 93–4. It is this revolution which is of interest to Lloyd, partly for the reason that his study concerns the wider field of rational discourse in Greek society, whether medical, philosophical, or mathematical; and his description of what was distinctive to the Greeks in the first quote is a good one. In the second quote Netz, looking at the collection of documents we have as evidence of how a community worked throughout the Greek period, concurs in seeing what was distinctive about Greek mathematics as a unified practice based on a 'toolkit' of argument about lettered diagrams. However, any search for the origin of this particular revolution, apart from general sociological speculation of the type outlined above, takes us back to the most unreliable parts of the commentators' story—to Thales (early sixth century BCE) and Pythagoras (around 500 BCE). Proclus claims, for example, that 'old Thales' proved six results, one being that the base angles of an isosceles triangle are equal (1970, p. 250/195). In other words, if the triangle ABC has sides AB, AC equal then so are the angles B, C.

This would certainly be 'revolutionary' if we had any reliable evidence (which, as far as Thales is concerned, Proclus is not). One should not discount the value of myth and late propaganda as historical source material. But it is, in the main, source material for 'what later Greeks thought about their origins' rather than for 'what Thales did'. The idea that such a fact, which Thales' predecessors could well have considered obvious, needed proving, and the attempt, whatever it might be, to construct a proof, would have marked Thales' geometry off from any earlier ideas of what geometry (traditionally, the measurement of land . . .) was about.

Speculation about what Thales did, which was once an acceptable part of Greek mathematical history, is now generally discounted as serious history (except as a metaphor, perhaps, for example, by Michel Serres, 1995, p. 105). The same is true, if anything more so, for Pythagoras, who Proclus claimed founded 'pure' mathematics ('transformed mathematical philosophy into a scheme of liberal education', 65/53). The scholarship of Walter Burkert in particular (see 1972), has established fairly conclusively that no mathematical discoveries can be soundly attributed to

Pythagoras at all.[8] That he did exist is a reasonable assumption, since he is referred to by Herodotus and Heraclitus not long after his time; but not much more can be said. With reasonable evidence that a revolution of some sort took place, we have no serious information on the state of affairs before or on what happened; only some idea of the situation some time after. Netz makes an interesting case for dating this 'first revolution' to the (rather late) time of Hippocrates, about 440 BCE:

> According to our evidence, mathematics appears suddenly, in full force. This is also what one would expect on *a priori* grounds. I therefore think mathematics, as a recognizable scientific activity, started somewhere after the middle of the fifth century B.C. (Netz 1999, p. 275)

His arguments are persuasive, but in the nature of things can hardly be conclusive.

However, as we have seen, there are more unusual things to be accounted for in Euclid, and in much of what followed, than the use of argument and diagram; and the introduction of such special features, which we could provisionally think of as the second revolution, is the one which has particularly attracted the imaginative historians. The evidence is still a mixture of gossip and inventive reconstruction, but there is more to it. Among the elements which need to be explained are:

1. Euclid's avoidance of numbers in describing lengths, areas, and so on.
2. His use—notably in book II—of a geometric language for manipulating areas where his Egyptian or Babylonian predecessors would have used something more like what we call algebra—see the discussion of proposition II.1 in the Introduction.
3. His theory of 'ratios' which are intended to replace numbers in contexts where these may not be fractions (area of circle to square on diameter, for example, see below).

Clearly, the most economical hypotheses to explain all this would be: first, that Euclid's practices were the result of a 'second mathematical revolution'; and second, that this second revolution was—on Kuhn's model—the result of serious problems which arose in the original practice of rigorous mathematics which made it impossible to proceed. Both of these theses, originally put forward around 1910, are still widely believed in a revised form; we now need to examine the arguments for and against them.

Exercise 5. *How would you prove Thales' statement on isosceles triangles, and what assumptions would you need?*

7 Drowning in the sea of Non-identity

"But betray me," said Neary, "and you go the way of Hippasos."

"The Akousmatic, I presume," said Wylie. "His retribution slips my mind."

"Drowned in a puddle," said Neary, "for having divulged the incommensurability of side and diagonal."

"So perish all babblers," said Wylie.

"And the construction of the regular dodeca—hic—dodecahedron," said Neary." (Beckett 1963, p. 36)

---

8. Nevertheless, you can of course still find long discussions of what Pythagoras did, on the Internet and even in 'general' histories.

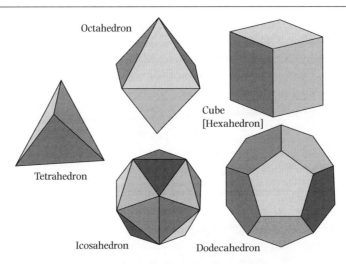

Fig. 3 The five regular ('Platonic') solids.

By 'the incommensurability of side and diagonal', Beckett means the fact, mentioned in our discussion of the *Meno*, that the ratio of a square's diagonal to its side, $\sqrt{2}$, is not a fraction. But his pub classicists most probably came across their story not in its Greek original source, but in its popularization in the twentieth-century history of mathematics, the 'secret' or 'scandal' of the irrationals. This story, in some more respectable form, is still widely believed. The basic 'fact' is that Pythagoras founded a sect of initiates whose secret knowledge was at least partly mathematical. He is said to have taught that 'all is number', where by number is meant whole number (1, 2, 3, . . .); and at the same time he or his sect attached religious/magical importance to the regular solids (see Fig. 3).

However, you cannot construct the regular solids, at any rate those which involve pentagons, without bringing in irrational ratios—see Appendix B. Both this, and the problem of the side and the diagonal, suggest a difficulty or, to put it more strongly, a 'scandal' for Pythagoras's supposed programme; because if 'all is number', then the ratio of the side to the diagonal should be the ratio of two numbers.

There are stories in various late writers that this was kept secret by the Pythagoreans (supposedly because it was a problem, though this is not explicitly stated), and that the secrecy was broken. Iamblichus, a commentator of the late third century CE (seven centuries after the events), refers to Hippasos of Metapontum as a member of the Pythagoreans who was expelled and drowned at sea (or some similar fate) for revealing a secret—in one version, the construction of one of the solids, and in another, the nature of the rational and irrational. How far you accept this story, as Beckett's characters did, depends on your estimate of Iamblichus as a source, and he does not go out of his way to inspire confidence. The next step for twentieth-century historians was to deduce that the secret or scandalous nature of the irrationals for the Pythagoreans extended to the Greek mathematical community in general; and that this accounted for their avoidance of measurement.

The definitive version of this story was due to Hasse and Scholz in the 1920s. Speaking of a 'crisis of foundations' for Greek mathematics, they used it to explain Euclid's use of proportions:

Given that the Greeks were born geometers, as they are usually held to be, it must be concluded with certainty that after such a foundational crisis they needed to construct a purely geometric mathematics. In such a mathematics we should naturally also come across a theory of proportions, in which no arithmetical parts remain. (Hasse and Scholz 1928, p. 13)

It is not accidental that they were writing at the time of Russell's paradox, when (modern) pure mathematics was experiencing just such a crisis. Extrapolating backwards, they read the anxiety of the ancient Greeks as a version of fin-de-siècle angst about mathematical certainty. Although the Hasse-Scholz thesis was criticized in the years which followed its appearance (notably by Freudenthal 1966) and never became the only accepted view, it has survived well, partly because it does explain the problems referred to above, and because it is easy to adapt and revise. The most recent sustained attack is by David Fowler (1999). Fowler rehearses many arguments which are now standard: that Plato and Aristotle, who are the nearest to contemporary sources, refer to the irrational without in any way suggesting that it was a problem; that the subject is not even mentioned by Proclus, who is supposed to be summarizing an earlier history; that the time of the supposed 'crisis' is also a time when many major mathematical discoveries were made, apparently without trouble; and that Iamblichus is notoriously inconsistent and unreliable. (In passing, this basic difference of opinion drives home the point about the difficulty of arriving at conclusions about the period.) Fowler contrasts Aristotle's hard-headed assessment (in one of many allusions to the problem):

A geometer, for instance, would wonder at nothing so much as that the diagonal should prove to be commensurable. (*Metaphysics* 983a, in Fauvel and Gray 2.H6)

with Pappus' (third century CE) apparent confusion:

the soul ... wanders hither and thither on the sea of non-identity ... immersed in the storm of the coming-to-be and the passing-away, where there is no standard of measurement. (*Commentary on Book X of Euclid's Elements*, I.2)

and concludes that 'the discovery [of the irrational] was no more than an incidental event in the early development of mathematics' (1999, p. 362.) While his thesis certainly advances some useful arguments, it depends strongly on a particular reconstruction of how Greek mathematics worked in the pre-Euclid period; while it fails to deal with a feeling that what is unusual in the Euclidean approach (as detailed above) must have come from *somewhere*. And despite all arguments against it, the discovery of incommensurability is still often thought to be *the* revolution in Greek mathematics, as opposed to the simple introduction of deductive method; two of the contributors to Gillies' (1992) consider it as such.

Exercise 6. *What is 'regular' about the five solids in Fig. 3?*

Exercise 7. *Why are there no others?*

8 On modernization and reconstruction

Eudoxus's general theory of proportions, which, from our vantage, amounts to a theory of real numbers, resolved the anomaly that the discovery of several incommensurables had introduced into Greek mathematics. (Calinger 1999, p. 110)

For example, the significant content of the proportion theory of *Elements* V is almost universally acknowledged to be due to [Eudoxus], in some form or another, though the explicit evidence for this is very tenuous indeed ... [A]lmost everybody says that Eudoxus' aim and achievement in Book V was to handle incommensurable ratios. I do not know how they can be so sure; for example, there is nothing to suggest this in the *Elements*. (Fowler, contribution to Historia Mathematica mailing list, 1999)

Today's history, while usually avoiding picturesque stories about drowning, still has to cope with the problem of presentism (referred to in the introduction) and that of reconstruction (see above, Section 4). Indeed, if we use the word 'problem', it is because both have their uses, and the historian's difficulty is to decide when they are the right tools to reach for in the box. The case of Euclid II.1 was already cited in the introduction; those who consider it to be equivalent to the distributive law for multiplication are, one might argue, guilty of presentism. If Euclid had wanted to state the distributive law, he was intelligent enough to do so. In this case, the problem is complicated by evidence that informal Greek mathematics, continuing in the Egyptian tradition, *did* use precisely such a translation—see, for example, Heron (Chapter 3). This means that we have two dividing lines to respect: between Euclid and his informal contemporaries on the one hand, and between Euclid and ourselves, on the other. All this is part of a proper respect for differing historical traditions. It makes the historian's work harder, but no one said it had to be easy.

A quite different example is provided by Euclid's theory of 'ratios' in book V. This complex treatment of a ratio can be successfully modernized so that Euclid's ratio (e.g. of the circumference of a circle to its diameter) translates as the modern concept of a 'real number' (e.g. π—we now think of the ratio as a number, and we take it for granted that it can be written as a decimal to as many places as we like). That this is possible has been taken to mean that the Greeks, specifically Eudoxus of Cnidus 'invented' the real number system over 2000 years before its development by Dedekind in the nineteenth century. I have given an example of confident statement (in a mainstream textbook) and scholarly doubt (in a listserv contribution) in the quotes at the beginning of this section. Apart from doubts about Eudoxus' role, the idea that he was concerned with what we call real numbers in any sense is unhistorical, and is now out of favour, although the reasons for doubting it are complex. Attacks on such ideas appear in the works of Knorr, Fowler, etc.[9] However, it is interesting that it is in the nature of mathematics that such translation can be done; it is not possible to make a similar translation of Aristotelian physics.

Question

See if you agree with this statement, and if you do, try to explain what features of mathematics favour the translation.

As for 'reconstruction', historians feel the need for it most particularly when a source refers to some mathematician's work without indicating how the work was proved. One then proceeds to present a plausible version of how the proof must have gone, with the hope of throwing some light on the state of mathematics at the time, or of supporting a thesis about it. Here are two examples:

1. In Archimedes' *The Method* he states that the [volume of a] cone is the third part of the [volume of a] cylinder having the same base and height. (This again was known to the Egyptians, incidentally.) He attributes the discovery to Democritus 'though he did not prove it', and the proof to Eudoxus. Archimedes may conceivably have had access to the works of both Democritus and Eudoxus, supposing these to have been written down; in any case, the

---

9. One could also contrast the extremely complicated definition of 'equal ratios' in Euclid V with Dedekind's 'disappointingly' simple definition of a real number (Chapter 9).

challenge is to reconstruct what the two did. The main fact known about Democritus is that he was an 'atomist', that is, regarded the universe as made up of atoms; and this has led to a great deal of speculation on his possible use of the infinitely small. For Eudoxus there is a more accepted reconstruction, based on the supposition that his work was much like Euclid's—a supposition which in turn owes something to Archimedes's passing remark.

2. In Plato's *Theaetetus*, Theaetetus claims that his teacher Theodorus was drawing diagrams to show that a square of 3 square feet was not commensurable in respect of side with a square of 1 square foot. This relates to the problem in the *Meno* for the square of 8 square feet; the idea is that the square root of three is not a fraction. Theaetetus claims that Theodorus continued with 5, 7, up to 17, and then for some reason could go no further. This has given rise to a great number of reconstructions of the (geometric) method which Theodorus could have used which would have worked for 17 but not for 19. Euclid's proof of the same result[10] is of no help; it is in his extremely difficult book X, and would work for any number you like.

Again, it can be seen that mathematics is peculiarly susceptible to this kind of 'history'. A brilliant example is Fowler's (1999), which gives a plausible version of how Greek mathematicians thought of ratios in the fourth century BCE, with a great deal of mathematics and scholarly apparatus to back it up; aside from giving a great deal of information on what we know *for certain* (manuscripts, papyrus fragments, etc.) about mathematics in Plato's time, Fowler constructs a detailed model of how it could have worked. Reviewers have been respectful (see, for example, Berggren (2003)), but have usually expressed natural reservations.

9 On ratios

It seems necessary to insert something here to clear up what, in classical Greek terms, a 'ratio' was—even if the experts are not altogether agreed on it. Euclid's book V starts by saying that a ratio is something which two quantities may have (simplifying, if they are of the same kind, say both lengths, or both times, ...). He then sets out a complicated criterion for two ratios to be equal. This was much disliked, when it was not simply misunderstood, by his Islamic and medieval successors; it is popularly believed that Isaac Barrow in the seventeenth century was the first mathematician (after the Greeks?) to understand the theory.[11]

I have already given some simple examples. The ratio of the diagonal of a square to its side is one; the golden ratio (see Appendix B) is another. A third is the ratio of circumference to diameter. And so on. Here, for the record, is the definition—it is a good example of what is really difficult in Euclid, and we shall leave it to you to spend some time thinking about its interpretation. (Or look at the commentary in editions of Euclid, websites, etc.)

Definition V.5. Magnitudes are said to *be in the same ratio*, the first to the second and the third to the fourth , when, if any equimultiples whatever be taken of the first and third, and any equimultiples whatever of the second and fourth, the former equimultiples alike exceed, are alike equal to, or alike fall short of, the latter equimultiples respectively taken in corresponding order. (Fauvel and Gray 3.C3, pp. 123–4)

---

10. That is, if x is a whole number but is not a square, then its square root is not a fraction.
11. For the medieval theory, see in particular Murdoch (1963). In particular, Murdoch claims that the eleventh-century Islamic mathematician al-Jayyānī *did* have a correct understanding of the Euclidean theory (the only writer between the Greeks and Isaac Barrow?).

This complex definition underlies the statement (see Section 3) that circles are in the same ratio as the squares on their diameters, for example. You can do much more with ratios, and you need to. For example, if you increase the length of a rectangle by one ratio and its height by a second, then you increase the area by a third, and so you have a relation of ratios which is similar to multiplication. (But again, maybe we should avoid confusing them.) In the section 'Plato and the *Meno*', we saw that if side and height ratios were both what we would call $\sqrt{2}$, the area would increase by 2. For a full discussion, look at at least some of book V and (if you can face it) book X, which introduces a whole classification of different kinds of irrationals, up to fourth roots or 'medials'. The exact reason for much of this theory is still debated, but the underlying theory is more or less understood.

A different take on ratios comes from astronomy, since it shows how they tie in with numbers. Suppose that Y is the length of a year (from spring equinox to spring equinox, say), and D is the length of a day. We need to know the ratio of Y to D (in our terms, the number of days in the year) to have an efficient calendar, which all ancient—and modern—people needed. It complicates matters that days are not all of the same length, but let us neglect that, as many of the ancients did. Our first observation is that the ratio is greater than 365; we find this by using 365-day years and observing that the date of the equinox gets noticeably earlier over a shortish period. We next introduce one leap year in every four (one extra day), and get a better result, a ratio of $(4 \times 365 + 1) = 1461$ to 4. After a few hundred years, we find this is too big, and so on. If we are astronomers, we continue, as the Babylonians and later Greeks did, to find more or less accurate approximations for the length of the year as fractions—sexagesimal or other kinds—and construct our calendars accordingly.[12] But if we are philosophers, we may well think that there is a *real* length of the year, and that this fiddling with figures is beside the point. This, perhaps, is the meaning of Plato's statement in the *Republic*, which is often ridiculed, that one should study ideal stars and not simply what one sees:

We shall therefore treat astronomy, like geometry, as setting us problems for solution', I said, 'and ignore the visible heavens, if we want to make a genuine study of the subject and use it to put the mind's native wit to a useful purpose. (Plato *Republic*, in F-G 2.E.3, p. 72)

Socrates even draws on the example of the calendar which we have been looking at for his argument:

[The astronomer] will think that the sky and the heavenly bodies have been put together by their maker as well as such things may be; but he will also think it absurd to suppose that there is anything constant or invariable about the relation of day to night, or of day and night to month, or month to year, or, again, of the periods of the other stars to them and to each other. They are all visible and material, and it's absurd to look for exact truth in them. (Plato *Republic*, in F-G 2.E.3, p. 72)

That the lengths of months were variable had long been known by Plato's time; that the lengths of days were too (since the four seasons had different lengths, for example), was perhaps a more recent discovery. And the reaction to these facts in the *Republic* is that they exhibit the failings of material stars and planets, as opposed to the ideal counterparts which were designed by their creator. If we put this together with the statement from the *Philebus* which opens this chapter, we could conclude that the ratio (e.g. of Y to D, above) is what the real mathematician uses in his kind of mathematics; while, in the other kind used by craftsmen, surveyors, and mere star-watchers,

---

12. This is, it has been pointed out to me, an oversimplification of the way ancient calendars were constructed. However, perhaps it can serve as a guide to thought.

recourse is made to fractions of all sorts as approximations. It is hard to go further, and a concept like 'ratio' shows the limits of the hermeneutic programme ('to understand an author better than himself'—Schleiermacher 1978, p. 112). The understanding which underlies the concept has been well hidden, in the nature of the texts we have; the best we can do is to try to understand the use and to guess at the ideas which lay behind it.

Appendix A. From the *Meno*

Soc. Mark now the farther development. I shall only ask him, and not teach him, and he shall share the enquiry with me: and do you watch and see if you find me telling or explaining anything to him, instead of eliciting his opinion. Tell me, boy, is not this a square of four feet which I have drawn?
Boy. Yes.
Soc. And now I add another square equal to the former one?
Boy. Yes.
Soc. And a third, which is equal to either of them?
Boy. Yes.
Soc. Suppose that we fill up the vacant corner?
Boy. Very good.
Soc. Here, then, there are four equal spaces?
Boy. Yes.
Soc. And how many times larger is this space than this other?
Boy. Four times.
Soc. But it ought to have been twice only, as you will remember.
Boy. True.
Soc. And does not this line, reaching from corner to corner, bisect each of these spaces?
Boy. Yes.
Soc. And are there not here four equal lines which contain this space?
Boy. There are.
Soc. Look and see how much this space is.
Boy. I don't understand.
Soc. Has not each interior line cut off half of the four spaces?
Boy. Yes.
Soc. And how many spaces are there in this section?
Boy. Four.
Soc. And how many in this?
Boy. Two.
Soc. And four is how many times two?
Boy. Twice.
Soc. And this space is of how many feet?
Boy. Of eight feet.
Soc. And from what line do you get this figure?
Boy. From this.

Soc. That is, from the line which extends from corner to corner of the figure of four feet?

Boy. Yes.

Soc. And that is the line which the learned call the diagonal. And if this is the proper name, then you, Meno's slave, are prepared to affirm that the double space is the square of the diagonal?

Boy. Certainly, Socrates.

Soc. What do you say of him, Meno? Were not all these answers given out of his own head?

Men. Yes, they were all his own.

Soc. And yet, as we were just now saying, he did not know?

Men. True.

Soc. But still he had in him those notions of his—had he not?

Men. Yes.

Soc. Then he who does not know may still have true notions of that which he does not know?

Men. He has.

Soc. And at present these notions have just been stirred up in him, as in a dream; but if he were frequently asked the same questions, in different forms, he would know as well as any one at last?

Men. I dare say.

Soc. Without any one teaching him he will recover his knowledge for himself, if he is only asked questions?

Men. Yes.

Soc. And this spontaneous recovery of knowledge in him is recollection?

Men. True.

Soc. And this knowledge which he now has must he not either have acquired or always possessed?

Men. Yes.

Soc. But if he always possessed this knowledge he would always have known; or if he has acquired the knowledge he could not have acquired it in this life, unless he has been taught geometry; for he may be made to do the same with all geometry and every other branch of knowledge. Now, has any one ever taught him all this? You must know about him, if, as you say, he was born and bred in your house.

Men. And I am certain that no one ever did teach him.

Soc. And yet he has the knowledge?

Men. The fact, Socrates, is undeniable.

Soc. But if he did not acquire the knowledge in this life, then he must have had and learned it at some other time?

Men. Clearly he must.

Appendix B. On pentagons, golden sections, and irrationals

The folklore has it that the construction of the pentagon (and the five-pointed star, which goes with it) were known to the Pythagoreans. When, or which Pythagoreans, is unclear, but by the time of Euclid the key steps were the following:

1. You need to construct an isosceles triangle ABC such that angles B and C are twice angle A (Fig. 4) This is because these angles will then be, in our terms, 72°, and angle A is 36° which is right for the star.

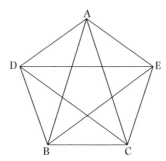

Fig. 4 Constructing a regular pentagon. The angles of the pentagon are 108°, so angles like ADE are 36°. Hence the triangle ABC has angles A = 36°, B = C = 72°. (Greek astronomers used degree measurements of angle, even if Euclid did not, so this description is not too modernized.)

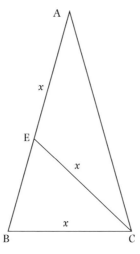

Fig. 5 The 'extreme and mean section' picture. CE bisects angle C, which is twice angle A. So the triangles ABC, AEC, ECB are all isosceles, so the lines marked x are all equal. The statement that EB : EA = EA : AB now follows by using similar triangles (CBE, ABC) and the various identifications we have.

2. To construct the triangle, you need to divide the side AB at E so that the ratio of BE to AE is equal to the ratio of AE to the whole of AB (see Fig. 5) This is called (by Euclid) 'dividing the line in extreme and mean ratio'. More fancifully in the Middle Ages it acquired the name of the 'golden section' or 'golden ratio'. If we want to use algebra: suppose AB has length 1, and AE = x so BE = $1 - x$, then

$$\frac{1-x}{x} = \frac{x}{1}; \qquad x^2 = 1 - x$$

which we would now solve to give $x = \frac{1}{2}(\sqrt{5}-1) = 0.618\ldots$. This is 'obviously' irrational to us (perhaps!), because of the $\sqrt{5}$ in it, but as usual we are not sure whether, or how, the Pythagoreans discovered the fact. Again, Euclid's proof is rather late in book X.[13]

---

13. Even this is simplifying. Euclid proves in book X that what he calls an 'apotome' is irrational, and in book XIII that the golden ratio is an apotome.

However, there is a simpler proof which is generally thought to be a probable reconstruction, as follows. If you have a ratio of honest whole numbers, say 42 to 15, then you can reduce it in steps, as follows:

$$\frac{42}{15} = 2 + \frac{12}{15}$$

$$\frac{15}{12} = 1 + \frac{3}{12}$$

$$\frac{12}{3} = 4$$

ending with an exact whole number. At each stage, you have a remainder less than 1; you invert this at the next stage, and get a whole number and a new remainder. This is Euclid's way of finding the 'common measure' or greatest common divisor for two numbers, in the above case 3, which divides both 42 and 15. The process stops because the denominators of the fractions get smaller, and so must finally reach 1. You cannot do this with the golden ratio, because of the way in which it is defined. The equation

$$\frac{1}{x} = 1 + x$$

which is equivalent to the one above shows that the 'fraction' procedure (invert the remainder, take the whole number part away, keep the remainder, and restart), never stops.

And indeed this can be stated geometrically, as even 'the Pythagoreans' could have done.

Exercise 8. *Explain why the fraction procedure cannot terminate.*

Solutions to exercises

1. I do not think it is quite obvious. Clearly the four sides are equal, since they are diagonals of equal squares. The other point to note is that all the triangles are isosceles (the diagonals bisect each other by symmetry, for example). So all their small angles are just half a right angle—as we would say 45°—and the angles at the corners of the figure, each being composed of two such small angles, are 90°. This makes the figure a square. We *have* used the result on angle sum of a triangle, which does not work in non-Euclidean geometry (see Chapter 8); but then the idea of 'square' becomes dubious anyway.
2. (Trying to use Greek language as far as possible.) 'Diagonal is commensurable with side' means that there is a line L such that the side S is a multiple, say q.L, and the diagonal D is also a multiple, say p.L. Now clearly (using arguments like the *Meno*), the area of the square on S is q^2 times the square on L, and that on D is p^2 times the square on L; and, by Pythagoras's theorem, the square on D is twice the square on S. So p^2 is equal to twice q^2, or, the square of the ratio of p to q equals the ratio of 2 to 1 (i.e. 2). Check that all these implications go both ways.
3. The usual proof (see, for example, Fowler for speculation on which proof the Greeks, in particular Aristotle, used), is the following. Suppose $(p/q)^2 = 2$; we can assume that p and q have no common factor. Then $p^2 = 2q^2$. It follows that p^2 is even, and so p is even.

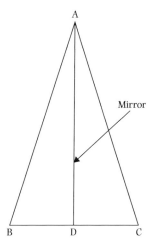

Fig. 6 Idea of proof by reflection of 'Thales' theorem'.

So $p = 2r$, say, and $p^2 = 4r^2$. Rewriting, $4r^2 = 2q^2$, so $q^2 = 2r^2$. Hence q^2 is even, and so q is; but this contradicts the supposition that p and q have no common factor.

4. This is vaguely formulated, and rather hard to make precise. It is clear from the proposition that the area of the parallelogram is equal to the area of the (unique) *rectangle* with the same base and height. To go further, and talk of 'multiplying base by height', you have to say what kind of numbers base and height are. The easy case is where the two 'have a common measure' (see Exercise 2); if one is $p.L$ and the other $q.L$, then it is not hard to show that the rectangle is pq times the square of side L. To go further, you need the general theory of ratios. In theory, this would allow you to show that a rectangle of sides $\sqrt{2}$ and $\sqrt{3}$ was equal to one of base $\sqrt{6}$ and height 1; however, it is difficult, to say the least.[14]

5. There are two questions involved; one is what is an acceptable simple proof, and the other is what proof might have been used by someone at a time when organized deductive geometry did not exist. Euclid's proof (I.5) is quite complicated, involving extra construction lines; it depends on I.4, that if triangles ABC and DEF have AB = DE, AC = DF, and the angles A and D equal, then they are congruent. In theory one could apply this to show that if ABC was isosceles (AB = AC), then it is 'congruent to itself', $\triangle$ABC = $\triangle$ACB. In practice it would be more natural to use properties of reflection; for example, that ABC is unchanged by reflection in a mirror which bisects angle A (Fig. 6). However, any of these ideas (congruence, reflection, . . .) are at the basic starting points of geometry, and one wonders what Thales and his contemporaries, if the story is true, would have considered a proof.

6 & 7. The property usually taken to define the 'regular' solids is that their faces are all regular polygons and all of the same kind (e.g. all squares); and that in a given solid S the same number of polygons meet at each point, or vertex as it is usually called. If we write (p, q) to denote the solid whose faces are p-sided, with q at each vertex, then we have:

---

14. Dedekind claimed in the nineteenth century (see Chapter 9) that he was the first person to have proved that $\sqrt{2} \cdot \sqrt{3} = \sqrt{6}$, and this is commonly accepted.

(3, 3) (tetrahedron), (3, 4) (octahedron), (4, 3) (cube), (3, 5) (icosahedron), and (5, 3) (dodecahedron). Each of p and q must be greater than 2; while the angles of a regular p-sided polygon are $\pi(1 - (2/p))$. (Why? Check examples, and prove the rule.) The sum of the angles at a vertex is therefore $\pi(q - (2q/p))$. This must be less than 2π for the result to be a solid (if it is equal to 2π, the polygons lie flat). From this, with a little algebra, $pq < 2p + 2q$, or $(p-2)(q-2) < 4$. It is now clear that the combinations of (p, q) which have been given are the only ones.

There is much more on this (covering star-polyhedra, 'semi-regular' polyhedra, and higher dimensions) in Coxeter (1963), an excellent introduction to the subject, with information on its later (post-Plato) history.

8. The process cannot stop, because when you invert x, the 'remainder' turns out to be x again. So you cannot (as you would with a ratio of integers, see our example) have a sequence of ratios with smaller denominators; from which it follows that x cannot be a ratio of integers.

Supplementary problem

I have left you one of the most interesting and typical problems in the history of early Greek mathematics as a 'research problem' to think about. This is known as *Hippocrates's quadrature of lunes.* It is a clever area calculation, supposed to have been worked out by Hippocrates of Chios before 400 BCE (and so, in a sense, our earliest serious result in Greek mathematics). You can find it in Fauvel and Gray; and also some account of the transmission line, which makes its status rather unreliable (the description comes from a text about 700 years later). Consider

1. the result;
2. how its status is evaluated by various modern historians—accepted without question, or grudgingly, or set on one side as dubious;
3. how you would assess the importance of the result, and of its authority;
4. the problems about how Hippocrates might have arrived at it, and why.

3 Greeks, practical and theoretical

1 Introduction, and an example

But, unlike Euclid, who attempts to prove musical propositions through mathematical theorems, Nicomachus seeks to show their validity by measurement of the lengths of strings. (Entry 'Nicomachus of Gerasa' in *Dictionary of Scientific Biography*)

The leading Greek geometricians were all master carpenters. Euclid, the author of the *Book of Principles*, was a carpenter and known as such. The same was the case with Apollonius, the author of the book on *Conic Sections*, and Menelaus and others. (Ibn Khaldūn 1958, II, p. 365)

The complaints of the previous chapter about the poverty of documentary evidence for Greek mathematics before Euclid need to be modified for the later period, say from 300 BCE to 600 CE. There is indeed a variety of material, but it is quite heterogeneous, and scattered in time and space. We are often vague about the dates of writers, and we have to guess about their communication; and still it seems that the survival of material is determined mainly by chance. Our first quotation describes the work of Nicomachus of Gerasa (Jerash, in Palestine), whose arithmetical, philosophical, and musical works were treated as important in the Islamic and European Middle Ages, and so survived although modern authorities consider him a desperately poor arithmetician. At least, as the quotation shows, he sometimes had a practical approach which was quite different from what we consider 'typically' Greek. The description of Euclid and Apollonius as carpenters runs contrary to all the information we have from other sources, but is it simply folklore? We have no way of knowing. How many ways were there, indeed, of being a Greek mathematician, and did they change over the 900-year period which we are considering? Did they interact? How did the Romans, generally portrayed as an uncultured master-race with no interest in science, contribute to the way mathematics was done, in the period when they dominated the Greek world (say from 100 BCE to 400 CE, when the 'Roman' part of the empire collapsed and the 'Greek' survived)?

For if often (e.g. with the Babylonians) we can say: 'At this time, mathematics was used in a different way from our own; the following methods were used, with the following ends in view', with the Greeks the situation is more complicated, and less well understood. The heritage which was passed on as important, particularly from the sixteenth century, was that of Euclid, Archimedes, and those who followed their models: axioms, theorems, and proofs. The ideology which went with it, which I have referred to in the last chapter as due (in part) to Plato, is that mathematics should not deal with the real world, or with applied problems. The reality is certainly more complicated; and even accepting that we have lost many records of carpenters, tax-gatherers, architects, and engineers—which we must suppose existed—there is quite a complexity and variety in what remains, even if some of it is what appears to us rather low-level mathematics.

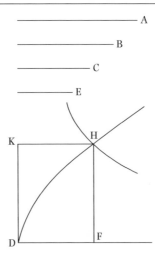

Fig. 1 Menaechmus's construction (from Fauvel and Gray p. 86). A and E are given, and it is required to construct B and C mean proportionals. The parabola is given by the rule (square on DK = rectangle DF by A); the hyperbola by the rule (rectangle KD by DF = rectangle A by E). DF equals C, and DK equals B.

A good example of the interaction of theory and practice is provided, surprisingly, by the classical problem of the duplication of the cube. The problem seems to have surfaced some time before 400 BCE in the form:

Given a cube C, to construct a cube D whose volume shall be twice the volume of C.

Using an argument similar to the one I have given in Chapter 2 (for Plato's 'Meno'), it is easy to see that, if the side of C is a feet, that of D must be $a\sqrt[3]{2}$ feet; and to generalize from doubling to increasing in any proportion. This was done quite early. The earliest solution, by Menaechmus is said to have involved the invention of the curves which we call conic sections. In modern notation, we would take a parabola whose equation is $y = x^2$ and a hyperbola whose equation is $xy = 2a^3$. They meet at $x = a\sqrt[3]{2}, y = x^2$ (see Fig. 1).

So far so good—see Exercise 1 for a check that this solves the problem. We have not said how Menaechmus defined the curves, and the sources do not either. In later times, they were defined in the first place as sections of a cone (e.g. the shadows which a lampshade casts on the walls or the floor); the information which is encoded in the equations I have given had to be proved, and was given its definitive form in Apollonius's difficult (late third century BCE) *Conics*, one of the major works in the 'classical' Greek tradition. However, there are doubts about whether such arguments were available to Menaechmus. The record which we have (which itself is late, see Fauvel and Gray 2.F.4) is related to the problem of two mean proportionals as in Exercise 2, and looks more like the coordinate definition which we would use. Knorr (1986, p. 62) gives a 'reconstruction' of how Menaechmus might have thought of it, with criticisms of earlier reconstructions by Heath and others.

We have a substantial amount of information on the cube-duplication problem, even if some of it is hard to interpret. One particularly interesting source is a supposed document by Eratosthenes (third century BCE), called the 'Platonicus', which only survives in quotation.[1] Eratosthenes gives

---

1. In Eutocius' (sixth century CE) commentary on Archimedes' *Sphere and Cylinder*.

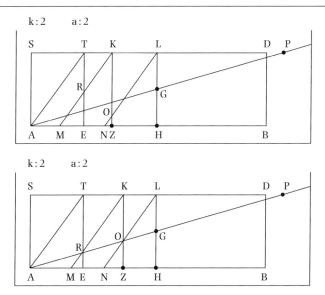

Fig. 2 The 'mesolabe' (Shown for the duplication problem). The triangular plate AET is fixed, while MZK and NHL move. To find two mean proportionals between AS and GL slide the two plates so that the meeting points R and O are in a straight line with A and G. (Try this at home.)

a folklore account of the origin of the problem (doubling the size of the altar at Delos). Though not reliable history, this at least has a practical appearance. He dismisses previous solutions, without describing them in detail, as 'impractical' or 'unwieldy'; and presents his own solution, by means of a machine called a 'mesolabe', or 'mean-taker'—which can construct any number of mean proportionals between A and B, if properly calibrated (Fig. 2).

This 'mechanical' solution in itself goes against the standard image of Greek geometry as purely abstract; and this is still further contradicted by Eratosthenes's claims that the method can be used in all sorts of ways:

We shall be able, furthermore, to convert our liquid and dry measures, the metretes and the medimnus, into a cube, and from the size of this cube to measure the capacity of other vessels in terms of these measures. My method will also be useful for those who wish to increase the size of catapults and ballistas. For, if the throw is to be increased, all the elements of these engines, the thicknesses, lengths, and the sizes of the openings, wheel casings and cables must be increased in proportion. (Eratosthenes quoted by Eutocius, in Fauvel and Gray 2.F.3.)

There seems to be some much stronger link between theory and practice here, though Knorr warns:

Eratosthenes was chiefly a man of letters, and one suspects that his vision of the practicality of a sensitive special-purpose instrument like the 'mesolabe' was rather overstated. Still, the ideology behind its invention seems genuine. (Knorr 1986, p. 212)

Knorr may underestimate the extent to which at least some Greek geometers thought of the problem as applied—the same motive for the construction appears in the undoubtedly military work of Eratosthenes's near contemporary Philo of Byzantium. Here then, without straying outside the boundaries of 'classical' geometry, we find a mechanical solution to a problem which is at least being promoted for its practical uses. Clearly—and this is the point which we shall investigate in this chapter—the nature of Greek mathematics is more complex than one might have thought. We shall look at some examples of the later tradition which do not entirely fit into the Euclidean mould, including Archimedes, Heron of Alexandria, and Ptolemy. All were tremendously influential in the later development of mathematics, and all raise interesting questions about the varieties of Greek mathematics which were (one might guess) competing for influence over the long historical period which concerns us.

At the same time, because of the length of the period, and the variety of the work produced, it would be impractical in a book of this kind to try to cover everything. In particular for the important work of Apollonius, Diophantus, and Pappus, you will have to look elsewhere.

The remarks on sources made in the previous chapter apply on the whole. The major historian who has recently concentrated attention on the late period is Cuomo, to whose works (2000) and (2001) we shall return in due course.

Exercise 1. *Check that Menaechmus's construction does give a line of length $a\sqrt[3]{2}$. How would you generalize it to solve the problem of increasing the volume of the cube by a factor m?*

Exercise 2. *Hippocrates of Chios (fifth century BCE) showed that the general problem (multiplying a cube by m) can be solved if, between two given lines A, B, with the ratio $B : A = m$, one can construct two 'mean proportionals' C, D; that is so that the ratios $A : C$, $C : D$, $D : B$ are equal. Why is this true?*

2 Archimedes

Archimedes is one of the most heroized figures in the history of science; but unlike Galileo and Newton, whose lives are available in minute detail, we know rather little about him. There is a growing literature on him; not so much 'biographical' as an attempt to understand him from his works. True, his life is better documented than that of any other Greek mathematician (with the possible exception of Hypatia), but that is not saying much. The chief sources tend to concentrate on a few memorable events—the 'Eureka story', his role in the siege of Syracuse, and his death at the hands of a Roman soldier. His works have always been seen as uniquely brilliant and difficult, and perhaps his portrait has been constructed to fit them; though unusually, there are letters introducing several of the writings which are 'personal' as not much else is in Greek mathematics. A late portrait of Archimedes as the absent-minded pure researcher is given in Plutarch's *Life of Marcellus*, and for whatever reason it has become influential. In line with a Platonic propagandist viewpoint, Plutarch (while crediting Archimedes with major military inventions), claims that such practical considerations were unimportant to him.

Yet Archimedes possessed so high a spirit, so profound a soul, and such treasures of scientific knowledge, that though these inventions had now obtained him the renown of more than human sagacity, he yet would not deign to leave behind him any commentary or writing on such subjects; but, repudiating as sordid and ignoble the whole trade of engineering, and every sort of art that lends itself to mere use and profit, he placed his whole affection and ambition in those purer speculations where there can be no reference to the vulgar needs of life; studies, the superiority of which to all others is unquestioned, and in which the only doubt can be whether the beauty and grandeur of the subjects examined, or the precision and cogency of the methods and means of proof, most deserve our admiration. It is not possible to find in all geometry more difficult and intricate questions, or more simple and lucid explanations. Some ascribe this to his natural genius; while others think that incredible effort and toil produced these, to all appearances, easy and unlaboured results. No amount of investigation of yours would succeed in attaining the proof, and yet, once seen, you immediately believe you would have discovered it; by so smooth and so rapid a path he leads you to the conclusion required. (Plutarch, in Fauvel and Gray 4.B.1)

One suspects that Plutarch had not read the works which he describes as 'smooth and rapid', since later generations have found them impressive but difficult. The geometrical core, which includes the *Measurement of a Parabola* and *On the Sphere and the Cylinder* carries on, with great ingenuity, from the harder parts of Euclid; we shall not deal with them here, but there are good extracts in Fauvel and Gray (see also Archimedes 2002). There is, however, more to Archimedes than these works suggest, and some of his other surviving works contradict Plutarch's image of the 'pure' mathematician. The *Statics* and *On Floating Bodies* are the most serious works of theoretical physics, outside the framework of Aristotle's thought, in the Greek tradition; and as such, they had a great influence in the Renaissance, particularly on Galileo—see Chapter 6. Further evidence of a mechanical tendency in Archimedes is provided by the strange document called the 'Method'. Extravagant claims have been made for this manuscript,[2] for example, that it contains a version of the calculus, and that the course of history would have been changed if it had not been 'lost'. There is no need for such exaggeration; *The Method* is, so far as we know, a very unusual work which had no imitators, and for good reason. In his introductory letter to Eratosthenes, Archimedes describes what he is doing, and why:

Seeing moreover in you, as I say, an earnest student, a man of considerable eminence in philosophy, and an admirer [of mathematical inquiry], I thought fit to write out for you and explain in detail in the same book the peculiarity of a certain method, by which it will be possible for you to get a start to enable you to investigate some of the problems in mathematics by means of mechanics. This procedure is, I am persuaded, no less useful even for the proof of the theorems themselves; for certain things first became clear to me by a mechanical method, although they had to be demonstrated by geometry afterwards because their investigation by the same method did not furnish an actual demonstration. (From Archimedes tr. Heath, in Fauvel and Gray 4.A9 (a))

The 'Method' referred to consists of measuring the areas of bodies (e.g. a segment of a parabola) by 'balancing' them against simpler bodies (e.g. a triangle), using a division into infinitely thin slices. (See Fauvel and Gray 4.A9(a) for an example.) Two things are striking here: first, the use of weighing as a guide to understanding, presumably inspired by the work in the *Statics*—this is the 'applied side' of Archimedes; and second, the insistence, in the letter quoted above, that this is not a proof, but that a proof has to be constructed once you have found the answer. (And, in some sense, that it clarifies why the answer is what it is.) I should stress that the fact that *The Method* is an applied work does not make it an easy read; if it had been, perhaps it would have been preserved and quoted more. To describe it as 'lost' is only partly accurate; someone in the ninth century, and various others before that, must have known it and thought it of enough interest to be worth copying. However, it had no influence on the later traditions, either through Byzantium or the Islamic world, so far as we know; and this although some Islamic mathematicians had a great respect for Archimedes and worked hard to reconstruct alleged works of his which they did not have.

In contrast, one work of Archimedes had tremendous influence, and still does. This was his *Measurement of a Circle*. It is very short—it is thought that it is only part of a longer work of which the rest has been lost; but what remained was found immensely useful by much more simple-minded mathematicians. The three theorems which it contains are worth quoting in full, as a typically Greek way of approaching what we would call the problem of calculating π:

Proposition 1. *The area of any circle is equal to a right-angled triangle in which one of the sides about the right angle is equal to the radius and the other to the circumference of the circle.*

2. Discovered by Heiberg in Istanbul in 1906, then lost again, but recently rediscovered, sold at Christie's for $2 m., and subjected to modern scientific reading methods.

Proposition 2. *The area of any circle is to the square on its diameter as 11 is to 14.*

Proposition 3. *The circumference of any circle exceeds three times the diameter by a quantity that is less than one-seventh of the diameter but greater than ten parts in seventy-one.*

It is clear that Proposition 2 is both wrongly placed (it depends on Proposition 3) and probably not as Archimedes stated it (it claims as exact what is recognized in Proposition 3 to be an approximation). This of course added to the confusion of medieval readers, who tended to go for the more usable Proposition 2; but at different times, all three parts were found useful.[3] The first states (in our terms) that the area A is $\frac{1}{2}rC$, where r is the radius and C is the circumference. The Greeks sometimes worried, as we would not, whether this implied the necessary existence of a straight line whose length was equal to the curved line C. The second states that $A = \frac{11}{14}(2r)^2 (= \frac{22}{7}r^2)$. The third also gives the approximation $3\frac{1}{7}$ for the ratio of C to $2r$ which is still used after over 2000 years, and was gladly taken as the 'right' answer by calculators who had no use for Archimedes' more precise formulation:

$$3\frac{10}{71} < \frac{C}{2r} < 3\frac{1}{7}$$

When we use the approximation $3\frac{1}{7}$ for π, we are therefore indebted to Archimedes, although we probably know nothing of his methods. These were interesting in themselves, however, as an example of how he calculated—again, a more down-to-Earth procedure than the Platonic model of mathematics would suggest. For the upper bound of $3\frac{1}{7}$, for example, he starts with a circumscribed hexagon (Fig. 3).

Archimedes assumes that any circumscribed figure has a greater perimeter than the circle, and proceeds to find successively smaller ones, by bisecting angles (Fig. 4); he derives the rules for the lengths of successive sides:

Rule 1: $A' : B = A : B + C$
Rule 2: $A'^2 + B^2 = C'^2$

These two rules make it possible to find the perimeter of polygons with 12, 24, 48, and 96 sides As aids in calculation, he (a) uses a fractional approximation for $\sqrt{3}$, whose origin is unexplained, but which is needed in the formula for the hexagon (see Exercise 3), (b) by successive applications of the rule gets a rather complicated fraction for the 96-sided figure, and (c) shows that this fraction is larger that $3\frac{1}{7}$. All this is a very interesting mixture of Euclid-style geometry and computation with ratios of numbers; the way in which the fractions are written and manipulated recalls the technique of the ancient Egyptians—unit fractions like $\frac{1}{5}$ rather than sexagesimals. There are repeated approximations to square roots which, while they seem correct, are not explained and so have been the basis for much speculation. All this is just what we claimed, perhaps prematurely (in the last chapter), Greek geometry avoided—the detailed engagement with numbers. This broad statement, true for Euclid, Apollonius, and the 'major' works of Archimedes, is, as we will see, not at all true for a variety of others. Is the 'Measurement of the Circle' intended as an aid for practitioners, or simply as an exercise in technique? We have no indication. And while Archimedes is always using the numbers (as a good geometer should) as *ratios*, not as absolute measures of length, the way is open for land-measurers to use them in other ways.

---

3. See Chapter 6 for Kepler's attempt to construct an infinitesimal version, around 1600.

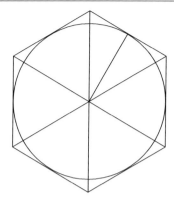

Fig. 3 Regular hexagon circumscribed about a circle.

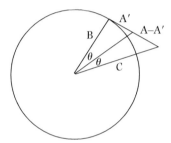

Fig. 4 Picture for Exercise 4.

Exercise 3. *Show that the perimeter of the circumscribed hexagon is $4\sqrt{3}r$.*

Exercise 4. *Rule 2 above is clearly Pythagoras's theorem. But where does rule 1 come from?*

3 Heron or Hero

Three centuries after Archimedes (probably in the first century CE) a very different mathematician left a number of works which were both accessible and popular. This was Heron, or Hero of Alexandria. (Because of translation problems, you may find either name used; I shall keep to the more usual 'Heron' in what follows.) His works are not easy to find, except in small extracts, but they are numerous and quite astonishingly diverse, dealing with theory and practice sometimes separately and sometimes together. That he was not despised, despite his practical bent and what some historians have seen as weak mathematical attainments, is shown by Pappus's description of his work—or that of his 'school', which in turn suggests influence.

The mechanicians of Heron's school say that mechanics can be divided into a theoretical and a manual part; the theoretical part is composed of geometry, arithmetic, astronomy and physics, the manual of work in metals, architecture, carpentering and painting and anything involving skill with the hands.
... the ancients also describe as mechanicians the wonder-workers, of whom some work by means of pneumatics, as Heron in his *Pneumatica*, some by using strings and ropes, thinking to imitate the movements of living things, as

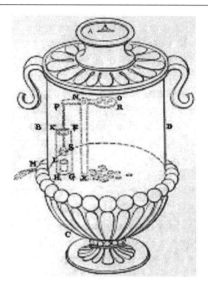

Fig. 5 Heron's slot machine.

Heron in his *Automata* and *Balancings*, ... or by using water to tell the time, as Heron in his *Hydria*, which appears to have affinities with the science of sundials.[4] (Pappus, in Fauvel and Gray 5.A.2)

The combination of quite classical geometry and detailed machine construction makes Heron interesting, unusual, and hard to classify. Predictably, his machines attract considerable interest on the Internet, notably the earliest description of a 'slot machine'; however unmathematical this may be, it is worth including to illustrate the variety of (some) mathematicians' interests:

Sacrificial vessel which flows only when money is introduced (see Fig. 5)

If into certain sacrificial vessels a coin of five drachms be thrown, water shall flow out and surround them. Let ABCD be a sacrificial vessel or treasure chest, having an opening in its mouth, A; and in the chest let there be a vessel, FGHK, containing water, and a small box, L, from which a pipe, LM, conducts out of the chest. Near the vessel place a vertical rod, NX, about which turns a lever, OP, widening at O into the plate R parallel to the bottom of the vessel, while at the extremity P is suspended a lid, S, which fits into the box L, so that no water can flow through the tube LM: this lid, however, must be heavier than the plate H, but lighter than the plate and coin combined. When the coin is thrown through the mouth A, it will fall upon the plate H and, preponderating, it will turn the beam OP, and raise the lid of the box so that the water will flow: but if the coin falls off, the lid will descend and close the box so that the discharge ceases. (Heron 1851, section 21, to be found at http://www.history.rochester.edu/steam/hero)

His geometry—the 'Metrics'—is both inside and outside the mainstream Greek tradition, giving rough rules for how to compute combined with Euclidean proofs on occasion. The most famous example, although not a typical one, is what has become known as 'Heron's formula' for computing the area of a triangle given its sides. The formula (see Appendix A) is very unusual in Greek mathematics in that it requires you to multiply four lengths and take the square root. While you could think of the product of three lengths as a volume, the product of four has no meaning in Greek terms—and Omar Khayyam was still dismissing such ideas a thousand years later. (In his *Algebra*; see Fauvel and Gray 6.A.3, p. 226.) All the same, the formula became widespread and popular in Islamic and medieval times, and a tradition claimed that it was originally due to

---

4. Pappus does not note that an accurate water-clock, compared with a sundial, would show the variation in the length of solar days; but this is not surprising, since elsewhere he criticizes water-clocks for their inaccuracy. The variation ('equation of time') is derived by Ptolemy purely on theoretical grounds—see below.

Archimedes (from whom Heron derived several other results).[5] If this is true, it shows Archimedes as innovative in a more striking way than had been thought.

What Heron's work shows, perhaps more than any other, is the existence of traditions which we do not otherwise know about—and this is a warning against easy generalizations on the nature of Greek mathematics. The quotation from Pappus does suggest that he was considered 'special' at least in his interest in machines, but it is unclear how much should be built on that; many of the machines which he describes, such as water-clocks, were certainly not of his invention. In parts of the *Geometrica* (which may not be his, but can be seen as work by his 'school'), he solves some quadratic problems by the Babylonian recipe; and this is taken by some scholars as a reason to fit his work into a genealogy which stretches (with notable gaps!) from the Babylonians through to the Islamic algebraists of the ninth century CE. However, he is also capable of advising intelligent guesswork, and using what appear to be quite new methods. The following example is again unlike our usual idea of Greek geometry.

In a right-angled triangle, the sum of the area and the perimeter is 280 feet; *to separate the sides and find the area*, I proceed thus: Always look for the factors; now 280 can be factored into 2.140, 4.70, 5.56, 7.40, 8.35, 10.28, 14.20. By inspection, we find 8 and 35 fulfil the requirements.

Note that the problem is not 'well-posed', by which I mean that it can have numerous solutions; as an equation, it is (setting the short sides equal to a and b)

$$a + b + \sqrt{a^2 + b^2} + \frac{1}{2}ab = 280$$

and of course the key point is that, for a 'nice' solution $a^2 + b^2$ must be a square, itself a favourite problem. It could be seen as an extension of the Babylonian problems which give you the side of a square plus its length; but it is considerably harder and more geometrical. Heron's solution continues:

For take one-eighth of 280, getting 35 feet. Take 2 from 8, leaving 6 feet. Then 35 and 6 together make 41 feet. Multiply this by itself, making 1681 feet. Now multiply 35 by 6, getting 210 feet. Multiply this by 8, getting 1680 feet. Take this away from the 1680, leaving 1, whose square root is 1. Now take the 41 and subtract 1, leaving 40, whose half is 20; this is the perpendicular, 20 feet. And again take 41 and add 1, getting 42 feet, of which the half is 21; and let this be the base, 21 feet. And take 35 and subtract 6, leaving 29 feet. Now multiply the perpendicular and the base together, [getting 420], of which the half is 210 feet; and the three sides comprising the perimeter amount to 70 feet; add them to the area, getting 280 feet. (*Geometrica*, in Thomas 1939, pp. 503–9; Fauvel and Gray 5.C.2 (c))

This calculation is completely enigmatic as it stands—we can see that Heron's answer is correct, but what is he doing? Thomas's translation gives a good explanation (due to Heath) related to the formula of Appendix A. In fact, if r is the radius of the inscribed circle, and $s = \frac{1}{2}(a+b+c)$ is half the perimeter, then the area is sr (see Appendix A) and so the sum of perimeter and area is $s(2+r)$. We can therefore 'look for factors', as Heron says, and guess $s = 35, 2 + r = 8$.

One could go on to ask how exactly, without algebra, such a procedure could have been hit upon. Such mathematics could be thought of, in the language of Høyrup (1994) as 'subscientific'—short on proof, although the numerical check is given; perhaps designed to display skill and virtuosity (or, like a crossword, to keep the mind active) rather than to be of any use. And yet the relation to the formula of Appendix A ties it in with 'real' geometry. Some time later than Heron (probably—the

---

5. In particular, he tended to use the 'Archimedean approximation' $3\frac{1}{7}$ for π.

dates of both are in dispute), Diophantus invented an 'algebraic' notation which might have made such questions easier; but although his works were studied and commented, they had no apparent influence on the solving of everyday puzzles for more than a thousand years.[6]

Exercise 5. *(a) Show that, in a right-angled triangle with short sides a, b and other notation as above, $r = \frac{1}{2}(a + b - c)$, (b) explain what Heron is doing in terms of this formula.*

4 Astronomy, and Ptolemy in particular

Mathematicians sometimes have difficulty in thinking about astronomy as a part of their subject; however, historically it is essential. Unified as a subject of study with mathematics from very early times, it has often provided work for mathematicians and motivation for much of their enquiries; the trigonometric functions (sin, cos, and so on) and the study of geometry on the sphere owe more to the study of the heavens, at least in their beginnings, than to the geography of the Earth.

There are considerable problems, touched on lightly in the last chapter, in setting up a mathematical astronomy. One could list some of them:

- find the length of the day
- find the length of the year
- before doing either of these, find a reliable way of measuring time
- find the path of the sun in the heavens, at a given place, at a given time of year.

It is worth taking time to think about these problems, and what instruments, observations, and/or calculations you would need to answer them. All of them are non-trivial, and all need to be understood with some finesse before we can begin to answer the subtler questions about the paths of stars, planets, and so on, let alone eclipses of the sun and moon. Necessarily, a great deal must already have been worked out for Plato (if it was his idea, as tradition has it) to pose the problem of accounting geometrically for the motions of the various heavenly bodies. The restriction classically put on these was that they must be 'composed of uniform circular motions'—we shall see later what this implied. The restriction had a practical advantage (circular motion is easy to work with) and a philosophical one (it corresponds to some ideal of perfection). The explanation, as I mentioned in the previous chapter, had to be descriptive, giving you the possibility of predicting where a body would be at a given time. This, incidentally, made it essential for astrology, which has almost always been a major concern for astronomers and by extension for mathematicians in general.[7] On the other hand, there was no call for a physical explanation of what force might make the heavenly bodies move; this theory, which is not easy to reconcile with the descriptive one, was given by Aristotle, and does not really concern us.

The major textbook of astronomy which has survived is the work of Ptolemy, who worked in Egypt in the second century CE. While he named it the *Mathematical Syntaxis* ('treatise'), from the time of the first Arabic translations it came to acquire the name 'Almagest' (= Arabicized Greek

---

6. Specifically, until the Renaissance (chapter 6). Hypatia among the Greeks and Qusta ibn Lūqa among the Arabs are known as students and commentators of Diophantus, and half of his work only survives in ibn Lūqa's translation.

7. Whether or not one believes in the influence of the planets' positions at a given time, the actual calculations which determine them are often quite hard mathematics. Ptolemy, in his *Handy Tables*, simplified the work so that the practising astrologer could look up the answer without reading his heavily theoretical *Almagest*.

'al-majistī', the greatest), by which it is usually known. Like Euclid, it was a standard textbook for over a thousand years, much commented and occasionally revised and criticized in the light of new theories but only losing its popularity in the seventeenth century as the theories of Copernicus, Kepler, and Newton came to form a solid alternative of a very different kind.

Ptolemy was not the first to develop the theory found in his book; the ideas on how the planets move were supposedly first framed by Apollonius (second century BCE) and dealt with in a textbook by Hipparchus (a little later); but their works, superseded by the *Almagest*, have not survived. What Ptolemy apparently added to Hipparchus was more accuracy and a simpler method of calculation. His book, however, though a major primary source and very interesting (translation by Toomer 1984), is not a straightforward read, and this also is worth thinking about. Where a mathematics textbook typically has one general subject, and starts from first principles to show you how to solve quadratic equations, or prove Pythagoras's theorem, or differentiate, an astronomy textbook has to set up a large and complex apparatus of general theory, observation, and particular verifications; and this is true of Newton's *Principia* part III (the explanation of the System of the World) as much as of Ptolemy. In the *Almagest*, Ptolemy first goes through the foundational assumptions: that the Earth is spherical and fixed in the middle of the universe, that the heavens move in circles around it, and so on. At this point, we have to set aside the fact that we 'know' that Ptolemy's system is wrong, since it is in fact an extremely good mathematical explanation of what is observed, and the mathematics is what should concern us. So we must accept these assumptions. We then observe that the stars have *two* motions: in a day they describe circles about the pole star, and in a year they describe circles about a different axis, defined by the path of the sun—the 'ecliptic' (see Fig. 6).

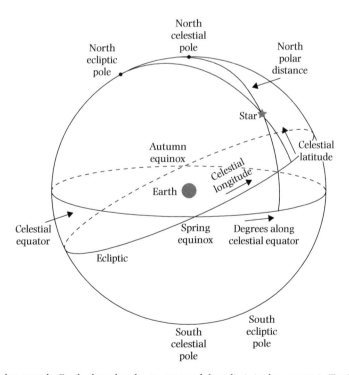

Fig. 6 The relation between the Earth, the poles, the equator, and the ecliptic in the geocentric (Earth-centred) model.

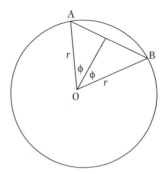

Fig. 7 The chord of an angle. If θ = angle AOB = 2ϕ, then the chord Crd(θ) = AB = $2r \sin \phi$ in our terms, where normally $r = 60$.

Again, we would say that one is 'caused' by the daily rotation of the Earth and the other by its annual rotation around the sun; but that is not how it looks to the naive observer.

Having got this far, Ptolemy needs to calculate the angle between the two axes, and this leads him into trigonometry or the theory of calculations about angles. The function which he uses is the 'chord' of the angle, which modern translators usually write Crd θ; it is the chord in the sector of the circle whose angle is θ, so Crd $\theta = 2r \sin(\theta/2)$, where r is the radius. The angle is written in our modern units (degrees, minutes, and so on), while the chord is written in units of length, in a circle whose radius is 60, using (Babylonian type) sexagesimal fractions, as for the angle (see Fig. 7). This again is interesting; we saw Archimedes calculating with Egyptian type fractions for the circle, and this may have been usual, but for setting results out formally in a table—which is what Ptolemy did—sexagesimals are clearly better. They point forwards towards modern decimal notation, as well as backwards to Babylon, and they were to be used in astronomy continuously until recently.

The table of chords (in intervals of half a degree) is worked out fairly quickly using some basic Euclidean geometry, with some results of Ptolemy's own. Square roots are often extracted, as they need to be, with no indication of how; but we might suppose that something like the method Heron uses for finding $\sqrt{720}$ (see Appendix A) is being used.

Next, (again this is typical of the variety of topics in the book) Ptolemy describes an instrument for measuring the angle between the two circles, the ecliptic and the equator. He derives the angle from his measurement, using the table of chords, as $23;51,20°$. We therefore think of the sun as moving around this circle at a uniform speed through the 12 signs of the zodiac (each of which takes up roughly 30°), in one solar year of around $365\frac{1}{4}$ days. We can then (one would think) find where the sun will be at any time on any day of the year; and in particular, which sign it will be in. The problem is that, as seen from the Earth, the sun does not move through the signs at uniform speed. The 90° from spring equinox to summer solstice, for example, takes two days longer than the 90° from summer solstice to autumn equinox. The sun is (or appears) slower in travelling through Taurus (May) than through Leo (August).

This 'anomaly', one among many astronomical anomalies which have to be explained, could be dealt with by supposing that the sun actually had a variable speed, and finding how it varied; this was Kepler's idea, arrived at around 1600, and it is what one would say today. But, both for ease in calculation and because of the general (Platonic) theory, Greek astronomy did not work in that way. Instead, Ptolemy puts forward two models of how the sun moves. They are, as he says, equivalent (you may try to work out why), but that is not really important, since the aim of an

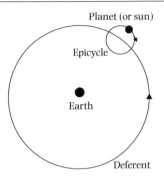

Fig. 8 The epicycle model. The planet moves clockwise round a small circle (epicycle) whose centre moves anticlockwise round a large circle (deferent) centred on the Earth. This explains why some planets (e.g. Mars) sometimes move backwards in relation to the stars.

explanation is not so much to say what 'really' happens as to give an accurate description from which you can make predictions. Again, the physics, as we would call it, is not considered. In the first model ('epicyclic'), the sun moves at constant speed round a small circle or 'epicycle'; and the centre of the epicycle itself moves at constant speed round the ecliptic, with the Earth as centre (see Fig. 8).

The second ('eccentric') supposes that the sun is travelling at uniform speed around a circle, but that the Earth is not the centre of the circle. It follows that the sun appears to be travelling more slowly when it is further from the Earth than when it is nearer (see Exercise 8). In Appendix B, I give the detailed working out by Ptolemy, from his observations, of where the centre of the orbit is in relation to the Earth. This is a detailed piece of Greek *numerical* mathematics, based on the geometric tradition. Is it practical? Very much so, in that it makes possible (the beginnings of) the calculation of where the sun will be. But it is supported by a formidable theoretical apparatus of results about chords in a circle, angles, and so on. Considered as a whole—and this is the justification for devoting so much time to his work—Ptolemy's *Almagest* gives more of an impression of the range and variety of Greek mathematics than any other text which we have.

Exercise 6. *Show that* $\mathrm{Crd}(60°) = 60$, *and* $\mathrm{Crd}(36°) = 60((\sqrt{5} - 1)/2)$.

Exercise 7. *How would you find* $\mathrm{Crd}(\theta/2)$ *given* $\mathrm{Crd}(\theta)$?

Exercise 8. *Explain the variation in the sun's apparent speed, on the eccentric hypothesis.*

Research problem. *Find the two reasons why the length of the day varies (a) as we would understand it, (b) as Ptolemy would have put it. (This is called the 'equation of time'.)*

5 On the uncultured Romans

With the Greeks geometry was regarded with the utmost respect and consequently none were held in greater honour than mathematicians, but we Romans have delimited the size of this art to the practical purposes of measuring and calculating. (Cicero, *Tusculan Disputations*, tr. Serafina Cuomo, in Cuomo 2001, p. 192)

The above quotation heads Serafina Cuomo's chapter on the Romans; and her recent book makes the first serious attempt to investigate and indeed question a view of their mathematics accepted since Cicero's time (first century BCE). Briefly, this is that the Romans in contrast to the Greeks made no contribution to mathematics, and that any works of theirs which contain or use it, such as Vitruvius' architecture, are trivial in comparison with the Greek achievements. The charge has some basis in fact, as Cuomo acknowledges—'I would be hard put to adduce a Latin equivalent of Euclid, Archimedes or Apollonius', she admits (p. 194). This seems like an understatement; where are the Latin equivalents of Heron or Nicomachus? True, these second- or third-rate mathematicians worked under Roman rule, and may have been, like St Paul, Roman citizens. However, they wrote in Greek, and in a tradition which was, and continued to be, overwhelmingly Greek. The Romans, with better access to the Greek classics than the ninth-century Arabs or the Renaissance Europeans, never seem to have felt the same need to build on their work and develop it. And while Cuomo also interestingly makes a historicist point about the class angle contained in Cicero's statement (ideas of 'the Roman' and 'the Greek' were marks of different kinds of prestige, while many users of numbers, and land-surveyors in particular, were seen as jumped-up technical upstarts), one is still left with an underlying feeling that it is an ideological statement based on good factual evidence. The most interesting part of her argument is a broader one, and does bear serious consideration: that the *practice* of mathematical methods of some sophistication pervaded the Roman world from top to bottom. Some of her examples, notably the charioteer whose tomb boasts that he

drove chariots for 24 years, ran 4,257 starts and won 1,462 victories, 110 in opening races. In single-entry races he won 1,064 victories, winning 92 major purses, 32 of them (including 3 with six-horse teams) at 30,000 sesterces . . . (CIL 6.10048 (Rome, 146 CE), tr. in Lewis and Reinhold 1990, pp. 146–7)

testify to the power of 'numbers' to impress rather than to the ability to do anything with them. However, her study of the practice of the despised land-surveyors (see also Dilke 1971), and of Vitruvius show how an appreciation, and application of classical geometry underlay their practice. Perhaps rather than decrying the 'low level' of geometry present in Vitruvius's architecture, we should think about the fact that it was a Roman, rather than a Greek, who bothered to write such a treatise; the architects of Greek temples were not, it would seem, given to exposition. We have different cultures (cohabiting in the same empire) with different ideas of what a book is for.

Similarly, the famous tunnel of Eupalinus in Samos, dated at 550–530 BCE, is often cited as an amazing example of very early practical Greek geometry; how did the builders of the tunnel, who started from the two sides of a mountain, contrive to meet so accurately in the middle? The answer is again that we do not know, and no Greek sources seem to have taken the trouble to explain how such a recurrent problem could be solved. The Roman surveyors, however, organized as a profession in which a discipline was transmitted by means of 'textbooks', both explained how they did it[8] and wrote instructions whose *foundation* is in their training in some derivative of Euclidean geometry.

This debate is only now beginning; the same applies to the doubts which Cuomo has cast on the idea that Greek mathematics was 'in decline' from (say) the time of Ptolemy, if not before. It is not so much a question of rehabilitating the Romans (awarding points to individuals or to civilizations for their excellence in mathematics should not be part of the business of history, though it often is). Rather, as we saw in Chapter 1 with the pre-OB periods, it is a question of looking at practices

---

8. See Cuomo (2001, p. 158) for a surveyor's account of how he helped the weeping villagers whose tunnel had manifestly gone badly wrong.

which have been dismissed as trivial or non-mathematical and seeing if they do in fact belong in our history—and if so, where.

6 Hypatia

Hypatia, the daughter of Theon the mathematician, was initiated in her father's studies; her learned commentaries have illuminated the geometry of Apollonius and Diophantus, and she publicly taught, both at Athens and Alexandria, the philosophy of Plato and Aristotle. In the bloom of beauty, and in the maturity of wisdom, the modest maid refused her lovers and instructed her disciples; the persons most illustrious for their rank and merit were impatient to visit the female philosopher; and Cyril beheld with a jealous eye the gorgeous train of horses and slaves who crowded the door of her academy. (Gibbon, n.d. chapter XLVII)

Hypatia was born in the later part of the Roman Empire, an era when women were not free to pursue careers. This was a time when orthodox belief effectively wiped out centuries of scientific discovery. Ancient Greek works were torched and scholars were murdered. Hypatia was the last proprietor of the Hellenic Age wonder, the Library of Alexandria. She is portrayed as a young adult facing the issues of a changing world. The reader will discover uncanny parallels to many current situations within the United States and, indeed, the world. Hypatia, a real, historically documented heroine, is a find for today's young adults who are searching for strong, non-fiction role models. (From review of a novel, 'Dear Future People', at www.erraticimpact.com/~feminism/html/women_hypatia.htm)

Mathematicians, like engineers and physicists, have very rarely been women—the rarity is far more serious than (for example) for poets and painters. As a result, the study of 'women mathematicians' faces a serious difficulty in even getting off the ground, since there has until the twentieth century been no continuing tradition from which to construct a history. The feminist historian (of whom Margaret Alic 1986 was a pioneer) therefore necessarily (a) points out the existence of numerous such women of whom we know little or nothing, and (b) attempts to weave the major lives of which we do know something into some kind of thread. The philosopher and mathematician Hypatia, who was stoned to death by a Christian mob in 415 CE, (this much is undoubtedly historical fact) is the most important early, perhaps the founding figure for the tradition, and the questions which surround her life and activity may illustrate the wider problem. For if one were to take any other figure from Greek mathematics as 'representative' of something (Nicomachus as a Palestinian, Ptolemy as an African, . . .), generalizations based on the little that is known of their lives would be hard, although the works at least give some basis for building theories. With Hypatia, the difficulty is the opposite one. Her life is unusually well documented, in the general context of Greek mathematics, as the result of friendly and hostile accounts by later Christian writers. Most particularly, her devoted pupil Synesius, bishop of Cyrene in Libya, wrote a number of letters to her and about her which give substantial detail about her life and teaching, if from a particular viewpoint. On the other hand, although her ability as a mathematician is well documented and at least the titles of some works have been preserved, there is no extant text attributed to Hypatia, no 'Hypatia's theorem', no discovery which tradition assigns to her. With Heron, as was noted above, we know the works but nothing of who he was; with Hypatia, it is the other way round.

On her life and her philosophy, for which the sources are good (Synesius does not appear to have been so interested in mathematics), Maria Dzielska's monograph (1995) is a recent excellent source. Dzielska begins with an account of the myths which have built up around her as an iconic figure since the seventeenth century, and which my opening quotes illustrate. She was a victim of Christianity and symbol of the death of the ancient learning at the hands of the new ignorance (Gibbon); a feminist icon and precursor of (for example) Marie Curie (Margaret Alic); a symbol of the imposition of European rule on Africa (Martin Bernal). Dzielska produces a convincing picture of Hypatia as an influential teacher of mathematics, astronomy, and neoplatonic[9] philosophy to a circle of initiates, both pagan and Christian.

> Around their teacher these students formed a community based on the Platonic system of thought and interpersonal ties. They called the knowledge passed on to them by their 'divine guide' mysteries. They held it secret, refusing to share it with people of lower social rank, whom they regarded as incapable of comprehending divine and cosmic matters ... Hypatia's private classes and public lectures also included mathematics and astronomy, which primed the mind for speculation on higher epistemological levels. (Dzielska 1995, p. 103)

She shows Hypatia becoming involved in a power struggle between factions in Alexandria in the years following 410 which led to a witch-hunt, and eventually to her death There was, indeed, a careful line to be drawn in late Roman times between the praiseworthy pursuit of geometry (*ars geometriae*) and the damnable art of astrology (confusingly, *ars mathematica*)—see Cuomo (2000, p. 39); and Hypatia was probably not the only scholar to be caught on the wrong side of the line. Dzielska further establishes, fairly convincingly, that her age at the time was about 60 (demolishing the image of a beautiful maiden cut down in the bloom of youth), and points out that her death was far from marking the end of learning in Alexandria, or the Greek world generally—or even of paganism.

> Pagan religiosity did not expire with Hypatia, and neither did mathematics and Greek philosophy. (Dzielska 1995, p. 105)

She also unearths a number of other references to women in the late Greek philosophical world, which show Hypatia's example to be not so unusual as had been thought.

This is helpful, but of course the historian of mathematics would like to have more, and here as so often we enter the world of more or less ingenious conjecture. As Gibbon states, her father was the mathematician Theon who has not been highly estimated in recent times ('a competent but unoriginal mathematician', Calinger 1999, p. 219). However, like many others of the period he studied and commented the difficult works of his predecessors, and edited the text of Euclid in a version which was almost the only one to survive. The titles of several of Hypatia's works—mainly commentaries, for example, on Ptolemy and Diophantus, are known from later bibliography, and Synesius, her student, was a philosopher not a mathematician; and by the time when Islamic scholars recovered and translated Greek works, none of them bore her name. The scholar who wishes to study her as a mathematician, supposing it possible, has to use a certain amount of imaginative reconstruction. Nonetheless, in line with the revival of neglected women in antiquity, she is given two pages in Calinger's general history (1999), and that common and convenient view which dismisses Theon's works as pedestrian and second-rate attempts to pick out the more interesting parts of them and ascribe them to Hypatia.[10]

Following the initial stage of 'recovery', where the aim was to point out Hypatia's status and relative neglect as a mathematician, Dzielska's work has been well received as perhaps marking the start of a second period in the study of women mathematicians, still rather in its infancy: an attempt to place them in a historical context, even when (as with Byzantine Alexandria) that

---

9. Neoplatonism was not simply a revival of Platonism, but had elements of mysticism; as such it played a semi-religious part in the late Roman empire.

10. For these and similar arguments see Knorr (1989), who also suggests that, since Diophantus's *Arithmetica* as it survives contains comments, the edition itself in the form we have it may have been prepared by Hypatia. Doubts are expressed by Cameron (1990).

context is remote and offers few opportunities for identifying role models. An assessment of her life and works, if any can be reliably ascribed to her, while sympathizing with her difficult, ultimately tragic situation, is not dependent either on approving her obvious ability and charisma, or on disapproving of the élitism, and that belief that the state would be better off if run by philosophers which she shared with other Neoplatonists.

Appendix A. From Heron's *Metrics*

[Introductory note. The standard translation (e.g. the one which you will find in Fauvel and Gray) has been changed so that there is as little as possible 'modernization' of Heron's language, and no explicit algebra. This is more difficult than one might imagine, since (a) Heron does think of lengths as numbers, and multiply them—this happens in the first part, (b) in the geometric proof the kind of straightforward and perhaps over-simple statements I have made about Euclid (areas of rectangles are just areas, not products of the lengths of sides etc.) are no longer clearly true, and it is possible that something like algebra, of an embryonic form, was in Heron's mind even if you cannot see it on the page. Prepositions like 'on' and 'by' indicate areas of rectangles or multiplication, and it is unclear which is being used. Most unexpectedly, 'the on ABC' means the product AB times BC, not the area of a triangle.]

There is a general method for finding, without drawing a perpendicular, the area of any triangle whose three sides are given. For example, let the sides of the triangle be 7, 8, 9 units. Add together the 7 and the 8 and the 9; the result is 24. Take half of this; the result is 12. Take away the 7 units; the remainder is 5. Again take away from the 12 the 8; the remainder is 4. And then the 9; the remainder is 3. Multiply the 12 by the 5; the result is 60. and this by 4; the result is 240. And this by 3; the result is 720. Take the side [= square root] of this, and it will be the area of the triangle. Since 720 does not have a rational square root, we shall reach a different [number] close to the root as follows. Since the square nearest to 720 is 729 and it has a root 27, divide the 27 into the 720; the result is 26 and two thirds. Add the 27; the result is 53 and two thirds. Take half of this; the result is $26 \frac{1}{2} \frac{1}{3}$ [This is the 'Egyptian way' of writing $26 \frac{5}{6}$.] Therefore the square root of 720 will be very near to $26 \frac{1}{2} \frac{1}{3}$. For $26 \frac{1}{2} \frac{1}{3}$ multiplied by itself gives $720 \frac{1}{36}$; so that the difference is a 36th part of a unit. If we wish to make the difference less than the 36th part, instead of 729 we shall take the number now found $720 \frac{1}{36}$, and by the same method we shall find a difference much less than $\frac{1}{36}$.

The geometrical proof of this is this (Fig. 9): *In a triangle whose sides are given to find the area.* Now it is possible to draw a perpendicular and calculate its magnitude and so find the area of the triangle, but let it be required to calculate the area without drawing the perpendicular.

Let the given triangle be ABC and let each of AB, BC, CA be given; to find the area. Let the circle DEZ be inscribed in the triangle with centre H [Euclid IV.4], and let AH, BH, CH, DH, EH, ZH be joined. Then the [rectangle] BC times EH is twice the triangle BHC [Euclid I.41], and CA times ZH is twice the triangle AHC, and AB times DH is twice the triangle ABH. So the perimeter of the triangle ABC times EH, that is, the [radius] of the circle DEZ, is twice the triangle ABC.

Let CB be produced, and let BF be made equal to AD; then CBF is half the perimeter of the triangle ABC because AD is equal to AZ and DB to BE and ZC to CE. So CF times EH is equal to the triangle ABC. But CF times EH is the side of the [square] on CF times the one on EH.

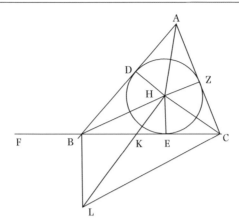

Fig. 9 Figure for Heron's theorem, Appendix A.

[This is the point at which we stop being able to think geometrically; the formula is that

$$CF.EH = \sqrt{CF^2 \cdot EH^2}$$

But while the left hand side is a rectangle, the right has arisen by taking the product of two squares and extracting the square root. From now on, numbers whose 'dimension' is 4 seem to come into the calculation.]

So the area of the triangle *ABC* by itself is the area of the [square] on *FC* by the [square] on *EH*.

Let *HL* be drawn perpendicular to *CH* and *BL* perpendicular to *CB*, and let *CL* be joined. Then since each of the angles *CHL*, *CBL* is right, a circle can be described about the quadrilateral *CHBL* [Euclid III.31]; therefore the angles *CHB*, *CLB* are together equal to two right angles [Eucl. III.22]. But the angles *CHB*, *AHD* are together equal to two right angles because the angles at *H* are bisected by the lines *AH*, *BH*, *CH* and the angles *CHB*, *AHD* together with *AHC*, *DHB* are equal to four right angles; therefore the angle *AHD* is equal to the angle *CLB*. But the right angle *ADH* is equal to the right angle *CBL*; therefore the triangle *AHD* is similar to the triangle *CBL*.

So as *BC* is to *BL*, *AD* is to *DH*, that is as *BF* is to *EH*, and interchanging as *CB* is to *BF* so is *BL* to *EH*, that is as *BK* to *KE* because *BL* is parallel to *EH*. And putting together, as *CF* to *BF*, so is *BE* to *EK*.

[**Note.** This is the rule which the translator, following the medieval Latin use, calls *componendo*: if $a/b = c/d$, then $(a+b)/b = (c+d)/d$ (Do you see that this works, and applies to the situation?)]

And so the [square on] *CF* to the [rectangle] *CF* by *FB* is as the *BEC* to the *CEK*, that is to the [square on] *EH*; for in the right angled triangle *EH* has been drawn perpendicular to the base. Therefore, the [square] on *CF* times the [square] on *EH*, whose side [square root] is the area of the triangle *ABC*, is equal to the *CFB* times the *CEB*. And each of *CF*, *FB*, *BE*, *CE* is given; for *CF* is half the perimeter of the triangle *ABC*, and *BF* is the excess, by which half the perimeter exceeds *CB*, and *BE* is the excess, by which half the perimeter exceeds *AC*, and *EC* is the excess, by which half the perimeter exceeds *AB*, since *EC* is equal to *CZ*, and *BF* to *AZ*, since it is equal to *AD*. So the area of the triangle is given.

Exercise 9. *Check through the calculation of the square root of 720. What method is Heron using for finding it?*

Greeks, Practical and Theoretical

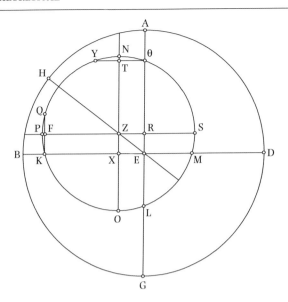

Fig. 10 Picture for Appendix B. Earth at E, sun on circle centre Z.

Appendix B. From Ptolemy's *Almagest*

(Ptolemy 1984, pp. 153–4)

Note. As already stated, lengths are in sexagesimals with the radius of the circle set equal to 60. The sixtieths are denoted by a small letter '*p*', corresponding to the degrees sign for angles.

In order not to neglect this topic, but rather to display the theorem worked out according to our own numerical solution, we too shall solve the problem, for the eccentre, using the same observed data, namely, as already stated, that the interval from spring equinox to summer solstice comprises $94\frac{1}{2}$ days, and that from summer solstice to autumn equinox $92\frac{1}{2}$ days. [Ptolemy then details his own 'very precise' observations in 139–140 CE, which confirm these figures, due to Hipparchus. These figures are all we need.]

Let the ecliptic be ABGD on centre E. In it draw two diameters, AG and BD, at right angles to each other, through the soltitial and equinoctial points (Fig. 10). Let A represent the spring [equinox], B the summer [solstice], and so on in order. [E is the Earth; the spring equinox occurs when the sun is in the direction of A (in Aries), and so on. The circle just drawn is the ecliptic as we see it in the heavens, and determines what we see.]

Now it is clear that the centre of the eccentre [i.e. of the eccentric circle] will be located between lines EA and EB. For semi-circle ABG comprises more than half the length of the year [187 days, as we have seen] and hence cuts off more than a semi-circle of the eccentre; and quadrant AB comprises a longer time and cuts off a greater arc of the eccentre than quadrant BG. This being so, let point Z represent the centre of the eccentre, and draw the diameter through both centres and the apogee, EZH.[11] With centre Z and arbitrary radius draw the sun's eccentre ΘKLM, and draw

---

11. The apogee is the point at which the sun is furthest from the Earth; from the picture, this is where the eccentre cuts the radius EH.

through Z lines NXO parallel to AG and PRS parallel to BD. Draw perpendicular ΘTY from Θ to NXO and perpendicular KFQ from K to PRS.

Now since the sun traverses circle ΘKLM with uniform motion, it will traverse arc ΘK in $94\frac{1}{2}$ days and arc KL in $92\frac{1}{2}$ days. In $94\frac{1}{2}$ days its mean motion is aproximately 93;9° and in $92\frac{1}{2}$ days 91;11°. [It covers 360 degrees in a year of $365\frac{1}{4}$ days, so slightly less than 1° per day; this is where these figures come from.] Therefore arc ΘKL=184;20° and by subtraction of the semi-circle NPO, arc NΘ + arc LO = 4;20°

So arc ΘNY = 2 arc ΘN = 4;20° and ΘY = Crd arc ΘNY = $4;32^p$, where the diameter of the eccentre is 120^p. [Remember that our tables deal with a circle of radius 60, diameter 120.] And EX = ΘT = $\frac{1}{2}$ΘY = $2;16^p$.

Now since arc ΘNPK = 93;9° and arc ΘN = 2;10° and quadrant NP = 90°, by subtraction, arc PK = 0;59°, and arc KPQ = 2.arc PK = 1;58°. So KFQ = Crd arc KPQ = $2;4^p$, and ZX = KF = $\frac{1}{2}$KFQ = $1;2^p$. And we have shown that EX = $2;16^p$ in the same units.

Now since EZ² = EX² + ZX², EZ = $2;29\frac{1}{2}^p$ where the radius of the eccentre is 60^p. Therefore the radius of the eccentre is approximately 24 times the distance between the centres of the eccentre and the ecliptic.

[This completes the first half of the calculation, showing how far the Earth is from the centre of the eccentric circle. It remains to find the direction of the line EZ so as to situate Z exactly; as you can see, this follows from the ratio of EX to EZ; we would use the tangent, but Ptolemy has to use the chord function again. The answer is that angle ZEX is 24;30°.]

Solutions to exercises

1. If (x, y) is on the two curves, then $x^3 = x.x^2 = x.y = 2a^3$; so $x = a.\sqrt[3]{2}$. The description of what Menaechmus did does not read quite like this—for a plausible version, see Knorr. If you replace 2 by m in the equation of the hyperbola, then you solve the problem of increasing the volume by m (and so the side by $\sqrt[3]{m}$), similarly.
2. Suppose C and D are constructed. Then (rules about ratios, think of them as fractions), A : B = (A : C).(C : D).(D : B). The three ratios in brackets are equal, so this is (A : C)³. If B : A = m, then (cube on B):(cube on A) is m^3. So (cube on C):(cube on A) is m.
3. Straightforward; the radius is the height of the equilateral triangle whose side is the side of the hexagon. So by Pythagoras's theorem, the side is $2/\sqrt{3}$ times the radius; and the perimeter of the hexagon is $6.r.(2/\sqrt{3}) = 4r.\sqrt{3}$.
4. This depends on Euclid VI.3, which you may not know. This says that in triangle ABC, if AD bisects angle A and meets BC at D, then AD : AB = BD : AB. (Look it up, or try to work out why it is true.) In our case, this gives (looking at the bisected angle at the centre of the circle in Fig. 4) A′ : B = A − A′ : C. Manipulating ratios (componendo, see Appendix A), A′ : B = A : B + C as required.
5. (a) (See Fig. 11) In the picture, CDOE is a square, so all of its sides equal r. Hence, AE = $b − r$, BD = $a − r$. But by the property of tangents, this means that AF = $b − r$ and BF = $a − r$. Hence, AB = AF + FB = $a − r + b − r$, and the result follows. (b) From the factorization, $\frac{1}{2}(a + b + c) = 35, \frac{1}{2}(a + b − c) = 6$; so $a + b = 41$. This is why we square 41, and get $(a + b)^2 = 1681$. But also the area sr equals $35 \times 6 = 210$, and this is (by a different

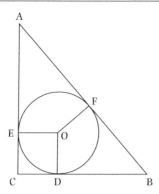

Fig. 11 The figure for Exercise 5; with BC = a, AC = b, AB = c, and the radius of the circle = r

calculation) equal to $\frac{1}{2}ab$. We now subtract eight times this number ($= 4ab$) from the 1681, and get 1, which must be $(a-b)^2$. Now we know $a+b$ and $a-b$, and the rest follows.

6. That Crd(60°) = 60 follows since the triangle of angle 60° is equilateral. For Crd(36°) we use Fig. 5 of the previous chapter. Here the angle at the vertex is 36°, and if the sides of the isosceles triangle are the radii (and so = 60), the ratio of the base to the side is what we have called the golden ratio, that is, $(\sqrt{5}-1)/2$. So Crd(36°) is 60 times this.

7. Without going into tedious detail, you would follow prescriptions similar to those of Archimedes in Exercise 4. You are dealing with inscribed polygons instead of circumscribed ones, but the essentials are the same.

8. Suppose that the sun is travelling at uniform speed (say $\theta°$ per hour) round a circle, and the Earth is at a distance a from the centre of the circle of radius r on which the sun is travelling. Then the $\theta°$ which the sun covers in an hour correspond to a distance $r\theta$, and so (roughly) to $(r/(r-a))\theta$ seen from the Earth at the nearest point, and to $(r/(r+a))\theta$ at the furthest point.

9. I shall not give the check. The method is to take a so that a^2 is near to b (in this case 720), and replace a by $\frac{1}{2}(a+(b/a)) = a+(b-a^2)/2a$. This is a 'standard' approximation taught in calculus if we write $b = a^2 + h = a^2(1+(h/a^2))$; $\sqrt{b} \simeq a(1+(h/2a^2)) = a+(h/2a)$.

4 Chinese mathematics

1 Introduction

Fu Xi created the eight trigrams in remote antiquity to communicate the virtues of the gods and parallel the trend of events in earthly matters, [and he] invented the nine-nines algorithm to coordinate the variations in the hexagrams. (Liu Hui, cited Shen et al. 1999, p. 52)

Mathematics is an important subject in the six arts. Through the ages all scholars who have participated in discussions on astronomy and calendars have to master it. However, you may consider it as a minor occupation, not as a major one. (Yen Chih-tui (sixth century CE), cited Libbrecht 1973, p. 4)

The usual warnings to avoid thinking of ancient mathematics in modern terms seem quite unnecessary in the Chinese case. Certainly the Chinese, like the Greeks, recognized a particular study called mathematics (*suanshu*). As a component of education, they seem often to have given it a rather subordinate role, as the second quote above shows. And yet, it was (as usual) essential for the standard preoccupations of irrigation, public works, and taxation. More particular to the Chinese, if still widespread, was the idea that the harmony of the universe is mathematically ordered, as the first quote expresses; guidance for future conduct can be gained from the 64 signs of the *Yijing*, or 'Book of Changes'.[1] Between abstract philosophy and low-level ditch-digging stood the essential practice of the calendar-makers and astrologers, who ensured—with variable success—that a complicated year ran smoothly enough and unlucky events in the heavens were accounted for. The earliest textbook, the *Zhoubi suanjing* (Cullen 1996), is an attempt to deal with these questions, and the problem of harmonizing the competing periods of days, months, and years is at least partly reponsible for the sophisticated number theory required for the 'Chinese remainder theorem'.

The relation of the various parts was a complex one, and yet Chinese mathematics is still often characterized as simply 'practical'. It is true that the bare classical texts often confront the reader as if that were their aim; but we have already seen in the Babylonian case that more may lie beneath the surface. Here, as an illustration, are two problems from the founding text, the *Nine Chapters*.

Now given a person carrying cereal through three passes. At the outer pass, one-third is taken away as tax. At the middle pass, one-fifth is taken away. At the inner pass, one-seventh is taken away. Assume the remaining cereal is 5 *dou*. Tell: how much cereal is carried originally?
Answer: 10 *dou* $9\frac{3}{8}$ *sheng*. (Shen et al. 1999, problem 6.27, p. 345)

Now chickens are purchased jointly; everyone contributes 9, the excess is 11; everyone contributes 6, the deficit is 16. Tell: the number of people, the chicken price, what is each?
Answer: 9 people, chicken price 70. (Shen et al. 1999, problem 7.2, pp. 358–9)

---

1. Although popular in the 1960s and often used in the West as an alternative to Tarot cards for fortune-telling, the *Yijing* ('I Ching') is a serious philosophical document, among other things.

Clearly the first question is 'easy', the second slightly harder, although the original reader may not have found it so. They belong in form to a very widespread tradition, which goes back to Egypt and Babylon. At first sight, only the specificity of the questions (the three passes clearly relate to a quite specific terrain and mode of tax gathering) reminds us to think about who was asking them, who they were addressing, and what was the aim. But aside from the particular organization of society, the well-known isolation of China meant that, while there were certainly cultural influences in both directions, they seem to have been rare and poorly documented. Furthermore, the ideas and aims of Chinese mathematics have elements which are hard to translate into our own terms. The questions and answers may be similar to those in other societies, but is that simply coincidence? How should the subject be studied?

The beginner may well be daunted. If we begin around 400 BCE (although there are no texts quite that early in date), and make the conventional end with the arrival of 'European' mathematics around 1600 CE,[2] we have an unbroken history of nearly 2000 years. Over this period, there are a succession of texts, similar in form—mainly lists of problems (like the ones above), with solution and commentary. The texts will have their own technical terms, which are not only exotic measures of length, but may refer to procedures within mathematics. Again, this helps historically in clarifying that we are not dealing with 'our' culture, but it does not help our comprehension. The non-Chinese reader is likely to know nothing of the Chinese language; it may be the second world language, but its script and structure make it inaccessible to most. Even transliteration can be a problem; while modern texts agree on using the now official 'pinyin' system, older ones will use some other one, so that the student should be warned that the mathematician who was formerly called Ch'in Chiu-shao is now Qin Jiushao (compare Peking and Beijing).[3] We shall try to provide some orientation on history and background; fortunately, not more than a minimum is absolutely necessary. For more, see the references in Section 2.

Added to these difficulties, classical Chinese mathematical texts can pose quite specific problems. Their language is compressed, so that the 'translations' which we have may be rather free adaptations. Some translators, in fact, (particularly Jock Hoe 1977) have tried to circumvent this by adopting a special telegraphic form of English which may help. Furthermore, they may be dependent on the specific calculator's skill of manipulating counting-rods, which for a long time was central to all work. This will have to be considered in its place, particularly in relation to claims argued forcefully by Lam and Ang (1992) that the use of the rods led to the Chinese invention of the place-value system and of decimal fractions. The questions usually asked, which extend some of these considerations, are:

1. What is specifically 'Chinese' about Chinese mathematics?
2. To what extent can similarities between Chinese mathematics and that of other cultures (Indian, Islamic) be attributed to cultural diffusion and to what extent are they independent? Specifically, one could consider the decimal system, Pascal's triangle, and methods for root extraction.

Despite its frequently mentioned restrictions, there is great diversity in Chinese mathematics, and this chapter can only discuss a part of it. Hopefully, you will be encouraged to read further.

Exercise 1. *Explain the answers to the questions from the* Nine Chapters, *given that 1* dou *= 10* sheng.

---

[2]. The later period has been neglected, and is still less studied; we shall consider it very briefly at the end of the chapter.
[3]. To help, the older form will sometimes be given in brackets, and quotes—for example Yijing ('I Ching').

2 Sources

The reader is *comparatively* well served by recent publications on Chinese mathematics. By this I mean that they are comprehensive and good, although the local library may have to be persuaded to invest in a copy. First among them is the classic work of Joseph Needham (1959). It has been claimed that the mathematical part of Needham's enormous work is the weakest, 'superficial and largely dependent on obsolete Western-language sources' being a recent judgment (Sivin 2004); but all subsequent scholars owe him an immense debt, and his emphasis on the social context is particularly valuable. Much more recent and very full is Martzloff (1995). This is thoughtfully structured, the first part on context and the second on content. Martzloff is scholarly, and anxious to give Chinese mathematics due credit; but he is equally cautious (some might say too cautious) about claims which rest on evidence which is scanty or late, or on conjectures about how things must have been. One could hardly wish for a fuller introduction, and it is highly recommended.

Of equal value—since the classic books are at the heart of Chinese mathematics—is the fact that several of these have been translated in the past 30 years, usually with a large amount of discussion and background material. This is most useful since there are no selections in Fauvel and Gray which deal with the subject. Certainly the most important is Shen et al.'s (1999) translation, with a very full commentary, of the fundamental *Nine Chapters on the Mathematical Art*, which fills the major gap; but also valuable are the translations of Cullen (1996), Lam (1977), Lam and Ang (1992), Swetz (1992), and the commentary of Libbrecht (1973).[4] The reader who can lay hands on some or all of these will be in an excellent position to form informed judgements. On the later, less studied period after 1600, Martzloff (1981) and Jami (1990) are both in French, but, if you can locate and read them, they provide a good opening on current research.

For a general history of China, an introduction—with much of interest concerning the history of science—is contained in Needham's volume I (1954). A more recent 'classic' is Fairbank (1992). Nathan Sivin has a good selected bibliography on Chinese science, including mathematics, online (Sivin 2004), and the new Chinese section of the St Andrew's website provides more detail on particular topics, for example, the *Nine Chapters* or individual mathematicians. Finally, there is a great deal of research activity, both in the standard publications (*Historia Mathematica, Arch. Hist. Exact Sci.*) and in specialist journals addressing Chinese science. Reference to some of these will be made where relevant.

3 An instant history of early China

The succession of dynasties, sometimes orderly and sometimes confused, which structures Chinese history is not 'general culture' as the European succession of states and empires is thought to be. What we need is a quick summary which is angled towards the main periods of mathematical interest, so far as we know them. Because the most important discussions concern the early period—up to about 600 CE—I shall cover that here, with brief inserts on the other key periods as we come to them.

---

4. There is still no translation of the key work on which this is based, Qin Jiushao's thirteenth-century *Shushu jiushang* (Computational Techniques in Nine Chapters); but perhaps one is on the way.

Dynasties

From earlier than 1000 BCE until the revolution of 1912 which both ended the monarchy and brought a new approach to 'Westernization', China (although initially smaller) was at least theoretically ruled by a king, later styled 'emperor', whose role in ensuring the harmony of the world and the social system was essential. As is usual, emperors succeeded one another in an orderly way on the whole, as a 'dynasty' (compare the dynasties of ancient Egypt, or the Tudors and Stuarts in England); but from time to time the succession was broken, one dynasty overthrew another or competing dynasties divided China between them. A not entirely rigid boundary, symbolized by the Great Wall, separated China and 'civilization' from the successive groups of threatening barbarians (Huns, Tatars, Mongols) outside; sometimes a successful barbarian conquest was followed by the adaptation of the conquerors to Chinese culture.

History, including the history of writing, calendar computations, highly skilled work in bronze, and some written texts containing basic mathematics, starts in the *Zhang dynasty ('Shang')*, before 1000 BCE. The dates are not certain, but the historical existence of the Zhang is not in doubt.

Zhou dynasty ('Chou'), c.1000 to 221 BCE

The form of Chinese writing, including the writing of numbers, was fixed. There must have been considerable development of mathematics during this period, particularly its later part, but virtually no documents survive—this is traditionally blamed on the famous 'Burning of the Books' by the first emperor of Qin. From about 500 BCE, under a system which Needham describes as similar to European feudalism, the country was divided into provinces ('Warring States') ruled by great lords and the emperor's authority was minimal. From this time date the main philosophical currents, Confucianism and Daoism ('Taoism'). These (together with Buddhism at a later date) had varying influences on scientific outlook. In Needham's view, Confucianism was socially orthodox and uninterested in science, and Daoism was the reverse. Interestingly, the founding work of Daoism, Lao-Zi's *Dao De Jing* (?third century BCE—estimates vary) provides one of the earliest references to the practice of mathematics:

Good mathematicians do not use counting rods. (Lao Zi, *The Dao De Jing*, cited in Lam and Ang 1992, p. 22)

Lao Zi's view of the most inventive tool of Chinese mathematics would turn out to be mistaken, as we shall see; one could compare the dogmatism of his near-contemporary Plato. In any case, he tells us that already in his time there were good mathematicians and more ordinary ones.

Qin dynasty ('Ch'in'), 221–207 BCE

A strong unifying and repressive government was instituted for the first time. This, although shortlived, has always been seen as a landmark in the central organization of the country. 'Feudalism' was ruthlessly eliminated. Nonetheless, society remained rigidly stratified into 'estates' with merchants, according to Confucian principles, at the bottom in prestige if not in power.

Han dynasty, 202 BCE–220 CE

The overthrow of the Qin was accompanied by a relative liberalization. The main features of subsequent Chinese society took shape, in particular rule by a bureaucracy recruited from the 'scholar-gentry' by examinations. According to Liu Hui (see later), mathematics was one of the six subjects required (the 'six gentlemanly arts'), but it did not long remain so.

The Chinese state now occupied its classical area—that is, as far as Guangzhou (Canton) in the south,[5] and the Great Wall was completed; as Needham suggests,

> to check the drift of Chinese groups towards coalescence with nomadic life, or the formation of mixed economies, at least as much as to keep the nomads out. (Needham 1954, p. 100)

The Han is the first major period for the history of mathematics, since the first works which were subsequently enshrined as classics were composed during the period. These include most importantly the *Nine Chapters on the Mathematical Art*, of which more will follow later in greater detail.

The 400 years which followed the Han dynasty were a time of division, conflict, and occasional unification. However, contacts with the outside world, in particular India, increased through the spread of Buddhism which was introduced in the first century CE. The remaining mathematical classics date from this period, including Liu Hui's commentary (third century CE) on the *Nine Chapters*, which transforms it from a collection of questions and answers (as in Section 1) to a mathematical text. It appears that the occupation of mathematician was respected and relatively flourishing, even if the work produced was of varying quality.

Sui dynasty, 581–618 CE

The Sui dynasty, like the Qin, was a successful unification which ran out of steam after organizing important canalization projects which helped to unify north and south. It marked a second point at which mathematics gained a place on the 'national curriculum', and the canon of 10 classical texts which students were required to study was fixed. In fact, a central mathematical school was set up, but it had few students; and as the official examinations remained exclusively literary/humanistic, it seems not to have lasted more than a few years.

4 The Nine Chapters

> Though it is called the *Nine Arithmetical Arts*, they can reach both the infinitesimal and the infinite. (Liu Hui's preface to the *Nine Chapters*, in Shen et al. 1999 p. 53)

> Whereas the Greeks of this period were composing logically ordered and systematically expository treatises, the Chinese were repeating the old custom of the Babylonians and Egyptians of compiling sets of specific problems. (Boyer and Merzbach 1989, p. 222)

Any consideration of Chinese mathematics has to start with the *Nine Chapters* or *Jiuzhang suanshu*, which dominated all subsequent work in much the same way that Euclid's *Elements* dominated Western (including Islamic) mathematics for the next 1500 years. It is not the earliest classic known—that place is held by the decidedly less mainstream *Zhoubi suanjing*[6] (Cullen 1996). The date and 'authorship' of this text is as uncertain as that of the *Nine Chapters*; Cullen considers it to have been a compilation from the early Han dynasty (second to first century BCE). It is a work

---

5. Sometimes a much wider area was covered, including Korea and northern Vietnam.
6. This title is not explicitly translated by Cullen; Needham translates it as 'The Arithmetical Classic of the Gnomon and the Circular Paths of Heaven'.

which combines the theory of heavens and earth with a certain amount of trigonometry. Being a manual for astronomers rather than a 'textbook', it did not have the same status as a founding work for mathematicians.

The comparison of *The Nine Chapters* with Euclid has been made so often that it is something of a cliché, which is not to say that it is without importance. In a now largely outdated discourse, exemplified by the quote from Boyer and Merzbach, a simple contrast was made between Euclid's use of proof and the axiomatic method as opposed to the supposedly basic practical orientation of the *Nine Chapters*. We shall see that the question is more complicated than that. Some initial points which can be made are:

1. Although much shorter than Euclid, the *Nine Chapters* is a substantial work, highly structured, with each chapter organized around a particular type of problem, and with short but full explanations for how the problems are to be solved. Like Euclid, the work appears to be the end of a process of development of which we have no record; the various methods described must have been worked out in the centuries which preceded the book's final compiling.
2. Historically, the *Nine Chapters* has always been supplemented by commentaries, most particularly that of Liu Hui, which add a theoretical element which is missing from the bare text. As the recent translation points out:

   > Liu was a unique mathematician, well-read in both science and literature, who wrote with great style, selecting appropriate phrases from historical and literary classics in his descriptions of the relevant scientific subjects, and showed succeeding generations how to solve problems and also how to justify and explain the rules used. The *Nine Chapters* would have remained a mere recipe book and not a complete classical mathematical textbook without Liu's work. (Shen et al. 1999, p. 5)

   The work of commenting has continued through Chinese history since Liu, indeed in a modern historicized style it is still ongoing. This (like the Euclid heritage) has had its positive and negative aspects; it has provided a tradition, but has also allowed generations of mathematicians to restrict their work within fairly narrow limits.
3. The detail given on the manipulation of counting-rods makes the book unique in its arithmetical specificity. At key points (e.g. on extracting roots), the text goes into the process of how to proceed with the rods with a precision which must have made clear to readers (if not always to us) both how they should apply the procedures and why they worked.

To illustrate some of these points, let us look again at the problem of the chickens cited in Section 1. In the Islamic world and Europe, the method was to become known as the 'method of double false position'. The brief exposition and answer of the problem (with several similar ones) is followed by the general rule.

The Excess and Deficit Rule. Display the contribution rates; lay down the [corresponding] excess and deficit below. Cross-multiply by the contribution rates; combine them as dividend; combine the excess and deficit as divisor. Divide the dividend by the divisor. [If] there are fractions, reduce them.

To relate the excess and the deficit for the articles jointly purchased: lay down the contribution rates. Subtract the smaller from the greater, take the remainder to reduce the divisor and the dividend. The [reduced] dividend is the price of an item. The [reduced] divisor is the number of people.

Liu's commentary. Let the bottom terms cross-multiply the top, [combine and] then uniformize by the common denominator ... [Lay down] the contribution rates. Subtract the smaller from the greater, this is called the assumed difference, which is taken to be the lesser assumption. Then combine the excess and deficit to be the determined dividend. Therefore dividing the determined dividend by the lesser assumption then gives the divisor to be the number of people, and reducing the dividend gives the item price. (Shen et al. 1999, pp. 359–60)

The first point to note is that the general rule is already, one might say, enough; a far broader account of what needs to be done than the pre-Greek texts which proceeded only by example. The explicit rules of procedure seem to lead to something like a matrix:

$$\begin{matrix} 9 & 6 \\ 11 & 16 \end{matrix}$$

with contribution rates at the top, and excess and deficit below. Cross-multiplying and adding gives the 'dividend' ($9 \times 16 + 6 \times 11 = 210$), while simply adding excess and deficit gives the 'divisor' (15). We are not told where to place these, unfortunately. These are not our answers, but they become what we want after dividing by a third number, the 'difference of the contribution rates', in our case $9 - 6$ or 3. Dividend then goes to price, divisor to number of people.

What does Liu's commentary add? In this particular case, not very much (there are better examples). The original has already laid down the basis of a technical language ('divisor', 'dividend', and so on); Liu's concern is to refine this, by introducing extra explanatory terms such as 'the lesser assumption'. He clearly feels, if the word 'therefore' means anything, that his scheme makes clear why the solution works. Yet it is not, in any sense, a proof.

In a carefully argued essay, Karine Chemla (1997) analyses one of Liu's more substantial commentaries—the one on addition of fractions, after problem 9 in chapter 1. Her argument is that in explaining that the commentary contains a 'proof' we risk simply finding what we are looking for. The sentences in the commentary which count as proof are only a part of the story; they break off at a point where, typically, Liu quotes a much more general idea, from the *Yijing*: 'Things of one kind come together'. What follows is a discourse on why *qi* ('homogenizing') and *tong* ('uniformizing')— the basic procedures in adding fractions—work in the way that they do. What kind of mathematics is it?

In other words, fractions with a common denominator can be added even if the numerators are quite different, while fractions with different denominators cannot be added even if the numerators are close to each other . . . Multiplying [the denominators] means fine division and reducing means rough division; the rules of homogenizing and uniformizing are used to get a common denominator. Are they not the key rules of arithmetic? (Shen et al. 1999, p. 72)

What Chemla is suggesting is that in attempting to correct an unhistorical judgement—'Chinese mathematics had no proofs'—one may fall into an equally unhistorical claim: that the closely argued commentaries of Liu are equivalent to 'proofs' in the Greek tradition. Arguably, they were not, and Liu's aim was a very different one: to explain for his readers how the parts of the *Nine Chapters* worked and came together as a coherent whole.

Exercise 2. *(a) In the general case of 'Excess and Deficit', suppose the price is x. If y people pay a each the excess is b; while if they pay a_1 each the deficit is b_1. What are the formulae for x and y, and what role do the 'dividend', the 'divisor', and the 'lesser assumption' play in finding them? (b) Use either the previous exercise or the method from the* Nine Chapters *to solve problem 7.3: Now jade is purchased jointly; everyone contributes $\frac{1}{2}$, the excess is 4; everyone contributes $\frac{1}{3}$, the deficit is 3. Tell: the number of people, the jade price, what is each?*

5 Counting rods—who needs them?

It follows that the Hindu–Arabic numeral system originated from the rod numeral system, which was developed centuries earlier. (Lam and Ang 1992, p. 148)

Most authors believe that counting-rods were manipulated on a special surface called the counting-board or chessboard, which would have been to rods what the frame and the bars are to the abacus. However there is no proof that such boards existed. (Martzloff 1995, p. 209)

What (in ancient Chinese mathematics) was done with counting rods was considered fairly well established before the doubts raised by Martzloff,[7] and the important claims for the numeral system made by Lam and Ang, following Needham and others, make it desirable to establish what we can. Not quite a way of writing numbers, nor simply a calculating tool, the rods were used to combine the two in a unique way which some specialists at least see as providing an approach to number-manipulation which was better than anything used before or since. Indeed, it seems that 'difficult' mathematics declined when, around the sixteenth century, the abacus replaced the rods as the instrument of calculation.

Numbers have been written in the Chinese script at least since the Zhou dynasty in a form which corresponds exactly to the words:

三	百	八	十	七
san	bai	ba	shi	qi
three	hundred	eight	ten	seven

which means: '387' (as is obvious). In particular, it is in this form that they are written in the classics such as the *Nine Chapters*. As Lam and Ang (1992, p. 14) point out, this means that, given the particular nature of Chinese writing, the usual distinction between writing numbers in words (e.g. 'three hundred and eighty-seven') and figures (e.g. '387') disappears.

This is quite convenient in itself. However, at some time in the Zhou dynasty the counting rods were developed as an aid to actually doing sums—one could conjecture, sums of the kinds needed by merchants or bureaucrats. A rod was

a round bamboo stick 1 *fen* (about 2.5 mm) in diameter and 6 *cun* (about 25 cm) in length... (Shen et al. 1999, p. 12)

(The exact dimensions, and the materials were more variable than this description suggests.) Placed in patterns, they could symbolize the numbers 1–9, in one of two forms or 'series'—either horizontal or vertical (Fig. 1).

There are references to how they were used in early classics, for example, the *Sunzi suanjing*,[8] but (presumably because the texts were supplemented by a teacher's instructions) they are not explicit. The usual explanation of their use, which follows the way in which they were used in Japan in the eighteenth century, is that:

1. Laid out along a single row of the counting board, the rods gave the decimal representation of a number, with an empty space denoting a 'zero': so '60390' was represented by Fig. 2 below.
2. By convention, vertical and horizontal types alternated, so that there was less room for confusion about where an empty space had occurred ('84' looked different from '804'—try to see why).

---
7. Who would concede that we know a great deal, but argues that it is not quite as much as we think.
8. This is the text, of very uncertain date (between the first and fifth centuries CE?) edited and translated by Lam and Ang (1992).

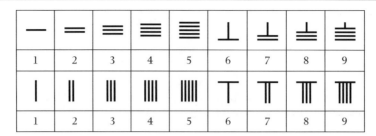

Fig. 1 Rod numbers.

Fig. 2 60390 as a rod number

3. As in our system, numbers read along the rows, while the various numbers belonging to a sum (multiplier, multiplicand, product) occupied different rows of the board.

To clarify, here is the rather simple operation of squaring 81, from the *Sunzi suanjing*. This is no more advanced than the *Nine Chapters*, in fact less so, but in places more explicit. The text clearly does describe *some* procedure with counting rods. The pictures in rod-numbers are the reconstruction in Lam and Ang (1992) of how the calculation would have been done, since the numbers in the text are *not* rod-numbers. In the first sentence, for example, 'nine' would be written 九, and '81' would be 八十一. The roman numbers refer to the diagrams below, showing the progress of the reconstructed calculation.

Nine nines are 81, find the amount when this is multiplied by itself. Answer: 6561.

Method: Set up the two positions [upper and lower] (i). The upper 8 calls the lower 8; eight eights are 64, so put down 6400 in the middle position (ii). The upper 8 calls the lower 1: one eight is 8, so put down 80 in the middle position (iii). Shift the lower numeral one place [to the right] and put away the 80 in the upper position (iv). The upper 1 calls the lower 8; one eight is 8, so put down 80 in the middle position (v). The upper 1 calls the lower 1; one one is 1, so put down 1 in the middle position (vi). Remove the numerals in the upper and lower positions leaving 6561 in the middle position (vii). (Lam and Ang 1992, p. 34)

The progress of this very simple example is illustrated by the rod-number diagrams (i)–(vii) (Fig. 3); you should translate these into 'Arabic' numbers for yourself. Note that the terms of the upper number are removed when they are finished with; and that the author takes it for granted that when you have put down the second 8 (stage v) you use basic rod addition to amalgamate it with the 648 you have already and get 656.

No one has come up with a better explanation of how the system worked. The first *written* records containing rod-numbers used mathematically date from the fifth to tenth centuries CE and the most coherent ones from much later again. In the meantime, the use of rod-numbers could have evolved. Martzloff's scepticism (it is no more than that) is based on the absence of evidence for two key assumptions: the use (a) of a 'board' to order the calculation, and (b) of blank spaces as a zero-equivalent at such an early date.

Let us, though, suppose the system granted, as it is widely believed to have been used and is a reasonable interpretation of the words in the *Sunzi suanjing*. The question of whether this

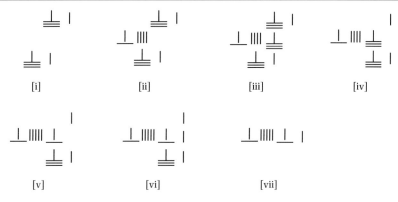

Fig. 3 The stages in calculating 81 × 81 by rod numbers.

constitutes an 'invention' of the decimal place-value number system, like other priority questions, then becomes serious, and needs some clarification. The counting rods, in some form, were certainly being used about a thousand years before our first certain record of 'Indian' place-value numbers (ninth century CE see next chapter). By that time, as we have seen, the Buddhists had established a fairly regular route for pilgrims between China and India, and in the seventh century the most famous of them, Xuan-Tsang, was defending his country's civilization to an Indian audience:

They have taken the Heavens as their model, and they know how to calculate the movements of the Seven Luminaries; they have invented all kinds of instruments, fixed the seasons of the year, and discovered the hidden properties of the six tones and of music. (Needham 1954, p. 210)

Although the shapes of Indian–Arabic numbers are quite unlike those of counting-rod numbers, Lim and Ang argue that the *idea* of decimal place-value computation must have been transmitted from China, probably to India, since the way of perfoming calculations in the earliest textbooks—in particular the Arabic texts of al-Khwārizmī and al-Uqlīdisī—is almost identical. Only the form of the symbols from 1 to 9 was changed, with the zero or dot being devised for the empty space.

Again, this is plausible, although the records which would establish it may never be found. However, as we have noticed, the counting-rod numbers, until the Tang dynasty (about 600 CE), retained the status not of a way of writing numbers but of a way of *working* with them. The early texts say what you must do, but nowhere do they even draw a picture of the counting-rod number, much less insert one into the text. The counting-rod numbers, in their early life, were 'fleeting' (to borrow Lam and Ang's term) and dynamic, there to be erased and transformed as the example shows.[9] Later, (perhaps from the tenth century CE—we canot be sure) they were written in mathematical textbooks to illustrate procedure; and in the thirteenth century a zero symbol was introduced[10] but even so they did not become the current way of writing numbers. If the idea of using a decimal number representation did diffuse to India, then it underwent an important change, since the system became something which (a) was written down and (b) became *the* principal representation of a number in everyday transactions. There may then have been diffusion, for example, of the zero symbol, in the other direction, from India or the Middle East to China. Assuming that the 'Indian' numbers did derive from the rods, they changed in becoming the

---

9. Interestingly, it seems that the early use of Indian numbers in the Islamic world involved erasure—they were traced on the 'dust-board' to which al-Uqlīdisī refers (see next chapter).

10. By Qin Jiushao, apparently (see later).

favoured medium for writing numbers, and in this sense we could think of a two-stage 'invention' of the place-value system. The reader who takes a little time to try out the method of counting rods might reflect that the *written* decimal place-value system which we have is not necessarily the best for all purposes.

On one other point priority is certainly established: the Chinese from an early period were quite happy with negative numbers, as Westerners were not. Liu is explicit on this; at the point where the *Nine Chapters* give a detailed and helpful 'Sign Rule'—'like signs subtract, opposite signs add'—he supplies a note on procedure:

> Now there are two opposite kinds of counting rods for gains and losses, let them be called positive and negative [respectively]. Red counting rods are positive, black counting rods are negative. (Shen et al. 1999, p. 404)

Martzloff speculates that this ease in dealing with signs may have arisen not simply from the manipulation of debts, but from the duality underlying Chinese natural philosophy:

> For example, astronomers imagined coupling the planet Jupiter with an anti-Jupiter, whose motion was deduced from the former by inversion; diviners practised a double-sided divination with symmetrically arranged graphics; not to mention also, of course, *yinyang* dualism. (Martzloff 1995, p. 200)

If Indians or Westerners 'borrowed' the idea of negative numbers at some much later date, they made more heavy weather of it.

Exercise 3. *Make your own set of counting rods and try to perform a simple multiplication on the lines of the one above.*

6 Matrices

So far we have only looked at the elementary parts of the *Nine Chapters*. This gives a wrong picture of early Chinese mathematics, which contained some sophisticated procedures—always framed in terms of straightforward problems with general explanation. Two in particular stand out:

1. The extraction of roots, a combination of counting-rod and geometrical arguments, which would lead to more general algebra.
2. The solution of systems of linear equations by an equivalent of what we call matrices.

Here we shall consider the second. Once again, to translate it into modern terms ('we are using matrices') is clearly a misrepresentation of the procedure of a Han dynasty mathematician using counting rods; and yet, the comparison of the methods is an interesting one, since we can see what elements there are in common. The subject is covered in the eighth chapter, 'Rectangular Arrays', or *fangcheng*; and the title in itself says something about the material. A large number of the problems concern different grades of paddy, and Liu comments, 'it is difficult to comprehend in mere words, so we simply use paddy to clarify'.

This is a fascinating remark, if we think of the question of abstraction. It almost seems as though Liu is undercutting the apparent concreteness of the *Nine Chapters* by claiming that he, at least, could use an abstract language ('mere words'). This would not, of course, be algebraic symbols, but given the nature of Chinese mathematics they could be rather similar, as we shall see. Perhaps we should think of the characters for 'low-grade paddy' (or medium, or high) as a more complicated version of the symbols x, y, and z.

Here is the text of chapter 8, problem 1, together with the 'Array Rule' which solves it.

Now given 3 bundles of top grade paddy, 2 bundles of medium-grade paddy, [and] 1 bundle of low grade paddy. Yield: 39 *dou* of grain. 2 bundles of top grade paddy, 3 bundles of medium-grade paddy, [and] 1 bundle of low grade paddy, yield 34 *dou*. 1 bundle of top grade paddy, 2 bundles of medium-grade paddy, [and] 3 bundles of low grade paddy, yield 26 *dou*. Tell: how much paddy does one bundle of each grade yield?
Answer: Top grade paddy yields $9\frac{1}{4}$ *dou* [per bundle]; medium grade paddy $4\frac{1}{4}$ *dou*; [and] low grade paddy $2\frac{3}{4}$ *dou*.

The Array [*Fangcheng*] Rule

[Let Problem 1 serve as an example,] lay down in the right column 3 bundles of top grade paddy, 2 bundles of medium grade paddy, [and] 1 bundle of low grade paddy. Yield: 39 *dou* of grain. Similarly for the middle and left column.

Use [the number of bundles of] top grade paddy in the right column to multiply the middle column then merge. Again multiply the next [and] follow by pivoting. Then use the remainder of the medium grade paddy in the middle column to multiply the left column and pivot. The remainder of the low grade paddy in the left column is the divisor, the entry below is the dividend. The quotient is the yield of the low grade paddy ...

The above quotation is enough (a) to compute the basic solution, the yield of low-grade paddy (the others can be found by substitution), (b) to show how the method is described in the *Nine Chapters*. For us, the description of what to do is unclear unless you have had the terms, for example, 'merge' and 'pivot', explained to you. The process begins as follows. First, imagine counting rods laid out (perhaps on a board) to represent the numbers used; in matrix terms:

$$\begin{bmatrix} 1 & 2 & 3 \\ 2 & 3 & 2 \\ 3 & 1 & 1 \\ 26 & 34 & 39 \end{bmatrix}$$

The textbook method today would be to get enough zeros in the matrix ('triangular form') by subtracting multiples of rows or columns from each other. This is nearly what the method described does. Here is Liu's explanation for the cryptic 'merge':

The meaning of this rule is: subtract the column with smallest [top entry] repeatedly from the columns with larger [top entries], then the top entry must vanish. With the top entry gone, the column has one item absent.

How does this work? First note that we have (from the method above) multiplied the middle column by 3 = number of bundles of top-grade. We now subtract the right column (the smaller) repeatedly from the middle (the larger). The stages are:

$$\begin{bmatrix} 1 & 6 & 3 \\ 2 & 9 & 2 \\ 3 & 3 & 1 \\ 26 & 102 & 39 \end{bmatrix} \rightarrow \begin{bmatrix} 1 & 0 & 3 \\ 2 & 5 & 2 \\ 3 & 1 & 1 \\ 26 & 24 & 39 \end{bmatrix}$$

The differences from today's procedure are fairly trivial. Most probably:

1. we would look for a column, for example, the left one, which could be subtracted from others without first having to 'multiply them up';
2. we would say we were subtracting twice column 3 rather than saying that we were subtracting it repeatedly.

These are details. The method, it could be said, has not changed even if questions about different grades of paddy are less frequent. There is, though, a more subtle difference. Confronted with such a question, a modern textbook would call the paddy yields x, y, and z, form three equations and *then* write a matrix to solve them; in the history of European mathematics, xs precede matrices by about 200 years. In the *Nine Chapters*, there is no intermediary between the paddy and the 'matrix'. There is indeed what one could call abstraction; but instead of our kind, which consists of replacing unknowns by symbols, it inputs data directly into a solution diagram. This particular kind of abstraction seems to have been peculiarly Chinese. It was clearly tied to the use of counting rods; and so, we could guess, it travelled less well than some other Chinese mathematical inventions.

Exercise 4. *Follow through the calculations, and check that they give the right answer; either by using matrices (if you know them), or using counting rods (if you can find or make them), or any other way.*

7 The Song dynasty and Qin Jiushao

There is no reason to doubt that the last half of the thirteenth century was the culminating point of Chinese mathematics. (Libbrecht 1973, p. 13)

In later generations scholars were very proud of themselves and, considering [mathematics] inferior, did not teach [or discuss] them. Only calculators and mathematicians were able to manage multiplication and division, but they could not comprehend square-root extraction or indeterminate analysis. In case there were calculations to be performed in the government offices, one or two of the clerks might participate but the position of the mathematicians was never held in esteem; their superiors left things to them and let them do as they pleased; [but] if those who did computations were only that sort of man, it was merely right that they should be disdained. (Qin Jiushao, *Shushu jiu zhang*, preface, in Libbrecht 1973, p. 60–1)

Subsequent historians have referred to the *Song dynasty* (960–1279) as a 'golden age' for culture in many respects, and for mathematics in particular. To mathematicians such as Qin Jiushao, who complained of their treatment as the nerds of the Chinese hierarchy, it did not appear so. The dynasty lost dynamism over a short period, and its territory shrank to the southern half of China, the north being controlled by a rival 'barbarian' dynasty, the Jin. In the thirteenth century which Libbrecht describes as the 'culminating point' the Mongols under Chinggis Khan fought a 50-year war and finally overthrew the Song rulers. They ruled under the name of the *Yuan dynasty* from 1260 to 1368. The outstanding mathematics for which the period is known is distinctly enigmatic. We have works from four apparently unrelated writers from the thirteenth/fourteenth centuries: the prolific but fairly elementary Yang Hui, and three more surprising mathematicians often sharing common concerns, but working in isolation, often with no official position to provide them with problems or support. Their work is both *innovative*, in that it is clearly different in kind from what has appeared before, and at the same time *traditional*, in that the models which it draws on are supplied by the classics.

1. *Li Zhi* (1192–1279), who lived in the north, worked under the Jin rulers and later the Mongols, and wrote the eccentric text called *Ceyuan Haijing* ('Mirror comparable with the ocean'), apparently dated 1248. This is entirely devoted to obtaining equations from problems of a geometrical type about a town whose plan is drawn at the outset (Fig. 4). The problem is always to find the town's radius; the answer is always 120.

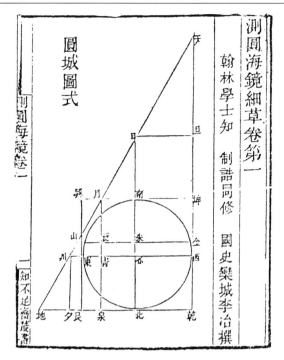

Fig. 4 Li Zhi's 'round town' form the *Ceyuan haijing*.

2. Qin Jiushao (c.1202–61), who worked in the south, and during a boisterous life (details on various websites, or in Libbrecht) wrote the long and semi-practical *Shushu Jiuzhang* (Computational Techniques in Nine Chapters). As well as material which can be found elsewhere, this provides the most advanced source for the 'Chinese Remainder Theorem': how to find a number n which leaves remainders $a, b, c, \ldots$ when divided by $p, q, r, \ldots$.[11] The *Shushu jiuzhang* is a complex work, organized around practical problems but often dealing with them in far-fetched ways. Among other things it illustrates the disturbed politics of the period by some of its questions: how to arrange soldiers in formation, how to find the distance of an enemy camp. At the same time, the mathematics introduced into the solution of the problems seems sometimes to pursue difficulty for its own sake.

3. Zhu Shijie (dates unknown, end of thirteenth century), another northerner, wrote two books which were printed but never seem to have been used for teaching. Like the other 'difficult' Song writings, they were probably soon forgotten. The very long (over 1000 pages) *Siyuan yujian* (approximately: 'Mirror trustworthy as jade relative to the four unknowns', see Martzloff 1995, p. 153) is, of course, again a collection of problems. However, it is very much more, since the problems lead to sums of series, high degree equations (again!), in fact a highly organized algebra whose ideas are similar to those of Li Zhi.

Since it is partly by chance that our key texts from the Song survived there may have been others. Many of the methods, and even the problems seem to have been common, and one wonders why.

---

11. There is no space to deal with this problem here, or the questions raised by Qin's treatment of it. His results were not rediscovered until the time of Gauss (1800)—and are in some respects more general than Gauss's work.

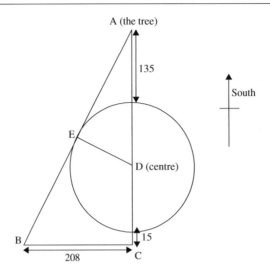

Fig. 5 Following Chinese convention, south is at the top. B is the point where the walker catches sight of the tree. Other lettering refers to the solution of the problem (later).

To give a particularly striking example: in his *Ceyuan Haijing* Li Zhi gives the problem:

135 *pu* out of the south gate [of a circular city] is a tree. If one walks 15 *pu* out of the north gate and then turns east for a distance of 208 *pu*, the tree comes into sight. Find the diameter of the town. (Ceyuan Haijing, problem 11.18)

Li Zhi solves the problem by an equation of degree 4, with the result 240 *pu*. The picture is shown in Fig. 5; clearly, its analysis requires more than basic geometry—the 'Pythagoras' theorem, called *gougu* by the Chinese, properties of tangents and similar triangles.

On the other hand, Qin Jiushao solves an extremely similar problem by an equation of degree 10 (see Libbrecht 1973, p. 134ff)—partly, it is true, by the artifice of using 'x^2' (as we would say) for the diameter. There are very strong reasons to suppose that Li and Qin never met or communicated—they lived as near contemporaries in mutually hostile parts of China. The two symbolize two kinds of mathematician: for Qin, the world consists of watchtowers constructed on the walls of cities (see Fig. 6), while for Li, people wander aimlessly round similar cities trying to catch sight of trees. Yet the mathematics is much the same. What is the explanation for this sudden eruption of a 'school' of mathematicians who, working apparently independently, produced work which is both original and in some ways related?

A part of the explanation is simple: it lies in our great ignorance. None of the writers, with regard to the work which seems most striking, claims to be innovating, and some refer explicitly to predecessors whose works are lost. The 'Golden Age' of the thirteenth century might therefore appear less golden if we knew more of the ages which had preceded it. So, for example, Yang Hui (like others) uses the 'Pascal triangle', but ascribes it to the eleventh-century writer Jia Xian whose works have not survived. Even the striking notation for polynomials called the '*tianyuan*' or celestial element method was apparently copied from an earlier lost writer. In Libbrecht's judgement:

[I]t is obvious that only a few names have been recorded, and that the greater part of Chinese mathematical works have been lost. (Libbrecht 1973, p. 18)

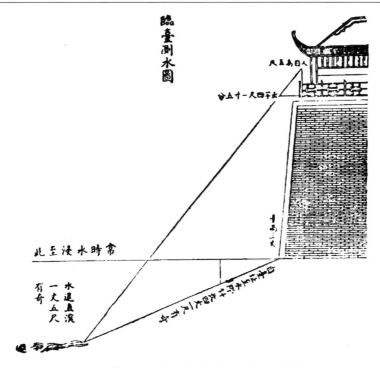

Fig. 6 Illustration from the *Shushu jiuzhang*, p. 167.

We have, as often in the early history of mathematics, an impressive part of a structure which has survived almost by chance while the rest is missing. Failing dramatic discoveries of the texts from the earlier period, we must be content with what remains, and try to understand the aim of this 'school' which was not a school. While our evidence for a 'culture of mathematicians' during the Song is sparse, the books do seem to aim at a specialist audience, whether it existed or not. Libbrecht emphasizes both the practicality of the *Shushu jiuzhang*:

In a splendid work on architecture, the *Ying-tao fa-shih*, there is a full description of materials and constructions, but what is lacking is plans for carrying out the work: the calculation of the building materials, the number of workmen, the provisions and wages. All this we find in Ch'in Chiu-shao's work. (Libbrecht 1973, p. 8)

at the same time as its 'advanced' nature:

In Ch'in's work all the basic operations (even the square root extraction) are taken for granted. For a beginner's textbook its problems are much too complex; it would be useful only for advanced students...It is possible that they [the Song texts] were unsuccessful substitutes for older books written in a less advanced phase of mathematical knowledge. But as mentioned earlier, none of the writers was a mathematics teacher; even Li Yeh [Li Zhi]'s work, which was printed in the thirteenth century, was never used as a textbook. (Libbrecht, 1973, pp. 8–9)

What can be said at the moment is that the work of the Song mathematicians, for the student who is prepared to overcome some formidable initial obstacles, is a really promising field of research. While much progress has been made in understanding them, there are still controversial points and open questions. The mixture of practical setting with the pursuit of difficulty apparently for its own sake, while not unique to the Chinese, could lead us to think again about what mathematics is 'for'.

Exercise 5. *Find the equation which solves Li's problem, and check that $x = 240$ is a root.*

A note on 'equations'

At various points in earlier history it has been necessary to exercise caution about using the word 'equation'—for the activities of the Babylonian scribes, for Euclid's book II, and so on. Still more have we tried to emphasize that, if we write a Babylonian problem in terms of *x*s, it is to help us read it and not to indicate how the authors thought of it. If this caution is noticeably absent in those who write about the Song mathematicians, this is because what they wrote down does *look* remarkably like an equation, even if it has no *x* and no '= 0'. To see this, look at Fig. 7, which reproduces the 'equation' for Qin's problem 6.2 (Martzloff 1995, p. 233ff). We translate this as

$$-x^4 + 15{,}245x^2 - 6{,}262{,}506.25 = 0$$

The coefficients are written in a column, from the constant term downwards $(1, 0, 15{,}245, 0, 6{,}262{,}506.25)$; below each one is its 'rank', a verbal/symbolic description of the power of the unknown to which it belongs. Positive (*cong*) and negative (*yi*) coefficients are distinguished by writing the relevant word by them. It is not an equation as we know it—but it can be seen as a convenient translation of one to the language of arrays which had been so successful in the *Nine Chapters*; and Qin's subsequent manipulations seem to be related to the *fangcheng* method. No more an equation than an array is a matrix, it is a clearly defined tool of equivalent sophistication. Perhaps we should still be cautious about translating it into our own terms; but we hardly need a dictionary to do so.

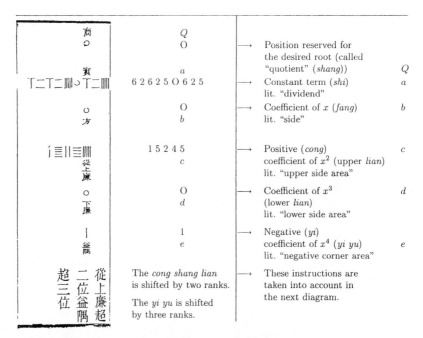

Fig. 7 The equation for Qin's problem 6.2.

8 On 'transfers'—when, and how?

Seek knowledge, even as far as China. (The Hadith)

Aside from the unusual nature of algebra in Song times—why was it being done, and what were its rules of operation?—we have what is sometimes an open question, sometimes an argument about origins. Several apparently similar techniques, most particularly the use of the binomial triangle ('Pascal's triangle') and of techniques for solving equations by approximation, seem to have appeared both in China and in the Near East around the same time, say the eleventh to twelfth centuries. We know of numerous contacts between China and the Islamic world, the earliest relevant one being the compiling of a calendar—a perennial headache for the government—by a Muslim named Ma Yize for the Emperor in the late tenth century. From then on, Muslim scholars seem to have been frequent visitors, until under the Mongols of the Yuan dynasty (thirteenth century) a large number of Muslim 'artisans' were settled in north China; and one Zhamaluding presented the Emperor Qubilai ('Kubla Khan', the son of Chinggis) with yet another calendar, and astronomical instruments. Arabic or Persian astronomical and mathematical books were also imported. Given all this, it seems to make for historical economy to suppose that some transmission of knowledge was taking place one way or the other. However, as with the question of 'counting rods' and the decimal system, it is more complicated than we might wish. For example:

1. There is no record of any mathematician of either culture referring to work from another. As Saidan notes:

   Al-Mas'ūdī... writes much about Hindu wisdom and learning, and refers to Chinese technology and social life, but never to Chinese science. The learned al-Bīrūnī does not refer to Chinese science in his *Chronology* nor does Ṣa'id in his *Ṭabaqāt*. (Saidan in al-Uqlīdisī 1978, p. 455)

2. The essential works in which the Chinese methods were introduced, such as those of Jia Xian (eleventh century), are lost and as Martzloff says:

   Our knowledge of this subject is very imprecise, since it is based on extracts from 13th century works such as those of Yang Hui or Zhu Shijie, accessible to us through the medium of 19th century editions. (Martzloff 1995, p. 17)

3. So far as we *can* draw a parallel between Chinese and Islamic algebra, they seem to have been very different pursuits.

   In non-Chinese algebraic manuals, this character [the equation] is treated with the utmost care and respect; it is studied in minute detail, in the search for the secrets of the algebraic formulae which unveil the results. This initially involves a study of equations of degree two, since these are the most docile... But in medieval China, the degree of equations is of little importance... Chinese equations are not exactly equations, but algebraic forms or schemes for extracting roots, which... consist of sequences of numbers to be operated on, as though one were extracting square, cube or *n*th roots. (Martzloff 1995, p. 261–2)

Given this, and given the specialized technical language of Chinese algebra, it is not surprising that the question of what could have been transmitted is a hard one. Let us consider a 'practical' problem which leads to an equation, from *Shushu jiuzhang*, III.1. The problem (Libbrecht 1973, pp. 97–9) is to find the area of a 'pointed field' (Fig. 8) when one knows the measurements shown. Again, the answer could be found trivially ($h_1 = 36, h_2 = 20$ by Pythagoras; area = $\frac{1}{2}(30 \times (36 + 20)) = 840$). Again, one assumes that Qin knew the answer in setting up the problem; and, as in the preceding question, that he knew it could be solved easily. Instead, he

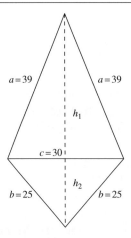

Fig. 8 The 'pointed field', from Qin Jiushao's problem.

takes the numbers

$$A = [b^2 - (c/2)^2] \times (c/2)^2; \qquad B = [a^2 - (c/2)^2] \times (c/2)^2$$

(in this case, $A = 90{,}000$, $B = 29{,}160{,}000$). He then shows that the area satisfies the equation

$$-x^4 + 2(A + B)x^2 - (B - A)^2 = 0$$

In fact, $x = \sqrt{A} + \sqrt{B}$, which is a root of the equation.

Feeding the numbers in, this equation becomes:

$$-x^4 + 763{,}000x^2 - 40{,}642{,}560{,}000 = 0$$

In the 'Western' world, which for our purposes at the time means Islam, the idea of applying such exotic methods to a simple problem would have been rejected out of hand. Equations involving only powers of x^2 were known, and had been dealt with by the simple method of treating them as quadratics in the variable 'square of thing' or x^2, but the Chinese notation with its negative signs would have posed difficulties.

What Qin does, and this is again odd if we suppose that he knew the answer, is to embark on a sophisticated approximation procedure. This makes sense if you are trying to find (e.g.) $\sqrt{2}$ to three or more decimal places, but for $x = 840$ it looks like overkill. The idea is to find the figures of the answer one at a time. If we know (say that the answer has three figures and the first is eight, then the equation $f(x) = 0$ becomes $f(800 + y) = 0$. Qin's method is to use a simple way of working out the coefficients of y in the new polynomial $g(y) = f(800 + y)$, which he demonstrates by a sequence of rod-number diagrams. The method which he used has been known since its 'rediscovery' in the nineteenth century as the Ruffini–Horner procedure.

It is now certain that some Chinese mathematicians (e.g. Qin) and some Islamic ones (e.g. al-Samaw'al, rather earlier) knew and practised this procedure, and the question of who might have borrowed from the other has become something of a crux in the question of what could have been transferred. It is not our intention to go further into the details of the procedure (see Martzloff 1995, p. 232ff and Libbrecht 1973, p. 180ff for the Chinese version and Rashed 1994,

p. 91ff for the Islamic one). Still less are we about to discuss the priority, on which the evidence is slender to non-existent; a recent evaluation is given by Karine Chemla (1994).

It seems almost inconceivable, however little mingling there was (e.g.) between Chinese and Muslims at the Mongol court that some knowledge was not shared, although it may have been at a fairly basic level. Given which, one might ask:

Why did Islamic mathematicians not learn anything of the 'matrix' method, and the use of negative numbers as routinely practised by the Chinese? Why did the Chinese not learn the formula for the volume of a sphere, the construction of the regular solids, and the properties of conics? The problem is as much to explain what was not transferred as to find what was.

A different question arises in response to the argument that the occurrence of the same idea in two cultures must imply copying. We could then ask (putting ourselves imaginatively into the situation of a medieval Chinese mathematician) how difficult we think its discovery might be. Pascal's triangle (the one with the binomial coefficients in, Fig. 9) is a case in point. At one level, it is a pattern of numbers which one could discover if one were playing with them idly. As Martzloff comments (1995, p. 91) on Needham's assertion that it was transferred from China to India: what was the triangle, in a given culture, being used for? Ancient Indians used something like it for problems involving combinations, Pascal for probabilities. In Samarkand and Beijing it seems to have been more an aid towards root-extraction, via the relation of the nth row in the triangle to the coefficients in $(a + b)^n$. But this is not so difficult that one has to suppose its discovery to have happened only once. Sometimes at an elementary level, the same ideas occur with the same

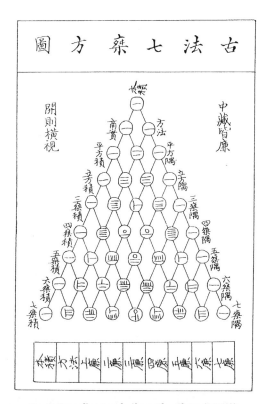

Fig. 9 Pascal's triangle (from Zhu Shijie (1303)).

numbers as examples (cf. Høyrup 1994); then one can suppose that there is a common source, though there is usually not enough evidence to deduce what the ultimate source was.

9 The later period

The *Ming dynasty, 1368–1644*, a Chinese dynasty which overthrew the Mongol rulers, saw a second period of 'decline', in that while mathematical work was still done, the emphasis was on commenting the classics and the innovations of the eleventh to fourteenth centuries were gradually forgotten (the works were lost, or found too difficult). Whether this was a result of the introduction of the abacus, as some writers have suggested, or because of a renewed value given to the literary as opposed to the practical arts, the description of the Song and Yuan as a 'golden age', and of the Ming as a period of stagnation are a commonplace (and were even recognized at the time). This was dramatically changed by the arrival of the Jesuit missionary Matteo Ricci at the end of the sixteenth century. An able scientist himself, Ricci saw (as others have after him) the key to conversion in the exploitation of Western science and technology, which were just entering the period we call the 'scientific revolution' (see Chapter 6). Once again, the calendar was seen as a way in:

We should change the Chinese calendar, this would enhance our reputation, the doors of China would be more open to us, our position would be more stable and we would be freer. (Tacchi Venturi 1911–13, II, p. 285)

Besides constructing the new calendar, Ricci with his Chinese assistants translated Euclid and other Western works, simplifying as they went along. The Jesuits successfully predicted eclipses, which the old calendar had been failing to do, and introduced logarithms not long after they had appeared in Europe. The fall of the Ming and their replacement by new outsiders, the Manchu *Qing dynasty* (1644–1911), did not basically interrupt this programme.

Was the result a victory of the new European methods over the classical Chinese? Emphatically not. The Jesuits hoped that the certainty of Euclid's geometry could be related to that of their religion—they were not the first nor the last to make the equation. Naturally, a reaction set in, and there was a revaluation of traditional Chinese mathematics, helped by scholars who rediscovered and edited many of the classics from the seventeenth to the nineteenth centuries. From this period date such fascinating 'hybrid' figures as Mei Wending.

Euclid's geometry is completely transfigured in Mei Wending's three-dimensional figures, which take no account of perspective, and in his immersion in numerical computation[12] . . . At the same time Mei Wending rehabilitated ancient Chinese techniques such as the *fangcheng* [array] method for solving linear systems. (Martzloff 1995, p. 25)

Again, there is much more work to be done on the history of Chinese mathematics *after* Ricci, precisely because it remained in tension between a vital, indeed increasingly strong tradition and the Western mainstream. The final victory/assimilation to Western mathematics came, naturally, with the fall of the Emperor in 1912.

---

12. It has to be said that many other writers in the Middle East and in Europe had 'transfigured' Euclid's geometry in different ways by this time.

Solutions to exercises

1. (a) We would probably call the original cereal x. Then after the three taxes have been exacted, there is left $\frac{2}{3} \cdot \frac{4}{5} \cdot \frac{6}{7}x = \frac{16}{35}x$ (which we are told is 5). Hence $x = \frac{175}{16} = 10\frac{15}{16}$ dou; and $\frac{15}{16}$ dou=$\frac{150}{16}$ sheng =$9\frac{3}{8}$ sheng. The *Nine Chapters* of course does not use 'x', but proceeds inversely, telling you to multiply the remaining 5 by 3, 5, and 7 and divide by 2, 4, and 6 ('the remainders'). The result is the same.

    (b) The *Nine Chapters* solution is given above in Section 4. We would call the number of contributors x (say) and the price y, and arrive at simultaneous equations:

$$9x - 11 = y \tag{1}$$
$$6x + 16 = y \tag{2}$$

It is then usual to subtract, and get $3x = 27$, so $x = 9$; from which y follows by substitution.

2. (a) We have:

$$ay - b = x \tag{3}$$
$$a_1 y + b_1 = x \tag{4}$$

From which, subtracting, $(a - a_1)y = b + b_1$. Note that $a > a_1$, because it gives the excess! Hence,

$$y = \frac{b + b_1}{a - a_1}$$

Here, $b + b_1$ is the 'divisor', and $a - a_1$ (greater contribution minus smaller) is the 'lesser assumption'.

Substituting back (see 1 (b)), and simplifying, we get

$$x = \frac{ab_1 + a_1 b}{a - a_1}$$

and now $a_1 b + a b_1$, the result of cross-multiplying, is the 'dividend'. (b) In this case, the dividend is $\frac{1}{2} \times 3 + \frac{1}{3} \times 4 = \frac{17}{6}$; the divisor is $4 + 3 = 7$, and the lesser assumption is $\frac{1}{2} - \frac{1}{3} = \frac{1}{6}$. This gives 42 for the number of people, and 17 for the price. (The *Nine Chapters* has rules for dealing with fractions, but I have assumed we do not need them.)

3. No solution required.

4. We have the matrix as simplified, and now we simplify it further. Following the *fangcheng* method, multiply the first column by 3 and subtract the third column. After which, you multiply the new first column by 5 (the first non-zero entry in the second column) and subtract 4 × the second column. Result:

$$\begin{bmatrix} 0 & 0 & 3 \\ 4 & 5 & 2 \\ 8 & 1 & 1 \\ 39 & 24 & 39 \end{bmatrix}, \begin{bmatrix} 0 & 0 & 3 \\ 0 & 5 & 2 \\ 36 & 1 & 1 \\ 99 & 24 & 39 \end{bmatrix}$$

This (if z is the yield of low-grade) gives us $36z = 99$ from the first column, so $z = \frac{99}{36} = 2\frac{3}{4}$ as required. We now use the second column to find y, the medium-grade yield ($5y + z = 24$), and so on.

5. (See Fig. 5.) The equation on the St Andrew's website actually is not 'the answer', since it computes the radius (correctly, as 120). If we want x to be the diameter, as the question asks, then we must have radius $= \frac{x}{2}$. We use the similar right-angled triangles ABC, ADE. We have $AC = x + 150$, $BC = 208$, and $DE = x/2$ easily. $AD = 135 + x/2$, so by Pythagoras's theorem, $AB = \sqrt{208^2 + (x+150)^2}$. Since

$$\frac{AD}{DE} = \frac{AB}{BC}$$

$AD \cdot BC = AB \cdot DE$, and $(270 + x) \cdot 208 = (\sqrt{208^2 + (x+150)^2}) \cdot x$ (doubling to remove the halves). Now get rid of the square root by squaring everything; we get

$$208^2(270+x)^2 = (208^2 + (x+150)^2)x^2$$

or

$$x^4 + 300x^3 + 22{,}500x^2 - 23{,}362{,}560x - 3{,}153{,}945{,}600 = 0$$

I leave it to you to check that $x = 240$ is a solution. It would seem likely that Li knew the answer in the first place not from solving the equation but (like most textbook writers) because he had chosen the numbers to come out exactly; in this case so that the sides of the triangles are in the ratio $8 : 15 : 17$.

5 Islam, neglect and discovery

1 Introduction

It should be clear from the present chapter that the traditional view of the Arabs as mere custodians of Greek learning and transmitters of knowledge is a partial and distorted one. (Joseph 1992, p. 344)

A number of medieval thinkers and scientists living under Islamic rule, by no means all of them 'Moslems' either nominally or substantially, played a useful role of transmitting Greek, Hindu, and other pre-Islamic fruits of knowledge to Westerners. They contributed to making Aristotle known in Christian Europe. But in doing this, they were but transmitting what they themselves had received from non-Moslem sources. (Trifkovic 2002)

The history of Islamic mathematics is clearly a contested area, and recent history has if anything sharpened the divisions. The view which Joseph described as 'partial and distorted' 13 years ago lives on in some academic circles, as the quote from an admittedly right-wing anti-Islamic columnist illustrates. It is perhaps natural that in the current context even questions about algebra in Baghdad in the ninth century should be charged with political relevance, and voices on the fringe should perpetuate old myths. As far as the mainstream of historians is concerned, the points made by Joseph are almost universally conceded, as Katz' recent respected textbook makes clear:

Islamic mathematicians fully developed the decimal place-value number system to include decimal fractions, systematized the study of algebra and began to consider the relationship between algebra and geometry, brought the rules of combinatorics from India and reworked them into an abstract system, studied and made advances on the major Greek geometrical treatises of Euclid, Archimedes and Apollonius, and made significant improvements in plane and spherical trigonometry. (Katz 1998, p. 240)

The only quibble which could be made against this generous assessment is that Katz does not mention the difficulties which previous scholars have had in getting such reasonable claims accepted. The major obstacle has been the viewpoint, referred to by Joseph, which sees the Arabs as transmitters rather than innovators. Why is this? We saw in the last chapter that Chinese mathematics, obviously outside the Western tradition, could be relegated to the sidelines as a mere collection of isolated problems without coherence and without any idea of proof. With the mathematics which was developed in the Islamic world from the ninth to the fifteenth century CE, the problem is the opposite. The work could with some justice be seen as a part of 'Western' mathematics, looking back to the Greeks and forward to the European Renaissance, and the existence of influences is not in dispute. However, because it was a specialist field of study and the original texts were often inaccessible, it was possible to 'forget' the ways in which the Islamic writers transformed mathematics and to claim (as Trifkovic does) that they did nothing but pass it on.

To undertake a proper discussion of the history as it is now understood, it is useful to look briefly at the West, the Islamic world, and their changing interactions. (Historians have a problem about the choice between 'Arabic mathematics' and 'Islamic mathematics'. Neither is completely accurate for the mathematics practised in the Islamic world between, say, 800 and 1500CE. Since a choice must be made, we shall opt for the more inclusive 'Islamic'.) The understanding of Islamic mathematics in Western Europe has gone through a variety of transformations. In the early Middle Ages, from the eleventh to thirteenth centuries, it was highly regarded, for the good reason that the level of achievement was visibly more sophisticated. Those works which were found most comprehensible or useful were translated from Arabic into Latin as were the contemporaneous translations of the Greek classics into Arabic. By the Renaissance (say by 1550) for complex reasons, there had been a change of view, even though the West had not overall achieved the Islamic world's level of achievement, much less overtaken it.[1] The practice of translation from Arabic was less frequent, while the publication of original Greek texts and their translation, again into Latin, made possible a claim that the Moderns were the direct inheritors of the Ancients. Even though, as far as algebra and the number system were concerned, this was clearly untrue, it was a useful myth in constituting a Renaissance world-view which built on the classics as a source of legitimacy.

We shall see later how much the work of Viète, Stevin, Descartes, and their contemporaries owed to Islamic precursors; what is important for the moment is that it was not normal to acknowledge the debt. It is not excessively oversimplifying to say that the broad outlines of the Eurocentric history, which Joseph criticizes were laid down in the sixteenth century, and were the dominant version of history until relatively recently. And yet a number of important, often striking Islamic works have been published and studied in western Europe over the last 200 years. Their understanding, and their incorporation into a general history remained the preserve of specialists with no impact on the mainstream view. A better understanding of what Islamic mathematics was has had to wait for:

1. a political motivation—the demand for recognition from the Islamic world from the 1950s on[2];
2. unified research programmes, partly related to that politics, which rapidly deepened and expanded the work of study and translation in the 1950s and 1960s.

We shall have more to say about what material is and is not available in Section 2. The important change has been not so much an increasing accessibility of sources as an increasing consciousness of the achievements of the Islamic mathematicians. Twenty years ago,[3] Roshdi Rashed, one of the leading historical researchers, made much the same points as Joseph:

> The same representation is encountered time and again: classical science, both in its modernity and historicity, appears in the final count as the work of European humanity alone; furthermore, it is essentially the means by which this branch of humanity is defined. In fact, only the scientific achievements of European humanity are the objects of history. (Rashed 1994, p. 333)

New texts, new research, and persuasive arguments by respected scholars have largely allowed Islamic mathematics to take its legitimate place in the histories; and among scholars with any serious academic credentials one will no longer find it neglected or downgraded. The main problems in building up a proper picture are constituted first by the great gaps in our knowledge—which are, of course, also there for the cultures of Greece and China—and second by the sheer diversity of activity (arithmetic, algebra, classical geometry, astronomy, trigonometry, and much else) over

---

1. This case is argued by Rashed (1994, appendix 2). The general point is incontestable, although there is disagreement about the detail.
2. Said's influential book (1978), although quite unrelated to the sciences, played a key part in making academics more self-conscious about how they treated things 'Oriental'.
3. Rashed's book dates from 1984, although its English translation is 10 years later.

what is once again a dauntingly long historical period. It should be easy for the student to approach Islamic mathematics, like Greek, without prejudice and make a fair evaluation. Assuming this possible, one could, if only to fix ideas, pose some questions:

1. Can one give a unified description of 'Islamic mathematics', given the length of time and space and the variety of fields covered —indeed, should we even try to do so?
2. How would we evaluate the 'Islamic contribution' to the development of mathematical thought?

2 On access to the literature

One would naturally like to recommend, as a follow-up to the general agreement on the importance of Islamic mathematics, that the student could consult texts and histories and examine—for example—the questions raised above. Unfortunately, this is not yet the case; and here an accusation of 'neglect' can still be made, in that access to the relevant materials remains extremely difficult. If we start with secondary texts, that of Berggren (1986) is full, readable, and well-informed. It is, in our current situation, where any reader should start. Rashed's work (1994) is more specialist, aimed at the exposition of particular points in arithmetic and algebra; it is also expensive and less often stocked by libraries. And while Youschkevitch's rather older text (1976) is fuller than either of these and contains much which they exclude, it is (a) in French and (b) long out of print. The situation for the student entering the field could be worse, but it is not very good.

With regard to primary sources, what is available reflects a long and patchy history of translation by individual enthusiasts. The relevant section in Fauvel and Gray, though it contains some essential texts, is relatively brief; and while the works of Euclid, Archimedes, and other major Greek mathematicians can often be found in libraries and are reprinted, this is far from being true of the classics of the Islamic world. One initial problem is that there is no longer a canon of a few great writers, rather a large collection of texts whose differing contributions are still in process of assessment.[4] More translation is now in progress, but there are major gaps. To take just a few examples:

1. The earliest, founding book on algebra which underlies all subsequent work is (Muḥammad ibn Mūsā) al-Khwārizmī's *Ḥisab al-jabr wa al-muqābala* ('Algebra', lit. 'calculating by restoring and comparing', date about 825). This exists in a translation by F. Rosen, dated 1831 (*The Algebra of Muhammed ben Musa*, London, Oriental Translations Fund). It has been reprinted by Olms (1986), and is therefore in a better situation than most (useful extracts are in Fauvel and Gray).

2. Much later, but equally important, is the algebra of Omar Khayyam ('Ūmar al-Khayyāmī), dating from about 1070. This has been known about for a long time; while it was first translated in the nineteenth century by Woepcke (into French), there is a more 'modern' English translation

---

4. By an irony in the history of research schools, a large number of very interesting texts were translated into Russian by Youschkevitch and his group in the 1950s and 1960s. Even for the readers, whoever they may be, for whom Russian is an easier option than Arabic, they are not accessible in most libraries.

(Khayyam 1931). This, however, is long out of print and far from easy to find. Again, there are good extracts in Fauvel and Gray.

3. A more recent find is the startlingly innovative algebra text *al-Bāhir fi-l jabr* ('The Shining Treatise on Algebra') of al-Samaw'al (twelfth century). This has been extensively discussed, and good summaries of what is said in some key passages concerned with sums of series and with polynomials are to be found both in Rashed (1994) and in Berggren (1986). However, while there is a modern Arabic text dating from 1976 with introduction and some footnotes in French by Rashed, there is no translation, indeed there are no translated extracts. And the edition itself, published in Damascus, is not likely to be stocked outside specialist libraries.

4. Lastly, one of the most famous works, often referred to for its sophisticated calculations—in particular the use of decimal fractions—is al-Kāshī's *Miftāh al-ḥisāb* ('The Calculator's Key'), written in Samarkand in the fifteenth century. This has been known and studied for over a century. Besides several editions in Farsi (the work was popular in Iran), and a translation into Russian by B. A. Rosenfeld in 1956, there is a modern Arabic edition, published in Cairo in 1967, and again long out of print. I know of no English translation, or even of any plans for one; although again one can learn something of the work's unusual features from descriptions in Berggren (and Youschkevitch).

There is now some serious translation underway; and since the field is very large, it is bound to be selective. One could single out A. S. Saidan's version of the (recently discovered) arithmetic of al-Uqlīdisī, a fascinating work to which we shall return; and numerous translations into French by Rashed, notably the works of Sharāf al-Dīn al-Ṭūsī (1986), and of ibn al-Haytham (a large project, ongoing). These translators (and others), being active researchers, will necessarily be selecting those authors of most interest to them, so that the act of editing and translating is often part of the construction of a personal 'canon' of what the translator considers major works. However, in the impoverished situation already described, any such work is invaluable.

It could be argued that a serious research engagement with Islamic science should include the acquisition of the ability to read Arabic (which some readers may have anyway). This seems misconceived, insofar as the works concerned are considered as major historical texts. The time is past when the student was expected to be able to have the leisure to learn languages as part of a general liberal education, and while the specialist might need to read Euclid in Greek or the *Principia* in Latin, no one would expect it of the student on a history course. In any case, as already stated, modern Arabic editions are not easily available, and the deciphering of the difficult manuscripts which are still our primary sources (Fig. 1) is an advanced research skill comparable to reading Sumerian. If the major works of Islamic mathematicians deserve study on an equal footing with the classics of other times, then they should be equally accessible. Those who research the Greek classics are in a fortunate position, in that critical editions and translations have been made available by scholars who (a century ago) considered it an essential part of their work. A commitment to fair treatment for the Islamic classics is now driving a similar effort as far as they are concerned. In a spirit of optimism, one could hope for a significant part of this vast literature, together with a variety of analytical histories, to be readable by students in 20 years time. (And perhaps a start should be made with al-Kāshī, see item 4.)

A good recent bibliography of sources and articles (which omits Russian works, but is otherwise comprehensive) is by Richard Hogendijk at www.math.uu.nl/people/hogend/Islamath.html.

Islam, Neglect and Discovery

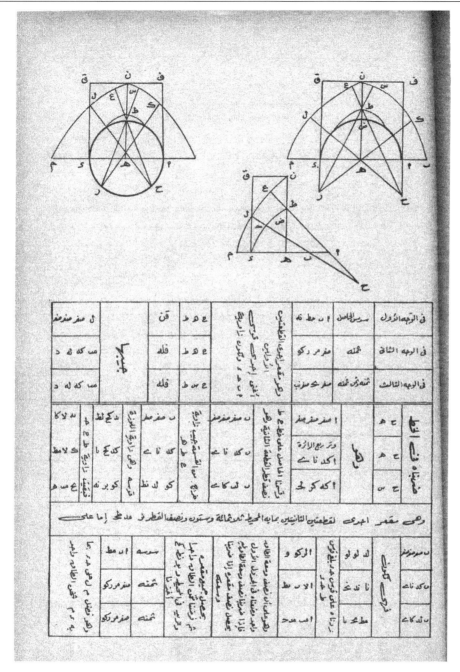

Fig. 1 MS, page from al-Kāshī.

And many of the out-of-print studies and aricles of the past hundred years are being printed as part of the vast series entitled *Islamic Mathematics and Astronomy*, by Fuat Sezgin (expensive, and rarely found in even the best libraries). The persistent student can find a great deal of material, but it may involve finding a friendly librarian, and possibly some expense.

3 Two texts

No curiosity that occurs, no strange method unheard of, no nice idea that is liked by them who hear it will be left out. These will be given and clearly explained, so that this book will contain everything the people enquire about. For indeed this arithmetic is often debated by people who enquire about its whys and hows. (Al-Uqlīdisī 1978, p. 36)

It is a characteristic of geometers that when you ask them a question on the division of figures or the multiplication of lines, they fall into confusion and need a long time to resolve it. (Abū-l-Wafā 1966, p. 115)

As an introduction to the nature and variety of Islamic mathematics, let us consider two texts from around the same date (tenth century CE). Both illustrate the problem of 'practical mathematics', which is raised by the two quotes above. For symmetry, one is a book, in print in an English translation, by an almost unknown writer; the other an untranslated book by an author of whom a fair amount is known. The first, which is relatively easy to find, is the arithmetic, or *Kitāb al-Fuṣūl fī al-Ḥisāb al-Hindī* (Book of chapters on Hindu reckoning), written by al-Uqlīdisī in Damascus in 951 CE (al-Uqlīdisī 1978). The book is one of the best sources on early arithmetic using the decimal 'Hindu' system, particularly since the earlier (earliest?) one written by the famous al-Khwārizmī has not survived in Arabic, and the various Latin translations seem all to have added and subtracted in different ways (see al-Khwārizmī 1992). On the other hand, while al-Khwārizmī was a notable scholar, nothing is known of al-Uqlīdisī's life at all. The name, which means 'The Euclidean' may indicate learning, but apparently people got this nickname for writing copies of *The Elements* for sale. (Tenth-century Damascus must surely have been unique as a place where copying the text of Euclid could earn you a living.) However, Greek learning makes no appearance in al-Uqlīdisī's text. It is long, detailed, and careful, and its world is that of street-corner calculators in Damascus who needed to work quickly and accurately, and who found that the new number system was ideal for their purposes. It was a competitive world—again this may appear strange—and one in which the partisans of one method of calculation would attack another. So al-Uqlīdisī defends his method, in phrases which are often quoted, as making it possible to carry out calculations among the distractions of street life:

Most scribes will have to use it because it is easy, quick and needs little precaution, little time to get the answer, and little keeping of the heart busy with the working that he [the scribe] has to see between his hands, to the extent that if he talks, he will not spoil his work; and if he leaves it and busies himself with something else, when he turns back to it he will find the same and thus proceed, saving the trouble of memorizing it and keeping his heart busy with it. (Al-Uqlīdisī 1978, p. 35)

The book is outstanding in its immediacy, and in the sense that al-Uqlīdisī has of his audience and what they need. Every rule is explained in great detail:

For example, we try to find the root of 576. We start from the six saying 'Is, is not, is', which falls under the five. We seek a number to draw under the five so that if we multiply it by its like, it exhausts most of the five. We find it 2. We insert it under the five, multiply it by its like and cast that out of the five. There remains one in place of five. We double the two in its place, shift the four under the seven, and seek a number to draw under the six so that if we multiply it by the four and by itself it will exhaust what is above it. We find it four. We multiply four by four, get 16, cast that out from above. We multiply 4 by itself and drop that from above; nothing remains. We halve the four which we have doubled. The result is 24. (Al-Uqlīdisī 1978, p. 76)

Clearly from the above, intelligence, numerical ability, and skill in following instructions are assumed; and there is no concession to a literary style once the initial points in defence of the book have been made. However, al-Uqlīdisī does take the trouble to explain his rules where he thinks it necessary. Why repeat 'Is, is not, is' to know where to start in root extraction? Why double the extracted root before shifting? These questions are answered in book III chapter 6 'Queries on Roots'. The book is in no way an advanced theorem–proof Greek text—but it makes no pretence to being that. It is a completely practical text on how to do arithmetic with Indian numbers, and the shadowy al-Uqlīdisī understands exactly what is required of such a book. We do not even know whether his text was popular—no other writers refer to it, and it seems to have survived by chance.

There is a great contrast in the comprehensive geometric text written about the same time in Egypt by abū-l-Wafā al-Buzjānī. Entitled *Kitāb fī mā yaḥtāju ilayhi al-ṣani'min a'māl al-handasah* (The sufficient book on geometric constructions necessary for the artisan), this has to date only been published in Arabic and Russian (Abū-l-Wafā 1966, 1979). It is therefore not a text easily available to the reader; but it has been considered important by Youschkevitch and Høyrup (who used the Russian version) and Berggren (who used an extract translated by Woepcke in the 1850s). We have done our best with the Russian text.

Abū-l-Wafā was at the other end of the scale from al-Uqlīdisī; a court mathematician and astronomer working in Baghdad who wrote (lost) commentaries on the classical works of Euclid and Diophantus and numerous other works on mathematics, astronomy, and other sciences. That he thought it useful to devote time to writing textbooks for artisans is the more significant. As ibn Khaldūn says, in the passage which immediately precedes the story of Euclid as geometer (which we quote in Chapter 3):

In view of its origin, carpentry needs a great deal of geometry of all kinds. It requires either a general or a specialized knowledge of proportion and measurement, in order to bring the forms (of things) from potentiality to actuality in the proper manner, and for the knowledge of proportions one must have recourse to the geometrician. (Ibn Khaldūn 1958, II, p. 365)

However, while the world of calculators who might have used al-Uqlīdisī's book is fairly easy to imagine from his text, the artisans who needed the 'Book on geometric constructions' seem more enigmatic. It is clear that abū-l-Wafā had in mind an actual audience, but he wished to raise the level:

[M]ethods and problems of Greek geometry ... and Abū-l-Wafā's own mathematical ingenuity are used to improve upon practitioners' methods, but ... the practitioners' perspective is also kept in mind as a corrective to otherworldly theorizing.
Interesting passages include Chapter 1, on the instruments of construction; and 10.i and 10.xiii, which discuss the failures of the artisans as well as the shortcomings of the (too theoretical) geometers. (Høyrup 1994, pp. 103, 312)

Indeed, the quote which opens this section is just such a criticism of geometers. As an example of abū-l-Wafā's method, here is the very classical construction of a regular pentagon (Fig. 2).

If someone asks how to construct on the line AB a regular pentagon, then we raise from point B a perpendicular BC [to AB] equal to the line AB. We divide AB in half at the point D, we describe with D as centre and radius DC the arc CE, and we extend the line AB to the point E. Then we draw arcs with each of the points A, B as centres and with radius equal to AE. They meet at the point G. We join the lines AG and BG. We have the triangle ABG, which is the triangle of the pentagon. (Abu-l-Wafā 1966, p. 71–2)

From this point on, the construction is easy (see Chapter 2, Appendix B); AGB is an isosceles triangle whose base angles are $72°$, and the isosceles triangles BFG and AHG which complete the pentagon have their short sides equal to AB. There is, as Høyrup remarks, no proof; and the

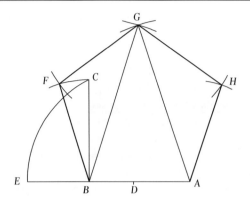

Fig. 2 Abū-l-Wafā's construction of the regular pentagon.

Greek 'We . . .' is mixed with the artisans' 'If someone asks you, . . . do'. And one wonders how often artisans might have needed to construct a regular pentagon. There is an evident wish to publicize Greek geometry and to extend its audience, as al-Uqlīdisī wishes to make propaganda for the Hindu numbers. 'Real' mathematics, expounded systematically in books, has suddenly entered a realm of popularization for practical men.

Exercise 1. *Explain al-Uqlīdisī's calculation. Why does one say 'Is, is not, is'?*

Exercise 2. *Justify abū-l-Wafā's construction.*

4 The golden age

The most venerable legal scholar Abū Bakr ibn Muḥammad al-Yafrashī told me in Zabīd the following story: It is related that a group of people from Fārs with a knowledge of algebra arrived during the caliphate of 'Ūmar ibn al-Khaṭṭab [634–644]. 'Alī ibn Abī Ṭalib—may God be pleased with him—suggested to 'Ūmar that a payment from the treasury be made to them, and that they should teach the people, and 'Ūmar consented to that. It is related that 'Alī—may God be pleased with him—learned the algebra they knew in five days. Thereafter the people transmitted this knowledge orally without it being recorded in any book until the caliphate reached al-Ma'mun and the knowledge of algebra had become extinguished among the people. Al-Ma'mun was informed of this and he made inquiries after someone who had experience in (algebra). The only person who had experience was the Shaykh Abū Bakr Muḥammad ibn Mūsā al-Khwārizmī, so al-Ma'mun asked him to write a book on algebra, to restore what had been lost of (the subject). (Brentjes 1992, pp. 58–9)

The above apocryphal story[5] of the arrival of algebra links the beginning of Islam with the beginning of mathematical knowledge among the Arabs. Significantly, however, it also introduces a 'group of people from Fārs' (Persia) who are responsible for its introduction; and it illustrates our ignorance of the first century of Islam, particularly in drawing attention to the oral tradition and the lack of writing. All other evidence which we have tells a different story: while the origins of Greek and Chinese mathematics are unclear and undocumented, Islamic mathematics begins

---

5. From the ease with which 'Alī learns algebra, the story appears to be Shi'ite in origin, but Brentjes gives no further information on this.

150 years later with an abundance of written texts from the ninth century CE, many of which have survived.

As with Chinese history, Islamic history can be structured by a succession of dynasties; however, after the earliest years this becomes confusing and it is simpler to give a broad outline. In fact, the world which was quickly conquered by the forces of Islam was large, and it hardly ever came under an undisputed single ruler. The conquest by those who accepted Muḥammad's new religion and message of Islam, is one of the most spectacular events of history, however interpreted; between Muḥammad's death in 632 CE and the end of the seventh century the whole of the Middle East, Egypt, North Africa, Spain, Iran, and parts of India and Central Asia were incorporated into the new state, under the rule of the khalif, first at Damascus and then at Baghdad. In the orthodox history of Islam, the period of 'Ūmar and 'Alī, the companions of Muḥammad mentioned above, was the golden age. Subsequent rulers, as always, fell off both in personal ethics and in their human rights record from the original standards, and the rulers who were remembered as good were (as in the Italian Renaissance) those who at least presided over a period of peace and promoted the arts and sciences. In this respect the Abbasid rulers of the early ninth century, particularly the khalif al-Ma'mun (813–833), were outstanding. Indeed, the history of Islamic mathematics, like Chinese, seems to divide naturally into two periods, an early one (say 800–1000) of quite concentrated activity, with a large number of mathematicians, working often in collaboration; and a later one of particular scholars, often very gifted, who, in times often of civil war or external attack, worked either in isolation or under the patronage of local rulers. There are signs that as early as the eleventh century al-Bīrūnī and Khayyam were looking back at the previous age and contrasting it with their own:

We have been suffering from a dearth of men of science, possessing only a group as few in number as its hardships have been many —persons who had recourse merely to a brief respite of time to concentrate on research and verification of facts. Most of our contemporaries are pseudo-scientists who mingle truth with falsehood ... In all circumstances we seek refuge in God, the Helper. (Khayyam 1931, p. 47)

Accordingly, when there was a revival, as in the Mongol court of the conqueror Hulaghu Khan (*c*.1260), or that of Timur's grandson Ulugh Beg at Samarkand (*c*.1410), scholars looked back to the period of al-Ma'mun and his 'House of Wisdom' at Baghdad as a model.

What was this 'golden age', and where did it come from? Early Islam was, as is well-known, tolerant particularly to Jews and Christians ('People of the Book'), and it is thought that much of the population of this empire were slow in acquiring the Arabic language and the Islamic religion, although both had advantages. Similarly, in the first 100 years the conquerors seem to have been unconcerned with the remnants of Greek learning which were cultivated by scholars—often refugees from Christian persecution—in centres like Harran in Turkey and Jundishapur in Iran.

The stage was set for a surprising union of cultures which took place in the late eighth and early ninth centuries. This was incidentally the period in which the religion of Islam took on most of its later form—the traditions with their injunctions about life and conduct, the legal system, and much else. The new Abbasid dynasty who ruled from Baghdad not only favoured trade, commerce, and public works (which, as usual, require mathematics at some level), but, particularly under al-Ma'mun, saw a value in pure research. In the context, this meant the discovery of the work of the Greek and Indian mathematicians, and its translation into Arabic. Scholars from what we can see as a melting-pot—Syriac, Greek and Arabic speakers, Christian, pagan, and Muslim—combined in the work of translation; and then immediately began to build on what they had translated. In fact, with such disparate sources, the idea that the Islamic work could be simple borrowing and transmission makes no sense; a synthesis was essential. This involved raising what appeared to be unanswered questions, and writing new books in more useful forms for practical ends (as the examples above illustrate).

In an article which we have already cited, which forms one of the most interesting theoretical discussions of early Islamic mathematics, Høyrup claims that this new synthesis marked a radical change in the use of mathematics comparable to the work of the Greeks discussed in Chapter 2.

[T]he break [which led to the acknowledgment of the practical implications of theory] took place earlier, in the Islamic Middle Ages, which first came to regard it as a fundamental epistemological premise that problems of social and technological practice can (and should) lead to scientific investigation, and that scientific investigation can (and should) be applied in practice. Alongside the Greek miracle we shall hence have to reckon an *Islamic miracle*. (Høyrup 1994, pp. 92–3)

You are referred to Høyrup's article both for his detailed arguments in establishing the nature of the new approach, and for his attempts to account for its origins. He considers and rejects a number of suggestions, finally opting for a description of the nature of Islam which he calls (perhaps unfortunately) 'practical fundamentalism'.

We shall return to the role of Islam as religion, philosophy, and way of life later. Let us now look at the interaction between new and old in the knowledge produced by the early mathematicians.

5 Algebra—the origins

I have established, in my second book, proof of the authority and precedent in algebra and *al-muqābala* of Muḥammad ibn Mūsā al-Khwārizmī, and I have answered that impetuous man Ibn Barza on his attribution to 'Abd al-Ḥamid, whom he said was his grandfather. (Abū Kāmil, cited Rashed 1994, p. 19, n. 3)

I have always been very anxious to investigate all types of theorems and to distinguish those that can be solved in each species, giving proofs for my distinctions, because I know how urgently this is needed in the solution of difficult problems. (Khayyam 1931, p. 44)

The word, in its derivation (from Arabic '*al-jabr*', usually rendered 'restoring'), suggests that what we call algebra begins with the Arabs. Like all other questions of origins, this can be disputed on various grounds; we have seen that the Babylonians knew how to solve problems which were equivalent to quadratic equations (Chapter 1). So what was so important and influential about the Islamic contribution? There is no better place to start than the original textbook by al-Khwārizmī. This was enormously influential both in the Islamic world, and in medieval Europe; abū Kāmil, as quoted above, illustrates the general agreement about al-Khwārizmī's priority, and his method and language survived with adaptations until the sixteenth century in Europe, when something more like our modern notation was introduced. Part of the text of his book (1986) is reproduced in Appendix A. This illustrates the core of the book, the treatment of quadratic equations, although a very large part is in fact given over to 'applications' to practical situations (e.g. inheritance), and to geometry. He *defines* 'roots', 'squares', and 'numbers', the three objects which enter into his algebra, in terms of *what you will do with them*; the definition is not so much conceptual as operational, and this itself throws light on how he is thinking.

A root is any quantity which is to be multiplied by itself, consisting of units, or numbers ascending, or fractions descending. (Fauvel and Gray 6.B.1, p. 229)

This may seem less than clear to us, but it enables a description—the first—of what a *general* quadratic equation is. Note that the 'root', or solution, is allowed to be a fraction although not worse.[6] You will still find this language, stretched to its limits, used in Tartaglia's rule for solving the cubic in the 1540s (see Chapter 6). There are six forms of the quadratic equation—this is dictated by the need for all numbers which are used to be positive. A typical one reads: 'Roots and squares equal to numbers'; some xs (as we would say) added to some x^2s equal some number. Al-Khwārizmī does not wish, like the Babylonians, to list particular cases and assume that you can deduce the general rule; he wants his statement to be general, but he does not have our version of a general symbolic language (which dates from the seventeenth century) 'a roots + b squares equal c numbers'. (Interestingly, although Diophantus's *Arithmetic*, which did use a more abstract notation, was translated relatively early into Arabic—ninth century, later than al-Khwārizmī—his methods were not adopted, any more than they were in the Greek world.)

If, in a parenthesis, we consider how one is taught to solve such an equation today, the commonest method is to give a simple literal formula, whether it is proved or not. Writing the equation $ax + bx^2 = c$, we deduce:

$$x = \frac{-a \pm \sqrt{a^2 + 4bc}}{2b}$$

which 'always applies'. The reason we can do this is because we can explain how to deal with several problems raised by the formula.

First, one, or both of the values we find may be negative numbers, which were first considered as possible solutions in India by Bhaskara in the eleventh century, and were still being argued about 400 years later; as we have seen (Chapter 4) this was found easy by the Chinese, but their attitude seems not to have been transmitted to the West.

Second, we have to be prepared to take the square root of any number we like. This raises two levels of problems; a 'naming' problem if the number is positive but not a square (say 5), which we shall see dealt with below; and a worse one—what are we talking about at all?—if it is negative (say -3). These were coped with at different times in more or less satisfactory ways, and a school mathematics course will similarly try to steer the student through them progressively.

Until the sixteenth century or later, though, no such general formula was considered, since even negative roots had to be dealt with separately if they were allowed at all. Hence the pattern which al-Khwārizmī set for dealing with equations case by case, as set out above. After describing the different cases, he moves on to the case 'roots and squares equal to numbers' mentioned above, and deals with the problem of abstraction by alternating the general statement with its application to the particular example 'one square and ten roots equal to thirty-nine dirhems'.[7] The solution goes back to Babylon ('you halve the number of roots, which gives five'), but has suddenly become general as well as particular. It is easy to see the reasons for the long popularity of al-Khwārizmī's text: he has grasped the idea of explaining the method through an example, as al-Uqlīdisī was to do in his arithmetic (and as was to become common practice in Islamic texts, and the European ones which derived from them).

---

6. Note also that despite al-Khwārizmī's role in introducing Indian numbers, they are not used in the algebra text, where numbers are always written out as words ('thirty-nine').

7. Why dirhems—a unit of currency? As one might say, '$x^2 + 10x = 39$ euros'. The implication is that there is a practical use for the sum; and there is some attempt to justify this later.

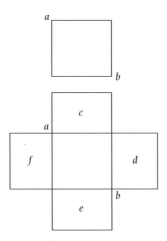

Fig. 3 Al-Khwārizmī's first picture for the quadratic equation.

'It is necessary', he continues, 'that we should demonstrate geometrically the truth of the same problems which we have explained in numbers.' Why is it necessary? There appear to be three requirements for the author:

1. to state what to do in general;
2. to illustrate it in particular;
3. to prove that it works.

Is it the weight of the Greek heritage which implies that 'proof' means geometry? One might suppose so, since the Greek texts were being translated when al-Khwārizmī wrote. In any case, the geometry looks nothing like Euclid, or even his more practically minded followers such as Heron. The picture (Fig. 3) compared with later proofs of the same method, is completely transparent; it is a good exercise to follow the proof through and see how verbal explanation and picture connect to give a convincing account of why the solution is the right one.

There has been considerable discussion of how 'good' a mathematician al-Khwārizmī was (the article in the *Dictionary of Scientific Biography* is dismissive). As already stated, the method which he set out was ancient, wherever he derived it; and his exposition, his examples, and his proof were (as the extract shows) at a fairly low mathematical level. However, this seems to miss an important point; such arguments assume that mathematicians deserve study only insofar as the work which they do is hard, while often this is not at all the case. (While Descartes was capable of hard work in mathematics, he disliked it, and his outstanding contribution, the coordinate representation of curves, is simple in the extreme.) What al-Khwārizmī did was to introduce a new way of thinking about the problem which brought together solution and proof in a major synthesis, involving both generalization and simplification. That the mathematics involved was *not* very difficult was an essential reason for the method's survival more or less unchanged over the next 600 years.

About 50 years later, Thābit ibn Qurra—who by general agreement *was* an able and interesting mathematician—wrote a text on quadratic equations. In contrast to al-Khwārizmī's treatise, it is a mere six pages. It was translated into German during the Second World War, and later into Russian;

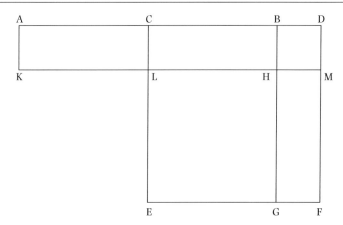

Fig. 4 The diagram for Euclid proposition II.6. The line AB is bisected at C (AC = CB), and BD is added. If now AK = BD, then 'the rectangle AD by DB' means the area of the rectangle ADMK; and this, together with the square on CB (which equals the square LHEG) is said to be equal (in area) to the square CEFD on CD. The proof is fairly obvious.

the chance of finding either translation in a library is slim.[8] However, it is a very interesting document. Thābit was one of the groups of Greek translators, and much of his prolific work expanded on Greek texts, commenting or dealing with problems which they raised. Here he uses his knowledge to draw on Euclid's proposition II.6 (for the case above described) and prove—in some sense—that the formula is the right one. Unfortunately, unlike al-Khwārizmī, he does not have an easy style, at least here.

Proposition II.6, in its particular form, says:

Let the straight line AB be bisected at the point C, and let a straight line BD be added to it in a straight line (see Fig. 4) I say that the rectangle AD by DB together with the square on CB equals the square on CD.

Those who believe that the results of book II should be interpreted as a form of algebra interpret this by saying: call AB 'a' and BD 'b'; then BC $= a/2$, and CD $= b + (a/2)$; the proposition says that:

$$(a+b)b + \left(\frac{a}{2}\right)^2 = \left(b + \frac{a}{2}\right)^2 \tag{1}$$

It is now on the whole thought unhistorical to claim that Euclid was thinking in such terms (see the remarks on this in Chapter 2). However, there is evidence that the Islamic translators of Euclid at some stage did come to use some sort of algebraic translation—after all, they now had algebra to help them. In the early tenth century the philosopher al-Farābī wrote that the rational numbers correspond to the rational quantities, and the irrational numbers to the irrational quantities (cited Youschkevitch 1976, p. 169). The distinction between numbers and lengths, which sometimes seemed so important to the Greeks, was being eroded, and in commentaries by Arabic writers on Euclids books V and X (which exercised them greatly) we can find many similar examples. Thābit does say that he is investigating the case 'square and roots equal to numbers'; but it is typical of his more hurried and more abstract approach that he gives no numbers as examples. You can find his argument in Appendix B. The end-result (of the extract) is that the root which we are looking for is 'known'—in classical geometric terms, it can be constructed.

---

8. Luckey's German version, with the original text and his discussion, is in Sezgin (ed.) 1999, pp. 195–216.

What Thābit does next is equally interesting; he goes through his method and shows, stage by stage, that it is the same as the method used 'in algebra'. The 'algebra' is the method described by al-Khwārizmī—without his geometric proof—and it seems reasonable on various grounds (short separation in time, the fact that they were near-colleagues, al-Khwārizmī's acknowledged status as 'founder') to suppose that it is precisely his book which is referred to. This 'dialogue' casts some light on different ways of thinking about geometry, numbers, and algebra in the earliest period of Islamic mathematics. It would seem that Thābit is saying: 'Anything which you can do by algebra, I can do by Euclid book II'. If so, there is some misunderstanding both of al-Khwārizmī's algebra (which is about numerical recipes for solving practical problems) and of Euclid (which is about something more abstract and quite different). More positively, we could see it as an attempt to harmonize the down-to-earth practice of algebra with Greek theory. Whether misunderstanding or harmonization, such a tension between theory and practice was to be of immense value in the further development of the Islamic tradition.

We have already entered the domain of (reasonable) conjecture about what the text means, in terms of the various ways tenth-century mathematicians thought about numbers and geometry. The problem is what Thābit means by 'is known'—the argument being that to say that the side (or its square) is known is to solve the quadratic equation. There are two competing interpretations of this. In *geometric* terms, it means simply that the line which represents the side can be constructed, which is certainly true. But what has been passed over is the *numerical* question of what happens when your answer is not a whole number, as it was in al-Khwārizmī's version. If the equation is 'square and two roots equal one', then the answer, whichever method you use to arrive at it, is (as we would say) $\sqrt{2} - 1$. Because Thābit is avoiding using numerical examples, he gives us no idea about whether such numbers are allowed *as numbers*, not as geometrically constructed lines. They have no name.

There is a useful word for 'having no name' in Arabic, which was variously applied: it is 'aṣamm', or 'deaf'. This was initially aplied to certain fractions; you can say the fractions up to one-tenth using one word, but after that you have to use phrases like 'one part of thirteen', and such fractions were 'aṣamm'. But in al-Uqlīdisī's arithmetic, the same word was applied to squares which have no roots; by extension (since if you are thinking for example, of a square of area 5, you are also thinking of its side) it denoted their inexpressible roots. This word translated, when the Arabic arithmetics were put into Latin, into the Latin word for 'deaf', 'surdus', used in the form 'surd' as recently as 50 years ago to refer to roots like $\sqrt{5}$. At some point a linguistic concept about numbers whose name you could not speak translated into a way round their unspeakability. They are still numbers, but numbers which need phrases rather than a single word to express them. Al-Uqlīdisī devoted some space to finding approximations for such square roots, in chapters which follow on the exact root extraction quoted above. His formulae were not new, but the use of Indian numbers makes the procedure more transparent. (A great deal has by now been written on this subject. A detailed and careful summary is Karine Chemla 1994.)

Exercise 3. *What kinds of combination of roots, squares, and numbers make an allowable equation, in the terms set out by al-Khwārizmī?*

Exercise 4. *Why is the algebraic formula given equivalent to Euclid's proposition II.6?*

Exercise 5. *How would (a) al-Khwārizmī's method and (b) Thābit's construction approach the equation 'square and two roots equal to one'?*

6 Algebra—the next steps

We have heard the great eastern mathematicians have extended the algebraic operations beyond the six types and brought them up to more than twenty. For all of them they discovered solutions based on solid geometrical proofs. God 'gives in addition to the creatures whatever He wishes to give to them'. (Ibn Khaldūn 1958, III, p. 126)

Not much later than Thābit's text, the Egyptian abū Kāmil wrote what is commonly considered the 'second-generation' algebra after that of al-Khwārizmī.[9] The work of al-Khwārizmī is explicitly referred to, and many of the examples are the same; but much else has changed. The simple geometrical diagram has been replaced by a reference to Euclid's book II (as it was in Thābit's text), but with numbers included. For the first time, as far as we know (and our knowledge is as usual limited), numbers have been introduced into Euclidean propositions as a matter of routine, and proposition II.6 is being interpreted more in the 'algebraic' sense referred to above. If this was done by the ancient Greeks, or by any of their successors, they were much more discreet about it than abu Kāmil.

However, what abu Kāmil did next was even bolder, as an innovation. Again, it may arise from the study of Euclid, in this case of his book X; but this is not made clear, and the language is completely different. He develops a set of rules—not complete, but useful—for calculating with roots, and uses them freely in many of his examples as if they were numbers. The result is an enormous expansion of the collection of equations you can solve, and of numbers you can name. Oddly, this appears not so much in dealing with whole number examples leading to square-root solutions (such as the simple one given above), as with examples where roots are part of the data of the problem. Here is the brief, but quite 'hard' problem 39:

If one says that ten is added to an amount, and the amount is multiplied by the root of five, then one gets the product of the amount by itself. For the solution, make the amount a thing and add ten to it to give ten plus a thing. Multiply by the root of five to give the root of five hundred plus the root of five squares equal to one square. Halve the root of five squares to give the root of one and a quarter. The root of the sum of the root of five hundred plus one and a quarter, plus the root of one and a quarter, equals the amount. (Abū Kāmil 1966, p. 148)

Notice that though the problem deals with numbers like $\sqrt{5}$, they are still expressed in words; there is no notation for them, and there will not be one for a long time (symbols for roots began to be used in the sixteenth century). For us, abu Kāmil's problem needs a considerable amount of 'unpacking'. In modern terms, setting x for the amount, it is:

$$(10 + x)\sqrt{5} = x^2$$

This abu Kāmil solves (roughly) by our usual prescriptions for quadratic equations again. In a slightly roundabout way he changes the left-hand side into $\sqrt{500} + \sqrt{5x^2}$. He halves $\sqrt{5}$ to give $\sqrt{1\frac{1}{4}}$, and arrives at the (correct) answer

$$x = \sqrt{\sqrt{500} + 1\frac{1}{4}} + \sqrt{1\frac{1}{4}}$$

All these numbers are still expressed in words, as is done in the above quote. The 'formula', if you like, is exactly the same as had been used by al-Khwārizmī; but the way in which it is applied has

---

9. This can be found in Levey's translation of Mordecai Finzi's medieval Hebrew version (Abu Kāmil 1966), again not easily. There is an extract in Fauvel and Gray 6.B.2., and a very interesting and complicated equation is discussed in Berggren, p. 110–111.

been vastly extended, without this ever being made explicit. The earlier author never said that numbers could not be square roots, and the later one never said that they could, but all the same the idea of what 'numbers' are allowed has changed.

It is easy, in the fairly open climate of discussion in Islamic mathematics, to find differences of approach such as those described above; and they are not confined to algebra. There are explicit arguments, for example, about the merits of those who were already being called (as they were by Pappus) 'the Ancients' (al-qudamā'):

> [Abū-l-Wafā'] said how much he appreciated the book, which he considered of great value, although he regretted that the author followed the way of the ancients in their use of the 'cutting diagram' and of compound ratios. He said that he had obtained, to determine the azimuth, elegant methods which were more concise and better. (Al-Bīrūnī 1985, p. 96)

However, to use this particular quarrel (about who was the first to find a formula for trigonometry on a sphere), or any other to divide Islamic mathematicians into 'schools' as has sometimes been done seems premature, and probably misguided. Saidan in his introduction to al-Uqlīdisī (1978) calls attention to attempts of earlier historians to distinguish those mathematicians who used Indian numbers from those who used sexagesimals (or 'astronomers' numbers' as they were called); and points out that it was common, especially in teaching texts, to use both, since the student might need both. As for Greek authority, it was universally recognized, and used as and when necessary together with more 'modern' methods. The case of Omar Khayyam (eleventh century) is particularly worth considering. In his algebra, he considers in detail the case of cubic equations. He was the 'eastern mathematician' referred to by ibn Khaldūn who had brought the number of types to more than 20 by introducing the various types of cubics (cubes and things equal to numbers, and so on). Besides being the natural next step after the well-understood quadratics, these had arisen in a number of special problems which he lists; a problem of Archimedes on cutting the sphere, trigonometric problems such as finding $\sin 10°$ given that one knows $\sin 30°$, and so on.

As has often been noted, he acknowledges that it would be desirable to find a solution in terms of a numerical procedure (what we would call a formula), as had been done for quadratics and as Tartaglia and Cardano were to do in the sixteenth century.

> When, however, the object of the problem is an absolute number, neither we, nor any of those who are concerned with algebra, have been able to prove this equation—perhaps others who follow us will be able to fill the gap—except when it contains only the three first degrees, namely, the number, the thing, and the square. (Khayyam 1931, p. 49)

Unable to achieve this, he followed the very Greek practice of drawing intersecting conic sections, just as Menaechmus had done for the simplest case $x^3 = 2$.[10]

On the whole, such a solution would have been acceptable to a Greek (supposing the problem to have been posed in the first place). Omar was in some ways particularly close to the Greek geometers in his outlook; he criticized ibn al-Haytham for using motion to prove the parallel postulate, and the algebraists in general for using the 'ungeometrical' powers of the unknown above the third. However, it may have occurred to him to ask a question which fits much better into the framework of the algebra we have been discussing above: namely, if you have constructed a solution (e.g. to $x^3 + x^2 = 3$) geometrically, what kind of a number have you found, and what can you do with it? There is a clue; when, in a different work, he considered the difficulties in Euclid's theory of ratios,

---

10. For an extract from Omar's work see Fauvel and Gray.

he came to a startlingly pragmatic conclusion. We should think, he says, of a quantity

not as a line, a surface, a volume or a time, but as a quantity which the mind abstracts from everything, and which belongs to numbers, but not to absolute and true numbers, for the ratio of A to B may often not be numerically measurable, that is to say one may not be able to find two numbers whose ratio it is ... This is how calculators and surveyors proceed when they speak of a half or other fraction of a supposedly indivisible unit, or of a root of five or ten etc. (Khayyam tr. Rozenfel'd pp. 105–6, cited Youschkevitch p. 88)

In other words, the calculators and surveyors are already using numbers on the assumption that they are the same as 'quantities'; that if you can construct a length, there is a number which corresponds to it (at least well enough). What is interesting is Omar's explicit suggestion that mathematicians could learn something from them.

Exercise 6. *Show that the equation given is equivalent to abu Kāmil's problem, and solve the equation.*

Exercise 7. *Use the formula* $\sin(3x) = 3\sin x - 4\sin^3 x$ *to find a cubic equation for* $\sin 10°$.

7 Al-Samaw'al and al-Kāshī

The *Calculator's Key* is an excellent guide to elementary mathematics, written to answer to the needs of a large public. Considering the richness of its subject-matter, and the clarity and elegance of its presentation, this work holds an almost unique place in the whole literature of the Middle Ages. (Youschkevitch 1976, p. 71)

It would take much more space to discuss all the varieties of mathematical practice which were undertaken in the Islamic world, their connections, and interrelations; although we shall return briefly to their views on Euclid in Chapter 8. However, in a specific attempt to investigate the themes of innovation, tradition, and continuity, let us consider two later mathematicians whose approach was very different, al-Samaw'al (1125–1180) and al-Kāshī (d. 1429). In both cases, there are acknowledgements of particular influences, neither is working in what we have called the Greek tradition, and both raise interesting unsolved problems about the aim and scope of their work. In particular, both present examples of what we might call excess, that is, calculation beyond what is necessary or useful and here we would differ from Youschkevitch's opinion above, with his references to 'elementary mathematics' and 'a large public'. In contrast to al-Uqlīdisī or abu-l-Wafā, they seem to be carried away by their subject. Why?

Al-Samaw'al appears the more straightforward case. His major work—of those which survive—is *al-Bāhir fi-l-jabr* ('The Shining Treatise on Algebra'). Written, it is said, when he was 19, it is a conscious attempt to strengthen and deepen the results of his predecessor al-Karajī, a century earlier. (In many respects al-Karajī had laid the foundation for al-Samaw'al's work, so much is undisputed; however, here we shall consider it in isolation.) The work is quite long and contains a variety of results (on systems of linear equations, for example), but it is most celebrated for the curious study of 'polynomials' (al-Samaw'al calls them 'composed expressions') in which:

1. the primary aim is not to find the 'thing'—it seems, in the main, to be treated as an abstract entity to be manipulated;
2. the powers of the thing considered are not only positive (in principle, as large as you like) but negative; what we would call $1/x, 1/x^2, \ldots$

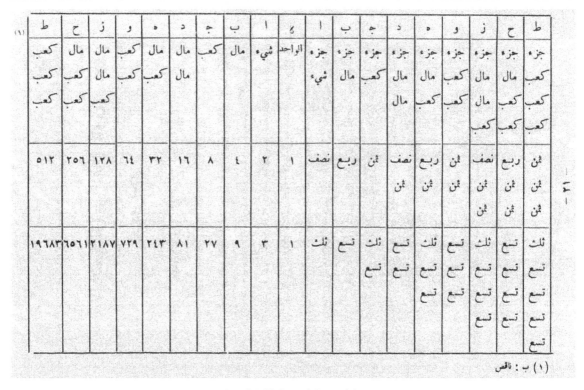

Fig. 5 Table from al-Samaw'al.

In al-Samaw'al's famous phrase, from his introduction, his aim is to proceed 'by operating on the unknowns by using all the arithmetical tools which the arithmetician uses to operate on known numbers'. In other words, you must—at least—be able to add, subtract, multiply, and divide your 'things' (xs and ys) any number of times. This leads to expressions which are complicated in our terms, let alone in twelfth-century notation, where '$1/x$' is designated by 'part of thing' and so on.

Al-Samaw'al wastes no time; by his fourth page he is giving a table of powers of the thing up to the ninth, which we would call x^9 and he calls 'cube cube cube' in the positive direction, and down to $1/x^9$, or 'part of cube cube cube' in the negative. The table (reproduced in Fig. 5) is an interesting mixture of notations. While the second row describes the powers in words ('square cube' etc.), the first row keeps track in a more rational way by using numbers going in both directions (expressed by letters of the Arabic alphabet), *including* zero. Underneath he gives the examples of powers, positive and negative, for the numbers 2 and 3. And here another notational problem; while the Indian numbers do very well to express $2, 4, 8, \ldots, 2^9 = 512$, the corresponding fractions have to be written in words starting with 'half' and ending with 'an eighth of an eighth of an eighth'. (In parenthesis, one notes that the ease with which the Egyptians, and the Greeks following them wrote unit fractions seems to have disappeared; changes in notation are not always for the better.) The power zero is correctly assigned to 1.

One has a sense, in the chapter on polynomials which follows, that al-Samaw'al is working at the limits of the notational possibilities which were then available, and trying to expand them

ISLAM, NEGLECT AND DISCOVERY

where he can. Sometimes an example (such as $(10/a^3)(a^2 + a) = (10/a) + (10/a^2)$) is set out and explained in words; sometimes a more general formula (such as $a(b/c) = b(a/c)$ is described by using a series of Arabic letters $a, b, c, \ldots$ to denote the 'unknowns' and the results of multiplying and dividing them. This in itself is not new—the use of letters to denote general numbers or quantities can be paralleled in Euclid—but in combination with the traditional algebraic language it gives the feeling (which Rashed expresses strongly) that we have something near to a 'new' abstract algebra.

The real coup is achieved when, after another 24 pages, al-Samaw'al sets out to divide two expressions (polynomials) according to the schema shown in Fig. 6. Before the table is set up, the problem is set out in words (with a few figures interspersed). Translated into our notation, it amounts to dividing

$$20x^6 + 2x^5 + 58x^4 + 75x^3 + 125x^2 + 96x + 94 + \frac{140}{x} + \frac{50}{x^2} + \frac{90}{x^3} + \frac{20}{x^4}$$

Fig. 6 Table from al-Samaw'al showing division.

by $2x^3 + 5x + 5 + (10/x)$. This is far from being the hardest such sum which will be tackled; in particular:

1. all the signs are positive;
2. the division has an exact result.

At this point the simple-minded reader might reasonably ask: what on earth did al-Samaw'al have in mind? The calculations which he undertook seem to be an end in themselves, a display of technical virtuosity on a theme which could have had no practical application, and which led nowhere. The example shown above, by the way, is by no means the end of the story; later, a division by 'six squares and twelve units' ($6x^2 + 12$) has no exact result. He therefore simply continues as far as possible, noting finally that any future coefficient can be determined by a formula. He is clearly on his way to understanding a particular form of infinite series. (The calculation is discussed in Berggren 1986, pp. 117–18.) Who the audience of his book could have been, and what they made of his work, remains at present a mystery; no subsequent algebraist refers to it. And on the face of it, such preoccupations give the lie to any easy characterization of Islamic mathematics as practical or down-to-earth.

A clue could be provided by a still more obscure work of al-Samaw'al, his recently discovered unpublished arithmetic. This is discussed at some length by Rashed (1994), where an extract is provided (untranslated), with a promise of future publication of the whole. In this text, according to Rashed, al-Samaw'al effectively introduced decimal fractions, using a schema very much like the one in Fig. 6; with ascending and descending powers of 10 (successive figures in the decimal expansion) taking the place of the powers of the unknown 'thing'. This of course has a much more useful appearance from our present-day viewpoint, although as Rashed concedes by writing the numbers in a table al-Samaw'al had not yet arrived at a simple and efficient notation.

Once the phrase 'decimal fractions' is mentioned, we have to deal with a long-running controversy over who was their originator. The question is interesting, but not because it really matters much any more. In textbooks from the 1950s or before, it was claimed that the invention was due to Simon Stevin (Netherlands, 1574), despite the fact that al-Kāshī's much earlier *Calculator's Key*, which used them extensively, was already known widely enough . There was no obvious line of influence from al-Kāshī to Stevin, and Stevin's was the first European discovery; it followed that he was the inventor.

Besides the obvious Eurocentrism of such a judgement, and the increasing evidence that al-Kāshī's work did influence western Europe via Constantinople and Venice,[11] this illustrates the whole problem of how one ascribes priority. The main interest in a mathematics textbook (medieval or modern), is to explain how you use a technique, not where the author obtained it; and this seems true even of Islamic writers who worked in a culture where citation of sources could be quite careful. Hence even where work is original, such originality may not be claimed, and this leaves the field wide open for historians (who may care more than is necessary) to argue about who is copying whom, and whether a writer really understands the method he is explaining. Al-Kāshī certainly did know what decimal fractions were; he has a technical term for them, and

---

11. This issue is discussed by Youschkevitch (1976, p. 75) and Rashed (1994, pp. 131–2).

uses them simply and with facility. In some sense, his introduction of them seems to be a claim to their invention—allowing that one does not always know who may have preceded one.

> We divided the unit into ten parts, we then divided each tenth into ten parts, and then each of them into a further ten parts, and then each of them into a further ten parts and so on, the first division being into tenths, and in the same way the second into decimal seconds and the third into decimal thirds and so on, so that the orders of decimal fractions and wholes are in the same relation as is the principle in astronomical numbering [i.e. sexagesimals].
> We call this 'decimal fractions'. (Al-Kāshī 1967 book 3, chapter 6)

From this (rather late) stage in his book, al-Kāshī sets out his results, where possible, in both forms, both sexagesimal and decimal. Whether his work 'diffused' to Stevin, whose notation was different and in some ways less user-friendly, is still unclear, though it appears increasingly a possibility.

But before al-Kāshī, as Rashed pointed out, stands al-Samaw'al, who also can claim a place as inventor; and before him there appears (according to Saidan) the still earlier tenth-century figure of al-Uqlīdisī, who seems to be using decimals in at least two places in his book. And in between these writers there may be many others of whom we know nothing. Rashed considers the claims of al-Uqlīdisī unacceptable; there is no sign that he was following a practice which he understood in a systematic way. On the other hand, he may have been one of a number of reckoners who had realized the obvious fact, as al-Kāshī states it: that, with Indian numbers as with sexagesimals, you could continue on the right as well as on the left, with your number (e.g. '5') having a smaller meaning the farther you went. This is what al-Uqlīdisī seems to be doing when, in one of his crucial passages, he performs a sequence of halvings on 19:

> For example, we want to halve 19 five times. We say: one half of 9 is four and a half; we set the half as 5 before the four; [remember that, Arabic being written right to left, 'before' means 'to the right of'] next, we halve the ten. We mark the units place. That becomes 95.
> Now we halve the five and the nine; we get 475. We halve that and get 2'375, the units place being thousands to what is before it, for if we want to say what we have got, we say that halving has led to two and 375 of one thousand...

A great deal of ink has been spilled over that single dash between the 2 and the 375; is it a decimal point, and why are there no others; and did al-Uqlīdisī understand the fact? A tentative conclusion might be that he did, to some extent, but that he would not dream of 'codifying' the idea as al-Kāshī did five hundred years later; he was a calculator, not a mathematician. Indeed, the illustration shows that the actual discovery of decimal fractions is not as much of a marvel as one might suppose. If you want to show your skill in Indian numbers by halving repeatedly, then you fall upon them almost naturally.

Having mentioned al-Kāshī in the context of decimal fractions, let us now turn to a broader appreciation of his work, and of *The Calculator's Key* in particular. In the long letter to his father published by Edward Kennedy in (1983), al-Kāshī gives a fascinating picture of the court of his patron Ulugh Beg. This may have been the 'end' of Islamic mathematics as far as our official histories go, but the society is far from being in decline; the atmosphere is one in which a sizeable and intensely competitive community of scholars strive to obtain the king's approval, primarily on the basis of their mathematical ability. Al-Kāshī, who was not given to false modesty (in Kennedy's classic understatement) makes it clear to his father that he has consistently come out best in all of these competitions, partly because of his skill in combining theory, calculating ability, and knowledge of the construction of instruments. It was for this unusually mathematically literate community of teachers and learners that al-Kāshī wrote *The Calculator's Key*, a very diverse collection of arithmetic, algebra, and geometry with results of the most various kinds. Unlike al-Samaw'al's book, this became something of a best-seller; the British Library, which is not strong on mathematics, has four manuscripts, two from the nineteenth century. His aims are stated at the outset, after a brief summary of his many achievements:

> Although some of these [methods] could not be discovered with the help of the six algebraic [forms] (i.e. al-Khwārizmī's six quadratic equations), yet in the course of this work I found numerous principles with whose help the groundwork of arithmetic is developed by the simplest means, on the easiest road, with the greatest profit and with the clearest exposition. I decided to write these principles and desired to clarify them so that they could be an instruction for others and a guide for the learned. Therefore, I have written this book and collected in it all which calculators may need, avoiding both the tedium of long-windedness and the excess of brevity. For the majority of methods I have drawn up tables, so as to simplify examination by the geometer. All the tables established in this book have been prepared by me and to me belongs all that is sweet and bitter in them, with the exception of seven tables ... (Al-Kāshī 1967, intro)

Indeed, the tables are a notable contribution to the work. We may already see a heavy dependence on the table for the exposition of complex calculations in al-Samaw'al; but in al-Kāshī they are everywhere, as he admits. There are the standard tables (multiplication, conversion from decimals to sexagesimals and back; sines, and so on); tables of currency conversion, of the properties of metals and other substances; tables of the areas of polygons, and more usefully (one might think), of different kinds of arches used in architecture (see Fig. 1). Almost always the numerical results are more accurate than they have any reason to be, and often they are given both in decimals and sexagesimals. As can be seen from the quote, al-Kāshī feels that they are an important contribution; he asserts his intellectual property in them, as well as an emotional relation (the sweet and the bitter). Most famously, beyond the 'static' tables, we have the 'dynamic' ones which show how you do a calculation. The reader is shown how to construct them, told in detail where to draw horizontal and vertical lines and make entries, so as (for example) to extract a root; and the often quoted example in which he extracts the fifth root of 44,240,899,506,197 in decimals can serve as a model.

This example (of a method which may be due to the Chinese, even if they did not carry it to such lengths—see Chapter 4; and which al-Samaw'al worked, if with less explanation, in sexagesimals) has been extensively discussed, in particular by Berggren (1986, pp. 53–63). When he comes to doing the same and more in sexagesimals, it is more a summary:

> In our treatise entitled 'Treatise on the circumference' [his calculation of π], we have found the roots of many numbers with many digits and adapted them in different ways. Anyone who wishes to know more can turn to this book. Furthermore, we present here an example of the extraction of a cube root and another example of the extraction of a cubo-cube [sixth] root, but, so as to avoid long-windedness in this book, we shall not here give an explanation of the process [as he did for the fifth root]. It is easy for anyone who knows how to do it with Indian numbers, as it was explained in Book 1.

At a certain point, we see, the tables, which are given, are a substitute for an explanation of the method.

To see al-Kāshī's style of exposition in a different context, an extract from the geometrical section of the *Calculator's Key*, on the regular solids, is in Appendix C, with the inevitable table which sets out all the measurements you may possibly need for them. Clearly considered an outstanding mathematician by his circle and beyond, al-Kāshī still appears something of an enigma. Given the obvious high culture of his milieu, one would like more information on what preceded it and what followed; and one wonders how far the sometimes obsessive accuracy of his calculations is motivated by the demands of practice, by competition, or by a pleasure in the activity of calculating itself.

Exercise 8. *(a) Look at the table for al-Samaw'al's polynomial division, and try to follow through the progress of the division, (b) show that the result of the division is $10x^3 + x^2 + 4x + 10 + (8/x^2) + (2/x^3)$.*

8 The uses of religion

Islam provides a whole set of fundamental values. Among these values one finds the uniqueness of truth, the lack of contradiction between revelation and reason...These values, among others, have without the least doubt pushed forth research and have fostered the creation of open scientific communities. (Rashed 2003, p. 153)

Allah is the ideal merchant. All is counted, everything reckoned...A more simply mathematical 'body of religion' than this is difficult to imagine. (C. C. Torrey, cited in Rodinson 1974, p. 81)

Earlier in this chapter it was suggested that the argument for the importance of Islamic mathematics, indeed its centrality in a tradition which links Babylonians, Greeks, and 'Moderns', is now established beyond argument. The idea that Islam itself played some role in the rapid development of the Abbasid period seems also undeniable; the question is, what was it? The argument (recycled in one of the quotes which opens this chapter) that many or even most 'Muslim' scientists were not Muslim at all is easily dismissed. Although a substantial number of important early figures belonged to tolerated non-Muslim religions, this had ceased to be true by about 1000 CE, and many leading mathematicians did more than simply conform, actively working in Islamic law or philosophy. If the Christians, Jews, and star-worshippers of the Fertile Crescent had it in them to create a mathematical revolution, one might ask, why did they have to await the advent of a new religion and social organization to do so? We could simply accept a sociological explanation (a new empire required scientific organization on a large scale—supposing that to be true); but this does not explain the specific value put on learning—which led to the Greek and Indian inputs—or the ways in which it was put to use.

We are unfortunately at some distance from ninth-century Islam, which was in many ways still in a state of flux. Either Rashed's characterization of Islam as promoting reason, or Torrey's more materialistic view of it as a kind of accountancy have germs of truth, and both were argued in the early conflicts of schools. Was there no conflict between the Qur'an and pagan learning or 'philosophy' (*falsafah*)? Had God decided everything and measured it from the beginning? Theologians discussed such points and competed for the favour of the khalifs.

For example, is what can be known in Arabic—the language of the Islamic revelation—different from Greek science and philosophy in part because of its linguistic home? Or does there exist a universal logic of thought that transcends (and is therefore superior to) particular expressions in use in a given culture? The ḥadīth, as yet one more category, already contain numerous admonitions about the value of knowledge, its reward and the duty to seek it, to gather and preserve it, to journey abroad in search of it. (McAuliffe (2001–), III, p. 101)

The general question of the relation of Islam to pagan and/or practical knowledge is a large one, and we have neither the space nor the ability to deal with it adequately. However, two points should be made:

1. Islam did certainly differ from Christianity (for example) in the value placed on knowledge, as the quote above illustrates; and the language of the Qur'an itself is strongly centred on appeals to reason:

   The Koran is a holy book in which rationality plays a big part. In it, Allah is continually arguing and reasoning. (Rodinson 1974, p. 78 (see also the following pages))

   (The reason in question, though, can hardly be equated with mathematical deduction; it is rather the deduction of our obligations to God from the beneficence of his works, and of ethical duties from basic principles.)

2. Høyrup's point, cited in Section 4: by the ninth century at least, Islam had become codified as a complete system of practice, organizing every sphere of human action; from which the needs not simply for knowledge in itself, but for knowledge to inform practice followed.

Rashed's very recent interview provides some starting points. By claiming that the values of Islam are specifically favourable to science, he raises the stakes, and makes some statements which even those who are quite committed to promoting better understanding of Islamic science might find difficult to accept. The whole interview is worth reading, since as a scholar he cannot only score good debating points but consider difficult questions such as the 'decline' of Islamic mathematics after the fifteenth century (how can it be understood and accounted for?). And he makes a more limited but important point, which has indeed been well appreciated, for example, by Kennedy (1983), that time has a particular value in Islamic observances which calls (one would think) for the application of science.

Science was an important dimension of the Islamic city. One element was the time-keeping (*miqat*) in the mosques. Astronomy was necessary to view the lunar crescent for religious purposes. It must not be forgotten that each of the large mosques had an astronomer associated with it... (Rashed 2003)

In fact, few religions have given practical mathematicians so much to think about as Islam, with its lunar months which start at the moment when the new crescent is visible, its carefully defined five prayer-times a day, and its fast which ends at dusk. Astronomers worked tirelessly on the improvement of their tables, developing the Ptolemaic and Hindu astronomy into a much more efficient instrument; but as early as the time of Thābit ibn Qurra, who wrote on the difficult question of the first visibility of the moon's crescent, they came to realize that their understanding of atmospheric phenomena always left some doubt about the key questions of what one could see.

The science of time was of course useful beyond a religious context, and similarly mathematics was important to the flourishing societies throughout the Islamic world insofar as it helped with commerce, surveying, architecture, and the various practical arts; and also in geography, the understanding of the known world. In this religion enters again, and the tenth-century universalist al-Bīrūnī can stand as a central figure, whose *Coordinates of Cities* made possible a general understanding of how the various widely scattered centres were related on the globe, using a well-developed understanding of geometry on a sphere. Both al-Bīrūnī and his modern commentators have claimed more; that such knowledge was essential for religious purposes, since to design the layout of a mosque (say in Seville) correctly it was essential to determine the *qibla*, the direction of Mecca where the faithful should turn for prayer. As he says:

[L]et us point out the great need for ascertaining the direction of the *qibla* in order to hold the prayer which is the pillar of Islam and also its pole. God, be He exalted, says: 'So from wheresoever thou startest forth, turn thy face in the direction of the Sacred Mosque, and wheresoever ye are, turn your face thither.' (Qur'an, Sura 2:150). (Al-Bīrūnī 1967, pp. 11–12)

The mathematicians may well have thought their knowledge essential; but mathematicians are not always as important as they think, and George Sarton pointed out in 1933 that many medieval mosques in North Africa and Spain have 'incorrect' alignments, despite the flourishing state of mathematics in those countries.

This problem has recently been cleared up, it appears, in a detailed study of legal writings and of the mosques themselves by Mónica Rius.[12] The answer is interesting for the light it throws on the status of mathematics: in fact, Islamic lawyers pointed out that the complex mathematical methods were (a) sometimes uncertain—particularly in the case of longitude—and (b) not accessible to the mass of the faithful, as they should be. They therefore allowed recourse to simpler definitions, which of course gave more 'approximate' directions for prayer. This is not to say that al-Bīrūnī and others were irrelevant; there must have been cases of mosques where the *qibla* was determined by mathematics. However, here, as elsewhere, its use could be contested and the idea that it was 'imposed by religion' certainly begins to seem simplistic.

This example can serve as a cautionary tale on the limits of the usefulness of mathematics, which was certainly important enough in the world of medieval Islam. As we shall see, Marxists tend to claim that mathematics is driven by the demands of society, and mathematicians, when it suits them, claim that they are doing vital and useful work. However, if much of the organization of Islam was favourable to science, there were certainly times and places when science could be dispensed with, even treated with hostility.[13] To make a parallel, Descartes, Pascal, and Galileo were no less good Christians than their predecessors. If they found that their religion could be harmonized with a rational and practical scientific outlook, the cause is perhaps to be found in the ideological climate, or what Marxists would call the relations of production. Accordingly, a particular difficulty in the statement with which this section opens is that Rashed seems to be treating Islam, as religion and philosophical outlook, as homogeneous in its positive effect on the sciences (at least during the medieval period). It will be interesting to see how other specialist historians react.

Exercise 9. *What would be necessary to know in order to determine the qibla? Given the necessary information, how would you do it?*

Appendix A. From al-Khwārizmī's algebra

(From Fauvel and Gray 6.B.1)

A root is any quantity which is to be multiplied by itself, consisting of units, or numbers ascending, or fractions descending.

A square is the whole amount of the root multiplied by itself.

---

12. *La Alquibla en al-Andalus y al-Magrib al-Aqṣà*, reviewed in *Isis* 94 (2003, p. 371).
13. Again, Rashed produces good exmples to show that an anti-science outlook cannot be equated with religious 'orthodoxy', but there were trends within orthodoxy which were opposed to science.

A simple number is any number which may be pronounced by itself without reference to root or square.

A number belonging to one of these classes may be equal to a number of another class; you may say, for instance, 'squares are equal to roots', or 'squares are equal to numbers', or 'roots are equal to numbers'.

[Al-Khwārizmī then deals with examples of these cases before continuing as follows.]

I found that these three kinds: namely, roots, squares, and numbers, may be combined together, and thus three compound species arise; that is, 'squares and roots equal to numbers'; 'squares and numbers equal to roots'; 'roots and numbers equal to squares'.

Roots and squares are equal to numbers: for instance, 'one square, and ten roots of the same, amount to thirty-nine dirhems'; that is to say, what must be the square which, when increased by ten of its own roots, amounts to thirty-nine? The solution is this: you halve the number of the roots, which in the present instance yields five. This you multiply by itself; the product is twenty-five. Add this to thirty-nine; the sum is sixty-four. Now take the root of this, which is eight, and subtract from it half the number of the roots, which is five; the remainder is three. This is the root of the square which you sought for; the square itself is nine.

[...]

[Geometrical demonstration]

We have said enough so far as numbers are concerned, about the six types of equation. Now, however, it is necessary that we should demonstrate geometrically the truth of the same propositions which we have explained in numbers. Therefore our first proposition is this, that a square and ten roots equal thirty-nine units.

The proof is that we construct a square of unknown sides, and let this figure represent the square which, together with its roots, you wish to find. Let the square, then, be *ab* [Fig. 3.] of which any side represents one root. Since then ten roots were proposed with the square, we take a fourth part of the number ten and apply to each side of the square an area of equidistant sides, of which the length should be the same as the length of the square first described and the breadth two and a half, which is a fourth part of ten. Therefore, four areas of equidistant sides are applied to the square, *ab*. Of each of these the length is the length of one root of the square *ab* and also the breadth of each is two and a half, as we have said. These now are the areas *c, d, e, f*. Therefore, it follows from what we have said that there will be four areas having sides of unequal length, which also are regarded as unknown. The size of the areas in each of the four corners, which is found by multiplying two and a half by two and a half, completes that which is lacking in the larger or whole area. Whence it is we complete the drawing of the larger area by the addition of the four products, each two and a half by two and a half; the whole of this multiplication gives twenty-five (Fig. 7).

And now it is evident that the first square figure, which represents the square of the unknown, and the four surrounding areas make thirty-nine. When we add twenty-five to this, that is, the four smaller squares which indeed are placed at the four angles of the square *ab*, the drawing of the larger square, called *GH*, is completed. Whence also the sum total of this is sixty-four, of which eight is the root, and by this is designated one side of the completed figure. Therefore when we subtract from eight twice the fourth part of ten, which is placed at the extremities of the larger square *GH*, there will remain but three. Five being subtracted from eight, three necessarily remains, which is equal to one side of the first square *ab*.

ISLAM, NEGLECT AND DISCOVERY

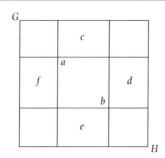

Fig. 7 Al-Khwārizmī's second picture.

Fig. 8 The figure for Thābit's proof. Compare Fig. 4 (Euclid II.6). ABCD (the way of writing it seems odd, but it is necessary for the statements to work) is the 'squares' of the problem, and rectangle DE (or BEGD) is the 'roots'. Their sum is the 'numbers', and is known AF = FE.

Appendix B. Thābit ibn Qurra

The first type is this: square and roots equal to numbers. The way of solving it with the help of the sixth proposition of the second book of Euclid's elements is as follows. We take for the square a square *ABCD*, and let *BE* be a multiplicity of units which measures a line, equal to the given number of roots. [So in the above example, *BE* is ten units.] We draw the rectangle *DE* [see Fig. 8]. Then it is clear that the root is *AB*, and the square is *ABCD*. In the domain of arithmetic and numbers, it is equal to the product of *AB* with a unit which measures a line. In this way, the product of *AB* with a unit which measures a line is equal to the root in the domain of arithmetic and numbers. But *BE* is such a number, equal to the given number of roots. And so the product of *AB* with *BE* is equal to the roots of the problem in the domain of arithmetic and numbers. But the product of *AB* with *BE* is the rectangle *DE*, as *AB* is equal to *BD*. So the rectangle *DE* is itself equal to the roots of the problem. And so the whole rectangle *CE* is equal to the square and the roots.

[The point of the repetitions seems to be that Thābit is carefully reminding the reader that we are working in a framework where numbers can be represented by lines, as they are in Euclid's arithmetic books; or by areas, if we make rectangles out of such lines, as happens in book X. He has now drawn a figure equal to (square and roots) which, unlike al-Khwārizmī's figure, is a single rectangle.]

But the square and the roots are equal to a known number. So the rectangle *CE* is known, and it is equal to the product of *AE* with *AB*, as *AB* is equal to *AC*. So the product of *EA* with *AB* is known and the line *BE* is known, as the number of its units is known.

In this way, the question leads to a known geometrical problem, namely: the line *BE* is known, it is produced to *AB*, and the product of *EA* with *AB* is known. But in the sixth proposition of the second book of the Elements it is shown that if the line *BE* is divided in half at the point *F*, then the product of *EA* with *AB* together with the square on *BF* are equal to the square on *AF*. But the product of *EA* with *AB* is known, and the square on *BF* is known. Hence the square on *AF* is known, and so *AF* also is known, and if from it is subtracted *BF*, which also is known, there is left the known *AB*, that is the root. If we multiply it by another equal to itself, we find the square *ABCD* is known. This is what it was required to show.

Appendix C. From al-Kāshī, *The Calculator's Key*, book 4, chapter 7

On the measurement of bodies with regular faces.

...

There are seven bodies. [Al-Kāshī considers not only the usual five but two semiregular solids (see Fig. 9) which have their faces all regular, and regularly arranged, but not all the same.]

The *first* contains four faces, which are equilateral triangles in the sphere, that is, it is the body bounded by four equal equilateral triangles. It appears as a pyramid with a triangular base, and is made up of four pyramids, whose bases are its faces, and whose vertices are at the centre. The measurement of this is as follows: take the square on the diameter of the circumscribed sphere, and find the root of two thirds of it, and also the root of half the square on the diameter, and the first will be the side of the base, and the second the height of the triangular side. If we multiply one of them by half the other, we find the area of one side. If we multiply this by two ninths of the diameter of the sphere, we find the volume.

Another way. We multiply the diameter once by 0 48 59 23 15 41 fifths, and we obtain its side, and another time by 0 42 25 35 3 53 fifths, and we obtain the height of the triangle. And do the rest as before.

[The main relation $s = \sqrt{\frac{2}{3}} \cdot d$, of the side of the tetrahedron to the diameter of the sphere, is to be found in Euclid XIII.13 and so 'common knowledge' among the savants at Samarkand; which is presumably why al-Kāshī feels there is no need to prove it. As has been said, his book is an exposition of methods, not of proofs, although from his other works we know that he could produce serious proofs when needed. As for the actual figures, in sexagesimals to 'fifths' ($1/60^5$, or roughly 1.2×10^{-9}), they follow from the standard method, which he has set out earlier, for extracting square roots; the first number is $\sqrt{\frac{2}{3}}$ and the second $\sqrt{\frac{1}{2}}$. It is interesting to compare the second figure with the Babylonian version on the Yale tablet (Chapter 1, Fig. 6), which has the

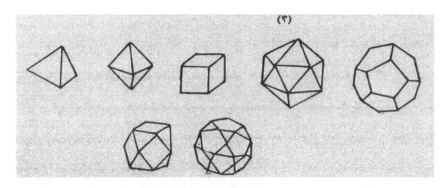

Fig. 9 Al-Kāshī's seven regular solids (the five 'platonic' solids of Chapter 2, and two 'semiregular' ones).

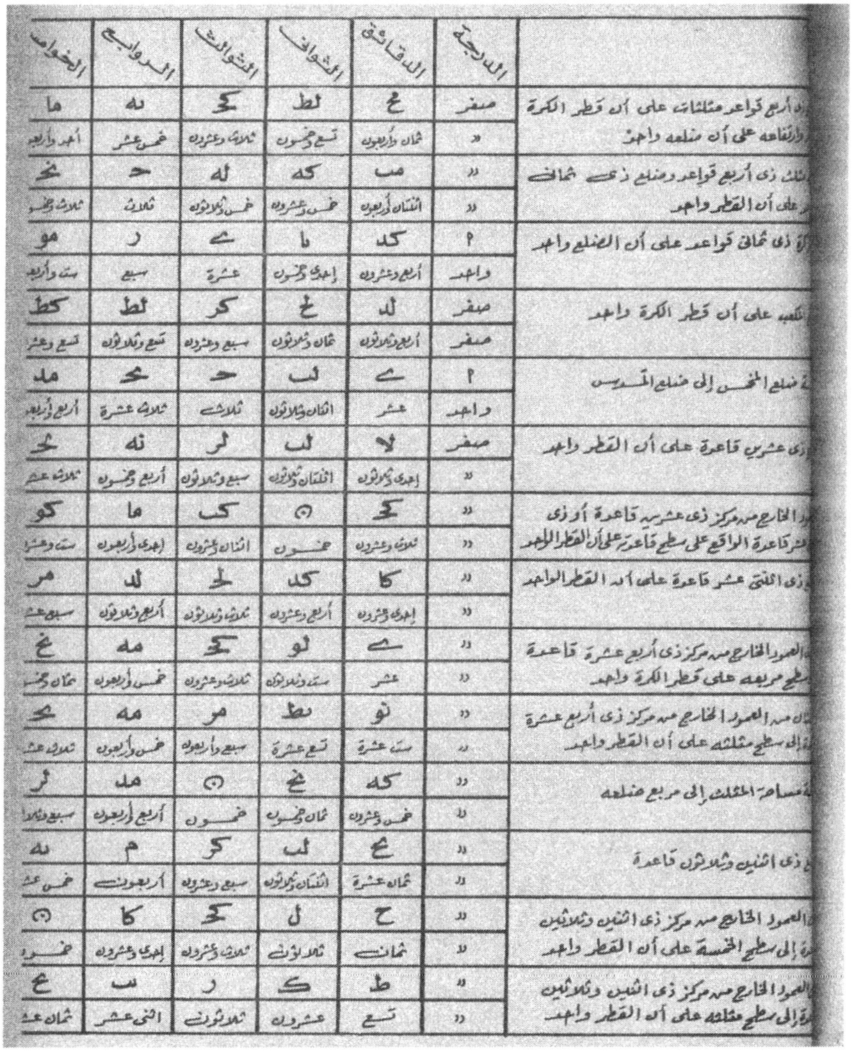

Fig. 10 Al-Kāshī's table. (Numbers are written using Arabic letters, cf fig. 5.)

value 42 25 35. Was the same method used?] Above (Fig. 10) is the table which al-Kāshī gives of the regular solids.

Exercise 10. *Take a tetrahedron with vertices*

$$(1, 1, -1), (1, -1, 1), (-1, 1, 1), (-1, -1, -1)$$

(half the vertices of a cube).

(1) Why is this a regular tetrahedron?
(2) What is the length of a side?
(3) What is the diameter of the circumscribed sphere? Verify the relation which al-Kāshī gives.
(4) What do his other statements mean, and can you check them?

Can you do any of this without using coordinates?

Solutions to exercises

1. Clearly 'Is' denotes an odd place (counting from the end); and the point is that your starting point is to look at the number up to the last odd place (e.g. 5 for 576, or 13 for 1369). The whole part of the root of this number—which is a single figure—gives you the first figure of your answer.

    You now have 2 as the root of 4, the largest square less than 5. You subtract its square (4) from the 5, and drop down the rest giving 176. You now double the two (4 again) and put it under the 7, so it is effectively 40. Al-Uqlīdisī's expression means that you are looking for an x such that $40x$ added to x^2 'exhausts' the 176 you have left. In other words, $(40+x)x = 176$. In fact this is satisfied by $x = 4$.

    The method is simply using the usual formula for $(a+b)^2$, with $a = 20$ and $b = $ the unknown x. If this is slightly confusing, try some other three- and four-figure squares. Then see how it generalizes to larger ones (it does).

2. As in Chapter 2, let us use algebra to simplify. Call the length AB 'a'. Then BC $= a$, BD $= a/2$, and so CD $= (a/2)\sqrt{5}$. Hence by construction, DE $=$ CD $= (a/2)\sqrt{5}$. So AE $= a((1+\sqrt{5})/2)$. This is the right length for the 'golden section' construction of Chapter 2; the triangle ABG whose sides are in the ratio $1 : (1+\sqrt{5})/2 : (1+\sqrt{5})/2$ has angles $36°$, $72°$, $72°$, and the construction proceeds as required.

3. Al-Khwārizmī gave six equation models, and these were always followed by his successors through the medieval and early modern period. There are three 'trivial' ones: roots equal to numbers, roots equal to squares, and squares equal to numbers; and three 'serious' ones, roots and squares equal numbers, roots and numbers equal squares, and squares and numbers equal roots. (The point is that all coefficients must be positive.) Again, because there must be a positive *solution*, the form (which we would think worth including) 'squares and roots and numbers equals zero' (e.g. $x^2 + 3x + 2 = 0$) is excluded.

4. AD is equal to AB+BD, or $a+b$. 'The rectangle AD by DB' in Euclid's language means the area of a rectangle whose sides are equal to AD and DB, so it is the product $(a+b)b$. Since C is the midpoint of AB, CB $= a/2$; while CD $=$ CB+BD $= (a/2)+b$. From this the statement follows.

5. (a) Al-Khwārizmī's method starts by halving the roots—result 1. Square this, result 1; add to 1 (the 'numbers'), result 2. Now our problem is to take the square root. If we can (call the result $\sqrt{2}$ as usual), subtract half the roots, that is, 1, and get the answer $\sqrt{2}-1$. (b) The line BE has length 2; and we must construct AB so that the square on AB and the rectangle AB·BE are equal to 1. We divide BE in half at F, so BF $= 1$. Euclid II.6 says that EA · AB together with the square on BF (i.e. $1+1 = 2$) equals the square on AF. So we construct a square of area 2 (compare the *Meno*!); its side is AF. Subtract BF (i.e. 1), and you have the result AB. This depends on the fact that you can construct AF, whose length is $\sqrt{2}$, geometrically without saying what the length is.

6. Call the 'amount' x. If 10 is added to an amount $(10+x)$, and the amount (i.e. the sum) is multiplied by the root of 5, we have $(10+x)\sqrt{5}$. This is said to be equal to the product of the amount (this word is being over-used) with itself; that is, to x^2. So, $(10+x)\sqrt{5} = x^2$ as stated. By the usual rules for quadratics: write it as $x^2 - \sqrt{5}x - 10\sqrt{5} = 0$. Then the solution is

$$x = \frac{1}{2}\left(\sqrt{5} \pm \sqrt{5 + 4.10.\sqrt{5}}\right)$$

Obviously for a positive solution we want the positive root, and a slight rearrangement of the expression puts it in the form given by abū Kāmil.

7. Since $\sin 30° = \frac{1}{2}$, setting $\sin 10° = y$, we obtain the equation $4y^3 + \frac{1}{2} = 3y$.

8. Rather than try to redo the division (which is a 'straightforward' long division of polynomials), consider the two tables shown in the figure. The first shows simply the dividend P arranged in columns according to powers, with coefficients $(20, 2, 58, 75,...)$; and below it the divisor $Q = 2x^3 + 5x + 5 + (10/x)$, shifted up by three places (so times x^3), ready to be multiplied by 10 and subtracted. The second table has the 10 in the cubes place of the top row (result); in the second row are the coefficients of $P - 10x^3 Q$; and in the third Q again, this time shifted up by only two places and ready to be subtracted again. The process concludes when al-Samaw'al finds that his final remainder $(4 + 10x^{-2} + 10x^{-3} + 20x^{-4})$ is an exact multiple $2/x^3$ times Q, and we can stop.

9. This is a slightly hard exercise in spherical geometry. We have to know: (a) our latitude, say $a°$, (b) the latitude of Mecca, say $b°$, and finally the difference between our longitude and that of Mecca, say $C°$. (Think of this as an angle of the triangle.) We then have a spherical triangle ABC (Fig. 11). The angle at the pole is C, and the two adjacent sides are a and b (degrees of latitude). The *qibla* is determined by the angle which the line from us to Mecca makes with North; the angle B in the figure. The 'sine formula' for spherical triangles:

$$\frac{\sin b}{\sin B} = \frac{\sin c}{\sin C}$$

would give us B if we knew c, since we know b and C. But we can get c from the 'cosine formula':

$$\cos c = \cos a \cos b + \sin a \sin b \cos C$$

(See Gray 1978, p 46)

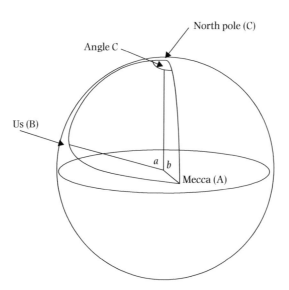

Fig. 11 Illustrating how you find the qibla (Exercise 9).

10. It is easy to check that the vertices given are distant $2\sqrt{2}$ from each other; which establishes (1), since the faces must be equilateral triangles, and also answers (2). You can find the centre of the sphere either by looking at the other half of the cube (the vertices are alternate vertices of a cube), or by finding the centre of gravity, obviously $(0, 0, 0)$. The radius is the length of the line joining this to a vertex, that is, $\sqrt{3}$, the diameter $2\sqrt{3}$. So $s : d = 2\sqrt{2} : 2\sqrt{3} = \sqrt{\frac{2}{3}} : 1$.

The 'height of the triangular side' is the height of an equilateral triangle of side $2\sqrt{2}$ in our model, that is, $\sqrt{6}$ (using $\sin 60° = \sqrt{3}/2$). The ratio of this to d is now $\sqrt{2} : 2 = \sqrt{\frac{1}{2}} : 1$. The statement about area (= half times base times height) is 'classical'. The volume is the area of the base (just found) times one third of the height, by the formula for the volume of pyramids. To find the height of the pyramid, note that the three points which are *not* $(-1, -1, -1)$ have centre of gravity $(\frac{1}{3}, \frac{1}{3}, \frac{1}{3})$. The height is the length of the line which joins this to $(-1, -1, -1)$, and it is easy to see that this is $\frac{2}{3}.d$. Hence al-Kāshī's 'two-ninths of the diameter'.

To prove it without using coordinates, look at Euclid XIII.13.

6 Understanding the 'scientific revolution'

1 Introduction

Philosophy is written in that vast book which stands forever open before our eyes, I mean the universe; but it cannot be read until we have learnt the language and become familiar with the characters in which it is written. It is written in mathematical language, and the letters are triangles, circles and other geometrical figures, without which means it is humanly impossible to comprehend a single word. (Galileo, *Il Saggiatore (The Assayer)*, tr. in Drake 1960, pp. 183–4)

It is sensible to begin any discussion of the scientific revolution with Galileo. The above quotation, which has been used to excess as a description of his position, at any rate serves to link Galileo's physics with the history of mathematics. It also illustrates the role which mathematics often plays in accounts of the scientific revolution—as a language whose use transforms science, not as an object of study in itself. As a result, those mathematicians for whom physics was not an obvious interest, like Cardano and Viète—whom we will discuss later—or who were better mathematicians than they were physicists, like Descartes, receive little or no attention in the history. Our version will necessarily have to be skewed in a different direction—to the development of mathematics itself, and to its interaction with physics; questions of the role of experiment and observation, which are central to the usual history, are not really important. There were exceptional changes in the way mathematics was done between, say, 1550 and 1700, some of which are discussed in this chapter. The most notable, the calculus, is the subject of the next chapter, but it is generally agreed that by the time Descartes's *Géométrie* was published (in 1637), a 'new mathematics' had come into being; and that the works of Viète, Stevin, Descartes, and others radically changed the way in which even ordinary practitioners worked. Much of this had some relation to the wider scientific revolution, but both the question of what was new and the question of origins need to be considered with due reference to the particularity of mathematics. So let us begin by posting, as major concerns for this chapter, some questions:

1. Was there a specific 'mathematical revolution' of the fifteenth to seventeenth centuries (say)? If so, what was its nature?
2. How far can developments in physics and mathematics in the period be 'disentangled', that is, to what extent do changes in one depend on the other?
3. To what external factors (if any) should we attribute any changes in mathematics which take place?

To ask our usual naive question: what is so important about the scientific revolution? Briefly, however it is defined it is central to the narrative of Western culture and how, for better or worse, it is viewed. And Galileo was only the most gifted among many contemporaries who, under the inspiration of Plato or Archimedes, saw the transformation in terms of the introduction of 'mathematical language', in a variety of senses, into the study of the natural world. True, mathematics had been present for a long time, in astronomy and optics and in Archimedes's statics, for example. However, Galileo's statement was explicitly expansionist (as well as being related to dynamics, his particular interest): you will have founded a proper science only if you have introduced the mathematics which is its hidden language.

2 Literature

In almost every preceding chapter, we have complained of the difficulty of locating sources, either primary or secondary or both. With the scientific revolution, we have the opposite problem. The major questions about the revolution (did it happen? what was its nature? what were its causes?) have been constantly debated, and have spawned a vast literature, which even specialists will find overwhelming. And the literature is easily accessible; the classic works of Duhem, Butterfield, Koyré, Dijksterhuis, and Kuhn are much easier to find in libraries than most of the other works I have recommended so far.[1] The same goes for many of the primary sources, most obviously Galileo's works. Help is at hand in the form of H. Floris Cohen's recent book (1994). Cohen is not particularly interested in the development of mathematics, and his account is if anything too painstaking, but he does describe the major currents in the history, the authors and texts who are worth further reading. Still, the literature continues to grow, and new theses and new material are continually coming to the forefront in the discussion. The reader should be prepared to keep a number of disparate, even conflicting ideas in mind (e.g. about the origin of Galileo's dynamics) at the same time—which is no bad thing for the historian.

About the mathematics specifically, the literature is much slimmer; even for key figures like Galileo and Kepler, the mathematical work generally takes second place to the physics. The most interesting and detailed discussion is hard to recommend: dealing specifically with the 'algebraic aspect' of the revolution, it is Jacob Klein's very dense and detailed book (1968), a translation of a German text of the 1930s. Despite its title, the key arguments of this book centre on what was revolutionary in the algebra of the period 1550–1650, and I shall be referring to them, but it is not easy to find, not very user-friendly, and (naturally) underestimates the Islamic contribution. All of these criticisms are frankly acknowledged by Klein in his author's note, but the problems remain. More recently, Hadden (1994), which draws on Klein for some key ideas, is less detailed and more polemical, but a relatively easy read.[2] The extracts in Fauvel and Gray are helpful, and there is a special section which throws light on early modern England—which we shall not deal with, but which is worth looking at.

Besides the problem of literature, we have a problem of timescale. Where do we start? The seventeenth century propagandists, of whom Galileo and Descartes were the most persuasive, tended to present their work as marking a clean break from a past of ignorance and sterile muddle-headed scholastic disputes. The Greeks—some of them—were precursors, to be sure, but no one else needed to be considered. This view was generally accepted as the history of science developed

---

1. As the key example of a 'revolution', the period is central in the writing of Kuhn and of his opponents, naturally.
2. See the review in *Isis* 86 (1995), pp. 642–3 for a criticism of this book's attempts to have it both ways on Marxism in particular.

as a discipline in the eighteenth and nineteenth centuries. One historian changed the situation completely: the reactionary catholic French physicist Pierre Duhem, writing around 1900. His key works, based on a careful study of French medieval writers, as well as of Leonardo da Vinci, aimed to show that there had been high-quality scientific activity from the thirteenth century on, and that the Church had played a decisive role in promoting it.[3] Furthermore, and this was his main contention, there had been no 'revolution'; Galileo's (physical) discoveries were already present in the works of the Paris school in the fourteenth century; and, typically, the history of science is continuous rather than catastrophic or revolutionary in nature.

[T]he mechanical and physical science of which the present day is so proud comes to us through an uninterrupted sequence of almost imperceptible refinements from the doctrines professed in the Schools of the Middle Ages. The so-called intellectual revolutions consisted, in most cases, of nothing but an evolution developing over long periods of time. The so-called Renaissances were frequently nothing but unjust and sterile reactions. Finally, respect for tradition is an essential condition for all scientific progress. (Duhem 1991, p. 9.)

The fact that many historians are still committed to some version of the Duhem thesis means that a serious account of the scientific revolution needs to start some time before; which, in turn tends to lead to an overload of often disparate information from a period of several centuries during which mathematics was used in a variety of different ways. Because the material for this chapter is rich, diffuse, and well, if unevenly, covered in various texts, we shall develop the story as a series of snapshots, or meditations on particular themes. To try for any degree of completeness would be to risk complete unreadability.

3 Scholastics and scholasticism

Although this is an instance of an unfounded mathematical formulation of a natural law that is not valid, Bradwardine's[4] argument is by no means destitute of historical significance. (Dijksterhuis 1986, p. 191)

The question naturally arises as to what the scholastics did with their interpretation of Eudoxus. What use can one make of the useless? (Murdoch 1963, p. 257)

Medieval science is now taken seriously, if often with the kind of patronizing despair expressed by Dijksterhuis and Murdoch. Thanks to the detail in Duhem's research, it is not necessary to agree with his more extreme theses (e.g. that the Church had helped scientific research by condemning Aristotle in 1277) to see that the work of the period preceding Galileo cannot be dismissed out of hand. However misguided his theories and specious his arguments, he contributed more than anyone else to changing our ideas of the scientific revolution by enforcing at the very least a serious consideration of Galileo's predecessors, even if the end result was (as with Koyré, for example, see (1978)), to conclude that there *was* a decisive break rather than a continuous evolution. The starting point of any thoughtful history is thereby pushed back; and where, before the sixteenth century, one chooses to start is likely to be determined by something

---

3. There is no consensus about how one uses the terms 'Middle Ages' or 'medieval'; and the problem is made worse by the fact that so much that is obviously 'Renaissance' which in some sense implies post-medieval, is also obviously 'medieval' in period—for example, the cathedral at Florence, or the work of Masaccio (both fifteenth century). Different ways of describing society may coexist more or less uneasily; my 'medieval' is roughly from 1100 to 1500, while my 'Renaissance', at least in Italy, is roughly from 1350 to 1600. To complicate matters further, the term 'early modern' is now academically popular for (roughly) the period 1500–1700.

4. Thomas Bradwardine, fourteenth century Oxford physicist-mathematician, whose study of motion has often been considered a precursor of Galileo's work.

other than 'revolution'. For mathematicians, it is likely to be the relatively early period when the major translations of Greek and Arabic texts were made (from Arabic into Latin, the universal language of culture in Western Europe), in the twelfth century. While this is commonly compared to the Arabic 'age of translations' three centuries earlier, the differences are as striking as the similarities.

1. In the first place, the Arabic translations were (on the whole) centrally organized around an institution—the 'House of Wisdom'—which was linked to the central political and religious power, the khalif. The western Christian world in which the translations were made was less centralized, and political and religious leaders, with a few exceptions, showed no particular interest. The Islamic translations were also widely diffused through the use of paper; Western libraries were smaller, literacy more restricted, and paper with all its cheapness and convenience only came into general use at around the time of the invention of printing in the fifteenth century.

2. More importantly, the 'caste of scholars' who had done the work of translation were not in a position to follow it up. We have, more than usually, a difficulty in identifying scholars as 'mathematicians', and the term is hardly useful before the fifteenth century. Apart from a scattered handful of specialists, most of those who studied mathematical questions (Albert of Saxony, Bradwardine, Oresme) should be considered physicists, philosophers, even theologians first with an auxiliary interest in mathematics—in some cases an intelligent one, but rarely interested either in practical problems or in following up the studies of antiquity. The difference was, most strikingly, in the lack of mathematicians interested in the more difficult work of Apollonius or Archimedes, for example. So, while in Baghdad we find the translators of Archimedes immediately taking up the problems which he failed to solve, or trying to understand his solution and work out an alternative, there is no sign that anything of the same kind was attempted in western Europe at all. Paul Lawrence Rose has pointed out the 'failure' (if one wants to pass judgment) of the scholastics to do anything useful with the major translations of Archimedes, by William van Moerbeke in the thirteenth century.

Why were Moerbeke's mathematical translations neglected? [True,] there are indications that Moerbeke was not at home with the mathematics of his subject. Yet the reason for the neglect lies not with the quality of the translation, but with the failure of medieval scholars to take up the tradition. Those responsible included scholastic philosophers who found a little Arabo-Latin Archimedes and a lot of Adelardian Euclid sufficient for their purposes. Equally to blame were the mathematicians including those who had perhaps encouraged Moerbeke in his project in the first place. (Rose 1975, pp. 80–1)

At this point, the reader may be feeling in need of an explanation of the word 'scholastic'. It is overdue, but the meaning is a complex one. In the first place, it refers to the tradition of teaching and study centred on the universities (Bologna and Paris were founded in the thirteenth century, then Oxford, Heidelberg, and others). A compromise between religious orthodoxy and admiration for Aristotle, as interpreted by ibn Rushd ('Averroes' in European translation) in particular led to an attachment to authority, both religious and 'ancient', and to logical arguments. The teaching and reasoning style is called 'scholastic'; its practitioners were 'schoolmen'. The arguments were of a particular kind (what were called *quaestiones*), in which a question was posed (e.g. could the sun be still and the Earth move?); the arguments on both sides were carefully set out and a series of objections had to be dealt with in a strictly defined format before a conclusion could be reached; in its later development this was the scholastic reasoning ridiculed by Galileo and Descartes, for example. Dijksterhuis makes the case against such methods forcefully:

> In fact, it had been one of the traditions of Scholasticism from the twelfth century onward to employ the so-called *sic et non* method, advocated especially by Abelard; its principle was that in dealing with a given subject all the opinions that had ever been pronounced about it and all the arguments that could be advanced for or against a certain view were enumerated and discussed as fully as possible . . . This method, of course, presented great advantages; it bespoke a striving after objectivity and it helped to prevent an idea, once it had been pronounced, from falling into oblivion again. It is, however, obvious that if the method were applied too thoroughly, the disadvantages would be bound to preponderate. (Dijksterhuis 1986, p. 167)

This method, moreover, makes some of the most interesting medieval work on mathematics appear peculiar in a unique way. The idea that a scientific question might be decided in this way by logical arguments ultimately derives from Aristotle. Its great virtue is that it encourages us to think of counter-arguments to the hypotheses to which we are committed, although the way of deciding between alternatives tended in the Middle Ages to depend on logic rather than what would now be called scientific method. And in mathematics, where we normally accept that there is exactly one right answer, it may seem quite contrary to the spirit of the subject. (What arguments could one produce against a method for solving quadratic equations? The question is worth thinking about.) It was not the method of Islamic mathematicians, even those who most respected Aristotle, so that the 'mathematics' of many of the schoolmen whatever it was worth, was genuinely a new area of enquiry. In the course of teaching mathematics in the faculty of arts, (which led to study in one of the advanced faculties of medicine, law, or theology), they frequently raised mathematical topics in the form of *quaestiones*, and tried to settle them by a form of debate.

The rational arguments of the schoolmen would not usually speak to today's rational understanding, as they rested in general on the basis of Aristotle's physics and logic (with a little Euclid), rarely went far beyond, and were often quite confused. For an example, we could consider Albert of Saxony's discussion of whether it is possible to square the circle; for this see E. Grant's sourcebook (1978), a good source for the schoolmen generally. It seems clear that Albert did not know, or did not understand the sophisticated methods of squaring by curves such as the quadratrix (for which see Knorr 1986), since he made no reference to them. His equipment consisted basically of some historians' references to circle-squaring, and Archimedes's *Measurement of the Circle*; the latter he misunderstood in the standard way to mean that the circumference was 22/7 times the diameter. He gives four arguments for squaring and two against, and then makes—again a typical scholastic trick—a distinction of five meanings which 'squaring the circle' could have. The distinctions are important in a scholastic argument, since clearly if you have conclusive arguments for and against a proposition, the proposition must have different meanings in the two cases. One argument for is simply that Aristotle said that it had been done by Antiphon and Bryson (which is not what Aristotle said in any case). The next introduces something new:

> If there could not be given a square equal to a circle, it would follow that there would take place passage from 'greater' to 'lesser', or from extreme to extreme, through all the means without ever arriving at 'equal' or 'middle'. But this is false. Therefore I prove the consequence. (Grant 1978, p. 171)

What Albert is doing here is a simple version of what we would call an existence proof—it shows by continuity that there must exist a square equal to the circle, while completely ignoring the problem which preoccupied the Greeks, that is, how you construct it. This argument was already known in Greek times, but Albert's presentation has something fresh about it. One could, critically, say that, however learned Albert was, he did not know very much about Greek geometry; it also shows that his ignorance, and his determination to proceed by what could be called common sense, led him to a new way of thinking about the problem. It is not strictly 'modern', but it is a break with ancient tradition.

Finally, he defines squaring the circle 'in the fifth way with respect to sense and to intellect' as the usual problem—to find a square whose area is equal to that of the circle. With respect to sense, because you cannot perceive the difference; with respect to intellect, because you can prove they are the same. Again we see the very specific nature of scholastic reasoning, and how odd it seems when applied to geometry. Albert 'proves' that this is possible, by using

Fact 1. Archimedes's result that the area equals half the radius times the circumference (well known, and often used);

Fact 2. Archimedes's 'formula' that the circumference is three and one-seventh times the diameter (used by Archimedes as an approximation, but as we have seen quoted at least from Heron's time onwards as if it were exact).

Although this *is* a mathematical argument, if a wrong one, the whole idea of settling the question by such a sequence of pros and cons seems to us exotic and 'unmathematical'; and it is easy to see why later generations were to consider the mathematicians of the Middle Ages, by and large, as simply confused. However, in Albert's favour, it should be said that the Greeks had failed to produce a conclusive 'answer' to the circle-squaring problem, and that the idea of posing the alternative—not to square it, but to prove that it could *not* be squared—was a new one, and (however poor his arguments were) pointed in the right direction.

This poses again the question of what might be regarded as revolutionary in science. We have a scientific practice which is unlike any that has preceded it, so it seems reasonable to describe the change (from Greek and Islamic mathematics, say, to that of Albert) as revolutionary; and if the revolution in some sense goes backwards, with a great deal of loss of content and sophistication, this is partly because the questions being studied are different. Science does not *only* progress—this is a modern myth, and later we shall see some alternative myths which were peddled in the sixteenth century. And without being completely relativist, it is clear that different societies have different ideas of what their object of study is. Our view that they are confused and/or wrong-headed should be tempered by an honest attempt to see what they were trying to do.

Exercise 1. *Given the two 'facts' above, how do you use them to square a circle?*

4 Oresme and series

Zénon! Cruel Zénon! Zénon d'Élée!
M'as-tu percé de cette flèche ailée
Qui vibre, vole, et qui ne vole pas!

Zeno, Zeno, cruel philosopher Zeno,
Have you then pierced me with your feathered arrow
That hums and flies, yet does not fly! (Valéry 1920)

The scholastic tradition in mathematics was, as we shall see, not the only one in the Middle Ages, but it was important. One of the best examples of new work produced by this approach is given by Nicolas Oresme (14th century). Oresme has been considered the originator of graphical (coordinate) representation of quantities before Descartes.[5] A particularly good example of his thinking, and of what the Scholastics could produce at their best, is given by his discussion of infinite series in his *Quaestiones super Euclidem* (Questions on Euclid).

The role of proportion in medieval thought was extremely important, both as a tool of elementary mathematics and as a philosophical theme; but the treatment of proportion in Euclid, particularly in book V (the 'Eudoxan theory'), was a constant problem on account of its difficulty. A detailed account of this theme (including the various mistranslations and misinterpretations in the medieval Euclid versions) is given by John E. Murdoch in 'The Medieval Language of Proportions'—see Murdoch 1963, pp. 237–271. The particular problem of what happened when—in modern terms—one took successive proportions $q, q^2, q^3, \ldots$ and added them had preoccupied Islamic mathematicians, because of its relation to the 'method of exhaustion'. The point is as follows. Euclid's proposition X.1 states:

If two unequal quantities be given, and if from the greater, greater than half be subtracted, and again from the remainder, greater than half be taken, and we continue successively in the same way, then it is at last necessary that there remain a quantity less than the lesser of those given.

In the Islamic tradition, the tendency was to ask: does it have to be 'greater than half'. This was answered by Naṣir al-Dīn al-Ṭūsī in his commentary: you do need (something like) Euclid's statement.

There is, then, underlying proposition X.1 the idea that you continue subtracting parts 'as long as you need to', and that at a certain point (if they are greater than halves) you can stop. However, it would seem that the scholastics were the first to consider the idea of taking an actual infinite sum; and the result was expressed most clearly by Oresme.

His text is given as Appendix A to this chapter; I have tried to doctor it as little as possible, so as to clarify exactly what he does say.

First, we should note that Oresme seems to have no doubt that you can *physically add* an infinite sequence of numbers. The numbers will be positive, as the techniques for dealing with negative numbers had not been developed, so some problems which arise in our general theory are absent. The sum may be 'infinite', whatever that means, or it may be a number; but he has no doubt that it exists. As far as I know, this is quite original. Since the days when Zeno (the 'cruel Zeno' of Valéry's poem) devised his paradoxes of the infinite in the fifth century BCE, there had been strong objections to treating a 'completed infinity' as opposed to a 'potential infinity' in Greek mathematics, which were spelt out by Aristotle. Indeed, Oresme deals with the argument from the authority of Aristotle before proceeding any further.

To consider what the extract says in detail, let us break the taboos on 'presentism', and translate his statements into modern language. The results are as follows:

1. A geometric series $a + ax + ax^2 + \cdots$ whose ratio x is ≥ 1 has an infinite sum; one whose ratio is < 1 has a finite sum. ['Second, it must be noted that...'.]
2. For example, $1 + \frac{1}{2} + (\frac{1}{2})^2 + \cdots = 2$; $1 + \frac{1}{3} + (\frac{1}{3})^2 + \cdots = \frac{3}{2}$; and generally, $1 + q + q^2 + \cdots = (1-q)^{-1}$, if $q < 1$. ['The first proposition is ... The second proposition is ...'.]

---

5. This is considered in detail by Dijksterhuis (1986, p. 266), who is *not* an uncritical supporter of the idea of 'revolution'; on the whole, his verdict is that Oresme's writings, however novel they were, cannot seriously be considered an anticipation of later ideas.

This is, at any rate, my interpretation of the way Oresme describes the summation of geometric series; it is hard to be sure not only because the language used is obscure but because, since there is no proof, we cannot see how the result was arrived at.

3. On the other hand, it is possible for a decreasing series to have an infinite sum, in particular the 'harmonic series' $1 + \frac{1}{2} + \frac{1}{3} + \cdots$ has an infinite sum. The proof is the usual one. ['The third proposition is . . .']. On this basis, Oresme is generally credited with discovering that what is now called the harmonic series (whose terms decrease to zero) has an infinite sum; and, unless an earlier candidate turns up, this seems right. He did understand that if you continued to add $(1/n)$'s you would get sums which were bigger than any assigned number. In other words, this is a particular instance where presentism gives a reasonably accurate version. We could say that here we have a fourteenth-century mathematician finding out facts about the convergence of series in a modern way.

Having established this to our satisfaction, we would still be left with some puzzling questions. To begin with, what exactly was Oresme trying to do, and why does the context of his work look so different? And, second, why did no one else deal with similar questions? Why were his results not reproduced for so long? Certainly there seems to be no record of a general acceptance that it is all right to use infinite sums, or of any similar use of them until much later.

The answer to all of these questions seems to lie in the framework of the discussion; the old-style scholastic *quaestio*. Unlike his early modern successors, Oresme was not concerned with series as the answers to problems in calculation. (Newton's *Method of fluxions and infinite series* is an obvious example of the later approach.) Rather, he wanted to know the answers to some questions about 'quantity', and the paradox—which is already present in a concealed form in the method of exhaustion, or in Euclid X.1—that an infinite number of finite quantities can have a finite sum. The Greeks would not have put it like this; the scholastics, for whom the infinite was attractive precisely because it was so fertile in contradictions and paradoxes, would. What, asked Oresme, are the conditions for an addition 'by proportional parts' to be possible? The question goes back to Zeno's paradox of Achilles and the tortoise. What was new about Oresme's treatment is that he gave precise conditions in terms of the ratios, and even summed the series. And, of course, that in going on to examine the possibility of a series which is 'by ratios of lesser proportionality'—decreasing—becoming infinite, he came up with the simple example of the harmonic series.

Exercise 2. *(a) What does Euclid's statement in proposition X.1, quoted above, mean? (b) Why can you not use proportions less than a half in general? (c) What has this to do, if anything, with sums of series?*

5 The calculating tradition

Forsooth, a great arithmetician
One Michael Cassio, a Florentine . . . (Shakespeare, *Othello*, Act I, Scene 1)

The claim of Duhem and his successors that the discoveries of the scientific revolution were in the main developments of earlier work by the scholastics has had the positive effect of drawing attention to what it was that the scholastics actually did. However, like most priority claims, it makes for bad history—because it focuses not on the work in its context with its proper connexions so much as on its place in an attempted genealogy; and in this instance the case can only be established:

1. by blurring the very important distinctions between the scientific aims pursued in (say) the fourteenth and seventeenth centuries;
2. by ignoring the lack of evidence of any transmission line (say from Oresme to Descartes).

Historians are rarely (never?) free of presuppositions, but many of them are now studying the medieval tradition for its own sake, as a particular historical tradition within mathematics. Much of the medieval work which was supposedly important for Galileo does not seem to have featured in his reading; and although there was undoubtedly a lively argument in progress about mathematics and its certainty in sixteenth-century Italian universities (conducted very much along Aristotelian lines, what is more)[6] it contributed much less to the shaping of mathematical *practice* than the two sources which Descartes identified—Greek geometry and Arabic algebra.

However, there was an alternative tradition, almost independent, with at least as much influence; that of the often very low-level practical calculators who were needed to teach the sons of merchants. Again we could compare the situation in Abbasid Baghdad, and again there seems to be an important difference: that in the Islamic world skilled mathematicians such as abū-l Wafā' wrote with such schools in mind, while the Western tradition seems to have been at a more basic level.

The works produced by such schools in Italy (where they were probably most important) has been studied in detail by Warren van Egmond (1980). The texts are referred to as 'abbacus books'; the title is misleading, since what we call an abacus, or counting-frame, was never used. The original text, and the most serious is one which you will often find referred to in histories, Leonardo of Pisa's *Liber abbaci* of the early thirteenth century. Leonardo was an unusually good mathematician whose distinguishing points are that he worked outside the university; that he had the good fortune to spend several years in the Arab world with his father, a Pisan merchant; and that he saw an opportunity to spread the useful things which he had learned, particularly the use of Hindu–Arabic numbers and algebra, to the practical men among whom he spent his later life. He was, in the context of the time, an intelligent student from a 'backward' country who received a good education in what was then the metropolis (North Africa), and did what he could with it when he returned.

The immediate influence of the *Liber abbaci* seems to have been the diffusion among the Italian merchants—who had an eye for what was directly useful, as the university men did not—of the most elementary parts of the Islamic tradition. We could, then, contrast two separate 'borrowings' from the world of Islam: the translations of learned works, Greek and Arabic, in the universities on the one hand, and the adoption of Indian numbers and simple algebra in the cities. These elements were taught in 'abbacus schools' using books, usually simplified versions of Leonardo's book and often in Italian to make them more accessible. On a smaller scale, similar works were produced in the various languages of western Europe—English, French, German, and of course their number increased dramatically after the invention of printing in the mid-fifteenth century. Van Egmond claims that 'nearly all the educated men of the renaissance gained their basic mathematical education in schools such as these, including, for example, such notables as Leonardo da Vinci and Niccolò Machiavelli' (van Egmond 1980, p. 8); a German printed version finds its way into that must-have

---

6. This not very enlightening controversy is documented in Rivka Feldhay's article: 'The use and abuse of mathematical entities: Galileo and the Jesuits revisited', in Machamer (ed). (1998, pp. 80–145).

Fig. 1 German arithmetic book from Holbein's 'The Ambassadors' (National Gallery, London).

catalogue of Renaissance things to own, the list of objects in Holbein's *The Ambassadors* (Fig. 1). He further claims that the introduction of these schools, around 1300, was commercially driven, a result of 'the commercial revolution of the thirteenth century'. This involved the increasing use of devices—some new, some perhaps adaptations from the Islamic world—such as banks, letters of credit, and bills of exchange. Later, we find more sophisticated accounting methods, leading eventually to the famous invention of double entry bookkeeping.

[The reader, if not an accountant, may well wonder what this important development which is so often referred to is. Briefly, it consists in the practice of entering every event (sale or purchase) twice, once as a credit and once as a debit; it was in use in Genoa in 1340 (but possibly earlier), and it was first properly expounded in texts in the fifteenth century, most famously by Fra Luca Pacioli, still considered as the 'patron saint' of accountants. For definitions see de Roover (1937), 'Aux origines d'une technique intellectuelle: la formation et l'expansion de la comptabilité à partie double', in *Annales d'histoire économique et sociale* 9 (1937). For its relation to the introduction of the zero, to perspective and much else, see Rotman (1987).]

Unlike the speculations of university men, the textbooks used in abbacus schools were justified solely by their supposed usefulness. Indeed, they did not even serve the purpose of creating a privileged caste, as in ancient Babylonia—solving equations was a skill, not a class marker. The calculating tradition is undoubtedly important, in contributing skills which were not obviously learned in the more formal context of universities. However, there seems little sign that in two hundred years the abbacus schools and similar institutions were responsible for innovation. Since the numerical requirements were relatively simple (no astronomy, for example), the kind of sophisticated approaches to number found in abū Kāmil, Khayyam, or al-Kāshī were not raised. The contents of the textbooks were often quite basic—the writing of numbers and how to calculate with them, a little geometry (measuring circles and triangles by approximate formulae); sometimes 'algebra' of the al-Khwārizmī kind and the extraction of square and cube roots were added. The problems addressed were pseudo-practical and generally solved by methods of false position which could be traced back to pre-Greek times:

1. A tree is 1/3 and 1/4 underground and above ground it is 30 braccia. I want to know how long it is altogether?
2. A man had a denaro and another came to him and he asked, 'I have one denaro. How much do you have?' And he replied as follows, 'I have so much that with the same amount and with one half of what I have and with a quarter and with your denaro it would be 100.' How much did he have?
3. How much does 87 gold florins 35 s. 6 d. earn in 2 years 7 months and 15 days at 10 per cent simple interest?

[I assume, but I may be wrong, that 12 d. make 1 s. and 20 s. make one gold florin. At any rate, the introduction of interest—which the Church condemned, and merchants used various devices to disguise—is a novelty in this mathematics, if in other respects it looks rather like the third dynasty of Ur.] (van Egmond 1980, pp. 22–3)

These questions (more are quoted in van Egmond's book) make clear the new input of merchants' needs into mathematics; but also (in my view) it was not so much for 'advanced' mathematics as for facility in training. Again the parallel with Ur III comes to mind. The 'abbacus schools' have come recently into prominence as a 'lowlier' form of mathematics than that of the universities; but it may be that claims for their influence on the major subsequent developments are overstated.

Exercise 3. *(a) Do questions 1 and 2 by the method of false position. How do you think you should approach question 3? (b) Assuming 240 pence to the pound, prove the neat calculation rule (from a problem in BL Add.MS): If the rate of simple interest is x pence per pound per month, then the annual rate is 5x per cent (that is, 5x pounds for every 100).*

6 Tartaglia and his friends

Let no man who is not a Mathematician read the elements of my work. (Leonardo 2004, vol. 1, opening admonitions)

It is around 1500 that the various developments sketched so far come together; the dividing line between university and informal mathematics is, at least to some extent, broken down; and the whole pattern of change becomes rather complex and difficult to classify. [For example, I shall omit completely (a) the very important subject of the effect of painting and perspective, which I recommend you to research if you are at all interested[7]; see Rotman (1987) and Field (1997); (b) trigonometry, an import from the Islamic world which was both theoretically and practically important.] Simplifying, we can trace two major threads: a rapid development in algebra and the general idea of 'number' on the one hand, and (later) the beginnings of a use of the infinitely small. Both are associated with the continuing problem of the Greek tradition; and in both cases we can see two important simultaneous and competing developments:

1. An increased familiarity with the works of the Greeks (including Archimedes in particular) through translation;

---

7. To open with a quote from Leonardo might seem, in contrast, to foreground painting; but Leonardo was interested in so many other practical pursuits that he can be considered rather as an example of the 'new model' of interested artisan.

2. A realization that the Greek writings were—depending on the author's particular take—too difficult, or too slow, or even mistaken, and that better methods could and should be found to solve the pressing new problems. The 'misunderstandings' of Euclid which dogged the medieval writers now change into something more creative: the invention of a method (a symbolic algebra, a primitive calculus) which masquerades as a true understanding, but is in fact something quite new.

The general solution of cubic equations (first half of the sixteenth century) is a good starting off point, because with it we leave the limitations of both the university and the abbacus school traditions. Although it was only one of several important developments around 1500, it illustrates a number of points about this period in mathematics. Briefly, the problem was to solve equations involving cubes of the unknown in the same way that, since al-Khwārizmī, quadratics had been solved—that is, by some sort of recipe. Omar Khayyam had hoped that such a solution could be found, but had to settle for his geometric constructions (Chapter 5).

The first point to note is that the history of the solution bridges the gap between university and non-university study. The first case was found by Scipione dal Ferro, a professor at Bologna; he did not publish it, but passed it on to his student Antonio Fiore. The general case, also not published, was found by Niccolò Tartaglia, a prolific mathematician working outside the university. He taught in Venetian schools, translated Greek texts—or sometimes passed off others' translations as his own—and wrote original works on algebra, the art of warfare and much else. Hieronimo Cardano, to whom Tartaglia revealed his 'secret' was again a university man, but a very unusual one, whose most celebrated work was in medicine and astrology; having allegedly promised not to publish before Tartaglia did, he 'broke' his promise and published in his *Ars Magna*.

This context of secrecy was very different from what we think of as research,[8] and was connected, at least in Tartaglia's case, with the chance of winning a reputation, and sometimes money, by competitions in which mathematicians set each other problems and tried to defeat each other. Clearly public knowledge of the method would ruin the contest.

The second point is that the mathematics itself is complicated and non-obvious, if all you have at your disposal is 1500-style algebraic methods. Tartaglia could probably justify his method in any particular case by calculation, but did he have the language for a general proof? Cardano gave a proof derived from Euclid, as an Islamic algebraist like abū Kāmil would have done. Tartaglia's well-known rhyme—in his version, a mnemonic to help him remember how to get the solution—goes as follows, for the case 'cube and things equal to numbers'. (Resisting the temptation to translate the sixteenth century mathematical rap song into verse, I will quote Fauvel and Gray's literal translation with its modern equivalents.)

When the cube and the things together
Are equal to some discrete number
[To solve $x^3 + cx = d$,]
Find two other numbers differing in this one.
Then you will keep this as a habit
That their product should always be equal
Exactly to the cube of the third of the things.
[Find u, v such that $u - v = d$ and $uv = (c/3)^3$.]

---

8. Although cases have occurred in more recent times—one could even mention Andrew Wiles's actions on the proof of Fermat's Last Theorem (see Chapter 10).

The remainder then as a general rule
Of their cube roots subtracted
Will be equal to your principal thing.
[Then $x = \sqrt[3]{u} - \sqrt[3]{v}$.]

The point about the solution of the cubic (which is never now taught in schools, and hardly in universities) is that it extended the simple reckoners' algebra beyond its capabilities, if not for any obviously useful purpose. One of the problems set by Fiore to Tartaglia in their 1535 contest sounds very much in the reckoners' tradition. However, it belongs in the category of problems which are practical only in appearance; one cannot imagine it being the answer to a merchant's needs.

A man sells a sapphire for five hundred ducats, making a profit of the cube root of his capital. How much is this profit?

This is the equation 'cube and thing equal 500', or as we would say, $x^3 + x = 500$.

How were such solutions written in the 1530s? Tartaglia's exposition in his published letter of 23 April 1539 to Cardano gives the answer, in a question which he seems to have chosen particularly to display his ability to deal with difficulties:

And if it were 1 cube plus 1 thing equal to 11, it would be necessary to find two numbers or quantities such that one is 11 more than the other, and that the product of the one by the other should be $\frac{1}{27}$, that is the cube of the third of the things, whence operating as above it will be found that our thing is ℞ u. cube ℞ $30\frac{31}{108}$ plus $5\frac{1}{2}$ minus ℞ u. cube ℞ $30\frac{31}{108}$ minus $5\frac{1}{2}$ and not other ... (Tartaglia 1959, p. 122)

The 'u.' in the above is for 'universal'; the whole means simply 'cube root'. '℞' is a common sign for 'root' at this time. We can recognize Tartaglia's solution, in our notation, as

$$\sqrt[3]{\sqrt{30\frac{31}{108}} + 5\frac{1}{2}} - \sqrt[3]{\sqrt{30\frac{31}{108}} - 5\frac{1}{2}}$$

And we can complacently note how much Tartaglia missed the use of brackets in particular, as well as many other improvements in notation which were introduced in the next century. In any case it seems that the arrival of formulae of this complexity meant that both the writing of algebra and the way in which numbers themselves were thought about needed radical change; and that is what happened. This, at any rate, is Jacob Klein's thesis:

While, however, the 'algebra' which has Arabic sources is continually elaborated in respect to techniques of calculation, for instance by the introduction of 'negative', 'irrational', and even so-called 'imaginary' magnitudes (numeri 'absurdi' or 'ficti', 'irrationales' or 'surdi', 'impossibiles' or 'sophistici'),[9] by the solution of cubic equations, and in its whole mode of operating with numbers and number signs, its self-understanding fails to keep pace with these technical advances. This algebraic school becomes conscious of its own 'scientific' character and of the novelty of its 'number' concept only at the moment of direct contact with the corresponding Greek science, that is, the *Arithmetica* of Diophantus. (Klein 1968, pp. 147–8)

It is probable that Klein did not know of the 'abstract algebra' of al-Karajī, al-Samaw'al, and Sharaf al-Dīn al-Tūsī; and that he did know that Diophantus was available to the Islamic world. He also seems to have given a lesser weight to the very influential introduction of decimal fractions, which made it possible at least to *think* of approximating roots, and even numbers like π, as closely as one liked. We have looked at the question of their 'invention' in Chapter 5; in Europe,

---

9. Each of these three pairs of Latin terms is the old equivalent of one of the modern English terms, at least approximately.

the most significant event was Stevin's propagandist work (*La Disme* of 1585). Here there *is* a possible debt to the Islamic world, specifically to al-Kāshī, but we are in need of further evidence. However, even given the various possible lines of transmission from Islamic mathematics, an analysis of what happened in the sixteenth century must take into account not just its 'influences', but its own particular momentum and early-modern ideology; Stevin was an early enthusiast for decimalization, who hoped to replace both 'astronomers' numbers' (sexagesimals) and the confused systems of measurement with which surveyors were faced.

> And the surveyor or land-meter ... is not ignorant (specially whose business and employment is great) of the troublesome multiplication of rods, feet, and oftentimes of inches, the one by the other, which not only molests, but also often ... causes error, tending to the damage of both parties ... (Stevin 1958, p. 395)

He also was responsible for producing tables of compound interest, in which again decimals simplified the task tremendously.

The new algebra, if we accept Klein's thesis, has a generally accepted 'starting point': the redraft by Bombelli of his algebra textbook of 1560 (published 1572). Having been shown a manuscript of Diophantus, Bombelli changed his emphasis to accord better with his ancient model, removing the traditional practical problems and replacing them by ones taken from Diophantus. This 'moment'—a change in the idea of number which overthrows many of the ancient Greek ideas in the interest of what is simple and practical—we could call a first mathematical revolution (to answer the first of the questions which we posed in section 1); the second is the gradual, equally un-Greek introduction of infinitesimal processes.

Exercise 4. *(a) Use Tartaglia's method to solve the equation 'cube and three things equal to four', or $x^3 + 3x = 4$. (Hint: You are given $u - v$ and uv; find $u + v$.) (b) Why do you not get the obvious answer 1? (c) Try to prove that x as given in Tartaglia's formulation is a solution of the general cubic equation 'cube and things equal to numbers' (1) by algebra and (2)—if you have the patience for it—by geometry, as Cardano did.*

Exercise 5. *Solve 'cube and thing equal 500' (as in the question of the sapphire and the ducats) by Tartaglia's recipe.*

7 On authority

> Behold, the art which I present is new, but in truth so old, so spoiled and defiled by the barbarians, that I considered it necessary, in order to introduce an entirely new form into it, to think it out and publish a new vocabulary, having got rid of all its pseudo-technical terms lest it should retain its filth and continue to stink in the old way ... And yet underneath the Algebra or Almucabala which they lauded and called 'the great art', all Mathematicians recognized that incomparable gold lay hidden, though they used to find very little. (Viète, *The Analytic Art*, in Klein 1968, pp. 318–9)

> It has become a matter of common usage to call the barbarous age that time which extends from about 900 or a thousand years up to about 150 years past, since men were for 700 or 800 years in the condition of imbeciles without the practice of letters or sciences ... but although the afore-mentioned *preceding* times could call themselves a wise age in respect to the barbarous age just mentioned, nevertheless we have not consented to the definition of such a wise age, *since both taken together are nothing but the true barbarous age* in comparison to that unknown time at which we state that it [that is, the true wise age] was, without any doubt, in existence. (Stevin, *Géographie*, quoted in Klein 1968, p. 187)

The question of innovation versus tradition was central to the major figures (and often the minor ones) in sixteenth-century science. If one considered the scholastic tradition a barrier to science, which of the Greeks did one call on to contest it? Copernicus claimed to be reviving an earlier theory of Aristarchus, Galileo drew particularly on Archimedes, Kepler was influenced by Plato and Pythagoras. In mathematics Aristotle was less important as a reference point, but the existence of a third tradition, that of practical algebra with its disturbing Islamic parentage made for a three-way contest; and many important textbooks start with explicit statements such as the above about where their authors stand. In the apparently very different field of (English) literature, Stephen Greenblatt[10] introduced the idea of 'self-fashioning', or what we might call the personal makeover, as a distinctive feature of the century:

Self-fashioning is in effect the Renaissance version of these control mechanisms, the cultural system of meanings that creates specific individuals by governing the passage from abstract potential to concrete historical embodiment. Literature functions within this system in three interlocking ways; as a manifestation of the concrete behaviour of its particular author, as itself the expression of the codes by which behavior is shaped, and as a reflection upon these codes. (Greenblatt 1980, pp. 3–4)

If we stop confining the narrow application of the word 'literature' to the writing which is called creative and allow the inclusion of algebra textbooks such as Viète's *Analytic Art*, Greenblatt's model provides a useful explanation of the projects of the new algebraists of the sixteenth and early seventeenth centuries—Tartaglia, Cardano, Bombelli, Viète, Stevin, and Descartes. (It is of course equally applicable to other scientists; Galileo notably was intensely aware, both as stylist and as self-presenter, of models to be adapted and avoided; and much of what Feyerabend (1975) presents as 'propaganda' could be looked at from this point of view.) The algebra texts actually solve equations in the author's favoured style ('the concrete behavior of the particular author'), they provide a model for others to imitate ('the expression of the codes by which behavior is shaped'), and, strikingly, they are given to programmatic statements which explain the author's attitude to the competing traditions and reasons for choosing a particular method or language ('a reflection upon these codes'). The statement which defines the author's innovation is also a self-portrait as the author would wish to be seen, as the extracts above show. And other aspects of Greenblatt's description of his self-fashioners apply easily to the mathematicians, in particular their social mobility (p. 7) and their need of an 'authority' and of an opposing 'alien' (p. 9); as Greenblatt points out (1980, pp. 3–4), 'One man's authority is another man's alien'. However, the authorities who shaped the mathematical discourse were (fortunately for them) unrelated to the great religious controversies of the day, so long as the geometry of the universe was not involved. For Viète, as his extract shows, the authorities were the ancient Greeks; while the aliens were the modern, barbarous, and filthy (one presumes Muslim) corrupters of the ancient art.

Viète was a notable innovator who invented the first fully coherent algebraic notation (to be superseded by the simpler version of Descartes, which we now use). His contradictory claims ('new, but in truth so old') are characteristic of modernizers of the time; renewal, as is implied in the term 'renaissance' has to be presented as rediscovery.[11] His book is hard to read, partly because of the notation (it is almost easier to read the traditional language of algebra which the Italians derived from the Arabs); and partly because he invented a new language of procedure, borrowing words from the Greek to describe his methods in solving problems, a language which no

---

10. The founding father of 'new historicist' criticism. See (1980).
11. The same strategy has been used up to the twentieth century for example, by T. S. Eliot—and no doubt after.

one afterwards adopted. Perhaps for this reason, you will find no extracts in Fauvel and Gray which show how he worked. Here, then, is one in its orginal form. (The 'standard edition' of *The Analytic Art* is a good example of the loss involved when an author's notation is updated; although it can be read to get an overall idea of Viète's project, the changes in terminology, such as '*BE*' for '*B* in *E*' make it both more readable and less interesting; one cannot see what innovations are specifically Viète's own.)

Book II Zetetic XVII. Given the difference between the roots and the difference between their cubes, to find the roots. [Try to read through this text to see what it means, if possible, before consulting the notes below.]

Let B be the difference between the roots and D solid the difference between the cubes. The roots are to be found.

Let the sum of the roots be E. Therefore $E + B$ will be twice the greater root and $E - B$ twice the smaller. [Why?] The difference between the cubes of these is B in E squared 6 + B cubed 2 which is consequently equal to D solid 8.

$$\text{Therefore} \quad \left\{ \frac{D \text{ solid } 4 - B \text{ cube}}{B\ 3} \right\} \quad \text{equals } E \text{ squared}$$

The squares being given, the root is given, and the difference between the roots and their sum being given, the roots are given.

Accordingly the difference of the cubes quadrupled, minus the cube of the difference of the sides, being divided by the difference of the sides tripled, there results the square of the sum of the sides.

If B is 6, D solid is 504, the sum of the sides $1N$, $1Q$ equals 100.

Notes. A 'Zetetic' is Viète's word for a method of finding out. In his notation the 'roots' are lines, so the sum of their cubes is a 'solid', which is why he calls it 'D solid'; his rule is that (as the Greeks prescribed) you must always keep track of the dimensions of quantities and not set lines equal to solids. For example, B and D solid are denoted by consonants, because they are known; while E is a vowel, because it is unknown. Numbers come *after* the letters, so that 'E squared 6' means what we would call $6E^2$.

What comes out of this, and many other examples like it in the *Analytic Art*, is not an outstandingly difficult result. It *is* a systematic treatment of algebra in which the objects being manipulated are *letters*, which stand not for natural numbers (as in Euclid's arithmetic), but for quantities, and in which the proof is not by geometry. In Viète's example, B is 6 and D solid is 504, so that E^2 is 100, and E is indeed a whole number 10. (Check this; and find the two roots.) But it is clear that a different choice (e.g. D solid $= 2, B = 1$) would lead to an 'irrational' answer, and that nothing in the method restricts answers to being whole numbers—or (to anticipate) to being numbers at all. It is this which leads Klein in particular to give Viète such a high value:

But above all—and it is this which gives him his tremendous role in the history of the origins of modern science—he was the first to assign to 'algebra', to this 'ars magna', *a fundamental place in the system of knowledge in general*. From now on the fundamental *ontological* science of the ancients is replaced by a *symbolic* discipline whose ontological presuppositions are left unclarified. (Klein 1968, p. 184)

Here, then, (if Klein is right) is the germ of Russell's 'Mathematics is the science in which we do not know what we are talking about'; and its extension to physics via the definition of 'occult' quantities, from Newtonian force to atomic spin, whose importance is not that they can be measured but that they can enter into equations. This is a great deal to ascribe to the work of a lawyer whose introduction of letters, if we are to believe his English interpreter Thomas Harriot, was inspired by the similar language of legal case studies.[12] It might be more reasonable to say that, following the rediscovery of Diophantus, such a transformation was 'in the air'.

Stevin by contrast appears as more practical, as can be seen from his biography (Dijksterhuis 1970). He was also more self-consciously innovatory. The quotation above shows his disregard for the Greeks, and his belief in a 'lost' programme of science from an earlier wise age. This was not completely eccentric, and was shared by a number of his contemporaries. Among the most important inheritances of the wise age, in his view, was the decimal system of writing numbers, and his role in promoting decimal fractions is undoubtedly related to that. When, in *La Disme*, he gives the result of a division by three as a decimal with (effectively) as many 3s after the point as you like, he has finally grasped a fact which seems to have eluded the Babylonians: the existence of repeating decimals and their necessity.[13] In his *Arithmétique*, he set out a deliberately 'controversial' view on numbers. The orthodoxy, transmitted in a confused way by the medieval schools from the Greeks, was that numbers (2, 3, 4, ...) were not magnitudes, that fractions or parts of a number were not numbers, and that 'one' was not a number since it was the origin of number. How widely these statements were believed in practice is uncertain, but Stevin enjoyed demolishing them, suggesting that those who denied that parts of a unit were numbers were 'denying that a piece of bread is bread'. He concludes by a statement of theses: one is a number (thesis I); there are no absurd, irrational, inexplicable, or surd numbers (thesis IV); and so on. Both Viète's and Stevin's viewpoint can be seen as contributing to the way that mathematics shapes our view of the world today; if we think of the law $E = mc^2$ as an essential equation irrespective of the values of E, m, and c, we are following Viète, while if we consider its use in telling us what happens when we substitute particular (computed) values of m and c, we are following Stevin.

Exercise 6. *Prove Viète's formula for the difference of cubes—it is of course easiest to modernize at least partly in your working—and deduce the formula for E squared.*

8 Descartes

I have constructed a method which, I think, enables me gradually to increase my knowledge and to raise it little by little to the highest point which the mediocrity of my mind and the short span of my life will allow it to reach. (Descartes 1968b, p. 28)

I have spent some time describing the ways in which a 'modern' outlook on numbers can be traced back to the late sixteenth century. The texts in which the work is done do not *look* modern, because they are written in a language which is in transition between that of the medieval world and our own. Descartes's *Geometry*, on the other hand, looks modern and is relatively easy to read—for us; his contemporaries found it difficult, because new. This is because he had the good fortune to invent the common notation of modern algebra (x, y for unknowns, a, b for constants; and $4xy$, for example, instead of Viète's '*A* in *E* 4'.) Of course, this could be looked at another way: if his terminology has stayed with us, it is because he had the intelligence to devise one which was clear and easy to use. As a result of this, and more specifically of his 'coordinate' representation for geometric curves, in the eighteenth century, historians of mathematics (French ones, in particular)

---

12. So, until recently, P. for plaintiff and D. for defendant—see any law book.
13. The Babylonians would have had to do more work, since $\frac{1}{7}$, the first Babylonian repeating decimal, repeats after six sexagesimal places, not after one.

considered Descartes *the* revolutionary who had freed them from bondage to the tedious methods of the ancient Greeks, by reducing hard geometric problems to simple algebraic ones. This is a view which is often now regarded with some suspicion, although Descartes himself promoted it:

> I have given these very simple [methods] to show that it is possible to construct all the problems of ordinary geometry by doing no more than the little covered in the four figures that I have explained. [That is, the figures which construct a sum, a product, a quotient, and a square root.] This is one thing which I believe the ancient mathematicians did not observe, for otherwise they would not have put so much labour into writing so many books in which the very sequence of propositions shows that they did not have a sure method of finding at all, but rather gathered together those propositions on which they had happened by accident. (Descartes 1954, p. 17)

It is noteworthy that Descartes here is *not* claiming to be rediscovering an ancient technique. In fact, the simplicity of his methods, he claims, is a proof that the ancients did not have them—or they would have found his results. It is sometimes claimed that he was unoriginal—the graphical representation came from Oresme, and the algebra from Viète. Descartes did acknowledge his debt to Viète, specifically defending himself against charges of difficulty by claiming (which he nowhere states in the *Géométrie*) that he supposed his readers to be familiar with the *Analytic Art*.[14] In any case, his project was different and specific: the relation of geometry and algebra. A standard modern textbook criticizes Descartes for not being more practical:

> Our account of Descartes' geometry should make clear how far removed the author's thought was from the practical considerations that are now so often associated with the use of coordinates. He did not lay out a coordinate frame to locate points as a surveyor or a geographer might do, nor were his coordinates thought of as number pairs ... *La géométrie* was in its day just as much a triumph of impractical theory as was the *Conics* of Apollonius in antiquity. (Boyer and Merzbach 1989, pp. 385–6)

This criticism is interesting, but, I think, misplaced. Coordinate geometry even today is not 'intrinsically' practical—even the statistician who studies whether points in a scatter graph lie near a straight line $y = ax + b$, let alone the geometer who wishes to picture the curve $y^2 = x^3 + x^2$ (Fig. 2) are not thinking as surveyors or geographers. On the other hand, for *some* practical tasks, the new ideas were very well adapted, as Newton and Leibniz were to understand. Galileo takes a great deal of trouble to establish using Apollonius' *Conics* that a projectile describes a parabola, a fact which follows very easily by finding its equation; and while Descartes does not deal with results of this kind (his physics was too different from Galileo's, and mostly confused), they are simplified and clarified by using the methods which are to be found in his book. To see this, and to see how, unlike Viète, he avoided the Euclidean heritage of formal definitions, propositions, and proofs, I have given the basic construction in which coordinates first appear as Appendix B. The idea is to draw a curve by using a simple-minded machine (a ruler which pivots, subject to constraints), and to find the equation of the curve. The description of the machine seems more complicated than it is in practice, and the derivation of the equation is not hard. At the end, the curve is said to be 'of the first kind', by which Descartes means a conic section; the reason being that its equation is of the second degree (quadratic) in x and y. Note that the use of machines for drawing curves could be seen as a typically practical Renaissance innovation; but like much else, it has a long heritage, both Greek (Eratosthenes) and Islamic, though neither is acknowledged by Descartes.

However, besides inventing a new method and a new notation, Descartes was introducing a new style of writing mathematics, which was also to have considerable influence. All previous books in

---

14. Letter to Mersenne, 1637 (Descartes 1939 t. II, p. 66.)

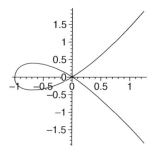

Fig. 2 Graph of a cubic curve.

Europe, even those of Stevin, had been formally set out, either on the Greek model (sequence of propositions and proofs), or on the model of the abbacus schools, which was also to some extent that of Diophantus, and of the Chinese (sequence of problems and solutions). It could be said that the same structure underlies the *Géométrie* (e.g. the extract I have given asks a question and solves it); but the whole is absorbed into a smooth narrative which appears to lead on without a break from one 'discovery' to the next, pausing for comments, explanations, or excuses for avoiding them:

> But I shall not stop to explain this in more detail, because I should deprive you of the pleasure of mastering it yourself, as well as the advantage of training your mind by working over it, which is in my opinion the principal benefit to be derived from this science. (Descartes 1954, p. 10)

The nearest approach to this scheme is Kepler's *Astronomia Nova* (see below), which purports to be an account of his struggles to discover the laws of motion of the planets. The latter, however, is a story of discovery, while *La Géométrie* is an account of how the reader should proceed. The novelty lies in the 'you' of the sentence quoted above: the reader can be addressed, not in the imperative ('Find . . .', 'Draw . . .') as a teacher addresses a student, but as an intelligent equal.

The fact that his contemporaries found the *Géométrie* difficult may help us, all the same, to guard against an 'unhistoricist' approach to his work. We read it from a perspective in which, on the whole, the translation from curves to equations and back is a familiar one. Descartes was making a major innovation, and clearly he did not explain it as well, to a seventeenth-century audience, as he hoped; its absorption took time, although only twenty years later young Isaac Newton was already (by some reports) finding Descartes more congenial than Euclid.

9 Infinities

Nature is an infinite sphere in which the centre is everywhere, the circumference is nowhere. (Pascal 1966, p. 89 (no. 199))

But let us remember that we are dealing with infinities and indivisibles, both of which transcend our finite understanding, the former on account of their magnitude, the latter because of their smallness. (Galileo 1954, p. 26)

Around 1600, more or less independently of the work in algebra, we see the first systematic use of 'the infinite' in European mathematics; by mid-century it was becoming frequent, and Pascal, a mathematical mystic, used it in a number of metaphorical statements (such as his famous wager about the existence of God), as well as in an early version of the calculus. The impetus seems mainly to come from physical applications, and from a recognition that infinities in some sense underlie Archimedes' work, although it may be necessary to be more careless than he was in what one allows. And indeed, in these early stages of what later becomes the calculus, there is a general sense of exploration, of trying out statements to see how they sound, and (by contrast with algebra) of a loosening rather than a sharpening of definition. While there are certainly traces in Stevin's work, the most interesting introduction is in Kepler's *Astronomia Nova*, which presents itself (only semi-realistically) as the account of the various false trails he followed until his final discovery of his famous planetary laws. (The account here is largely based on a detailed analysis by Bruce Stephenson (1987). Any analysis of such a complex text as the *Astronomia Nova* is contestable, but the broad lines seem persuasive enough.) Kepler was faced by a new problem almost from the beginning. Both Ptolemy and Copernicus explained the motion of planets—Mars in particular—as being composed of uniform motions in a circle around a point which was not the real centre (sun for Copernicus, earth for Ptolemy) but a point in empty space, the 'equant'.[15] The advantage of such a scheme is that it is relatively easy to calculate using uniform motion in a circle. A modern physicist would point out that this is 'unphysical', in that one is postulating a force linking the planet to the equant, where there is no matter. Kepler's version of this, which stemmed from his own ideas on planets, was that the souls which animated them could perfectly well perceive the sun (for example) and its distance and adjust their movement accordingly; but it was unreasonable to suppose them capable of perceiving the equant. He therefore had to suppose that motion was dependent on distance from the sun, and slowed down when the planet was further.

At this point we have two problems about velocity. The first is the old question of whether Kepler or his contemporaries could define, or even think of 'velocity at an instant', as we register at an instant that a car is travelling at 45 miles per hour. It has been suggested that Thābit ibn Qurra and al-Bīrūnī used such an idea (see for example Hartner and Schramm 1963), but it does not seem to have been in general use among Islamic astronomers. The first steps towards understanding what this might mean were taken by Galileo—see below. The second was the old problem that even velocity over a time interval was the ratio of quantities of different kinds—distance and time— and so unacceptable in a Greek framework. Kepler's way of avoiding these problems was to use the 'delay'—the time taken by the planet to travel along a small interval of its orbit—in circumstances where the intervals were approximately equal. He had to frame a hypothesis about how the delay depended on the distance, and he tried several; but in each case, he faced the problem of adding a large number of very small delays to arrive at the measurable time-intervals given by observations. Here, in chapter 40, he is introducing the difficulty, supposing that the planet moves in a circular eccentric orbit around the sun:

Since, then, the delays of the planet in equal parts of the eccentric are in the inverse proportion to the distances of those parts [from the eccentre], but the individual points are changing their distance from the eccentre all around the semicircle; I thought it would be no easy work to find out how the sum of the individual distances could be arrived at. For unless we had the sum of all, which would be infinite, we could not say what was the delay of each. And so we should not know the equation. For as the whole sum of the distances is to the whole period, so any part of the sum of the distances is to its own time.

[Kepler began by dividing the circle into 360 degrees, but found the calculation tedious; but then he had an idea.] For I remembered that once Archimedes also, when he was seeking the proportion of the circumference to the diameter, divided the circle into an infinite number of triangles, but this scheme was hidden by his proof by contradiction. Hence where before I had divided the circumference in 360 parts, now I cut the plane of the eccentric in as many lines from the point from which the eccentricity was computed. [Fig. 3]. (Kepler 1990, pp. 263–4)

---

15. Even this is a serious oversimplification of the system of epicycles and equants which both needed: points which were occupied by no real bodies, but whose rotation was a necessary part of the description of the 'phenomena', the observed motions.

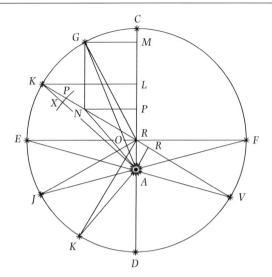

Fig. 3 Kepler's diagram from 'Astronomia Nova'.

This is a relatively early stage in the research; Kepler is still following a 'wrong' theory, and finds a result which fails to agree with his data; and even before this he has to adjust his method because his ideas on summing triangles do not work out. However, he has made a very bold statement about Archimedes: that his proofs by contradiction are a way of concealing infinite methods. This is not what the ancient Greek texts say (as far as we know), and it did not conform to the orthodox view of them in Kepler's time. (The *Method* in which Archimedes did use a sort of infinite process was unknown at the time (see Chapter 3).) This 'misreading' of Archimedes was useful to Kepler at the stage he had reached; and the idea that a circle could be thought of as a polygon with an infinite number of sides had already been used by the mystic Nicholas of Cusa, whose mathematical/theological thinking certainly influenced him. It recurs in his work on the measurement of wine-barrels, which I reproduce in Appendix C—there the recourse to the infinite is justified as simply being quicker.

10 Galileo

SAGR. But I, Simplicio, who have made the test can assure you that a cannon ball weighing one or two hundred pounds, or even more, will not reach the ground by as much as a span ahead of a musket ball weighing only half a pound, provided both are dropped from a height of 200 cubits. (Galileo 1954, p. 62)

And so, finally, we return to Galileo as innovator or revolutionary; he who (as in the passage above) overthrew the authority of Aristotle by appeal to experiment. As we have already suggested, Galileo, however committed he was to mathematics as the language of the universe, was mathematically on the conservative side. Despite learning from the artisans of the Venetian shipyards (like Tartaglia) and writing his major works in the vernacular (like both Tartaglia and Descartes), he continued to grind out propositions in the Euclidean style whose proof was by appeal to classical geometry. His two major works are valued as brilliant examples of Italian literary style; the *Dialogue on the Two Major World-Systems* and the *Discourses on Two New Sciences* are both cast in dialogue form, and full of artful reasoning and rhetoric. (Misleadingly, the Dover edition of the second work translates 'discorsi' as 'dialogues', which gives the two works the same (short) title. In the text here we refer to it as the *Discourses*.) His 'Euclidean' bent may seem strange, given the importance which he is often given as the inventor of 'instantaneous velocity', since that idea at least would seem to need the infinitely small (or some equivalent) to define it—as the ratio of infinitesimal distance to infinitesimal time, say. However, it relates to his extreme reverence for Archimedes (again) as a true scientist in opposition to Aristotle. The infinitely small is perhaps present as a subtext of his discussions of motion in the two texts. However, they tend to be concealed by a vague description of velocity (which was in some sense Galileo's favoured term) as a 'degree of swiftness'; we know what it is if it is uniform, as he often says, but that is not really the point. The following exchange shows how, and with how little clarity, the infinite was introduced:

SAGR:[16] A great part of your difficulty consists in accepting this very rapid passage of the moving body through the infinite gradations of slowness antecedent to the velocity acquired during the given time . . .

SALV: The moving body does pass through the said gradations, but without pausing in any one of them. So that even if the passage requires but a single instant of time, still, since a very small time contains infinite instants, we shall not lack a sufficiency of them to assign to each its own part of the infinite degrees of slowness, though the time be as short as you please. (Galileo 1967, p. 22)

It has been pointed out by those who favour Duhem's thesis that uniformly accelerated motion was already introduced in the fourteenth century (at Merton college Oxford), and that Galileo's key result (that if acceleration is uniform, the time taken to cover a distance is equal to the time which would be taken moving constantly at the mean speed) was also known. The crucial contribution which Galileo made was the observation—which was confirmed by his experiments—that *free fall* was uniformly accelerated. This, with the related deductions (e.g. that the path of a projectile is a parabola) set him quite apart from the fourteenth century discussions of uniformly accelerated motion *in general*, however much he may have drawn on them. For the Oxford men, speed was a 'quality', whose intensity could vary, and the difficulties about instants of time which worried Galileo's characters seem not to have arisen. The fact that Galileo did return repeatedly to the infinitely large and small, with varying degrees of sophistication, shows an increasing feeling that a defence was needed.

Galileo better than most others (probably Kepler in particular) realized the pitfalls of reasoning with the infinite. In the First Day of the *Discourses*, he devotes a long digression to the subject. Can one divide a continuum into an infinite number of pieces? Why can one use a limiting argument to show that a circle and a point have the same 'volume', although the circle is clearly much bigger? His classic example is the first instance of what, many years later, would be called a one-one correspondence between infinite sets. I have given it as Appendix D. The conclusion tends to a sensible caution: one cannot use the terms 'equal', 'greater', or 'less' for infinite quantities. A great deal of such caution—which derived partly from Galileo's respect for the Greek tradition—had to disappear for further progress to be made.

---

16. Both works are presented as 'dialogues' between three parties: Simplicio, who is the mouthpiece of Aristotelian views, Salviati, who represents (roughly) Galileo himself, and an intelligent arbiter Sagredo, who tends to raise objections, while seeing the force of Galileo's arguments.

Appendix A

(From Oresme, *Quaestiones super Euclidem*)

Next we inquire whether an addition to any magnitude could be made by proportional parts.

First, it is argued that it cannot be, since then it would follow that a magnitude could be capable of being increased to an actual infinity. This consequence contradicts what Aristotle says in the third book of *Physica* and also Campanus's statement [in his commentary on the Common Notions in Euclid I], where he distinguishes between a number and a magnitude, in that a number can increase indefinitely and not decrease indefinitely, but the reverse is true of a magnitude.

Proof of the consequence: From the fact that the addition takes place indefinitely it follows that the increase too takes place indefinitely.

Against this it is argued: anything that is taken away from one magnitude can be added to another. It is possible to take away from a magnitude an infinite number of proportional parts,[17] therefore it is also possible to prove that it can be increased by an infinite number of parts.

[This is the only 'argument against' the possibility of addition, and follows fairly closely the *quaestio* structure with its objection, reference to authority, and so on. Notice (and this agrees with the Duhem thesis, perhaps) that Aristotle's authority is cited in argument, but is not treated as conclusive—in fact, the counter-arguments override it. After this, Oresme goes in more detail into the mathematics at issue, beginning with some definitions of types of ratio, which I omit.]

Secondly, it must be noted that if an addition were made to infinity by proportional parts in a ratio of equality or of greater inequality, the whole would become infinite; if, however, this addition should be made [by proportional parts] in a ratio of lesser inequality, the whole would never become infinite, even if the addition continued into infinity. As will be declared afterward, the reason is because the whole will bear a certain finite ratio to the first [magnitude] assumed to which the addition is made ... [Here follow some definitions on fractions.]

The first proposition is that if a one-foot quantity should be assumed and an addition were made to it into infinity according to a subdouble [that is one-half] proportion so that one-half of one foot is added to it, then one-fourth, then one-eighth, and so on into infinity by halving the halves [lit. doubling the halves], the whole will be exactly twice the first [magnitude] assumed. This is clear, because if from something one takes away successively these parts, then [one is left with nothing, and so] from the double quantity one has taken away the double, as appears by question 1 [which was about subtraction]; and so by a similar reasoning, if they are added.

The second proposition is this, that if a quantity, such as one foot, were assumed, then a third were added and then after a third [of that] and so into infinity, the whole would be precisely one foot and a half, or in the sesquialterate proportion [this is the medieval terminology for the ratio of 3 to 2]. Furthermore, this rule should be known: We must see how much the second part falls short of the first part, and how much the third falls short of the second, and so on with the others, and denominate this by its denomination, and then the ratio of the whole aggregate to the quantity [first] assumed will be just as a denominator to a numerator. [This looks very obscure, but the meaning seems to be that, if your ratio is a fraction q so the series is $a + aq + aq^2 + \cdots$, then the 'falling short' is $1 - q$; and you invert this—exchange the denominator and the numerator—to get the sum $1/(1 - q)$, which is the ratio of the sum to the first term a. There is no proof.]

---

17. This is the substance of the previous 'question', which deals with the successive subtractions which exhaust a magnitude, as in Euclid X.1.

For example: in the above the second part, which is a third of the first, differs from the first by two-thirds, therefore the ratio of the whole to the first part or the assumed quantity is as three to two and this is sesquialterate.

The third proposition is this: It is possible that an addition should be made, though not proportionally, to any quantity by ratios of lesser inequality, and yet the whole would become infinite. For example, let a one-foot quantity be assumed to which one-half of a foot is added during the first part of an hour, then one-third of a foot in another, then one-fourth, then one-fifth, and so on into infinity following the series of numbers, I say that the whole would become infinite, which is proved as follows: There exist infinite parts any one of which will be greater than one-half foot and [therefore] the whole will be infinite. The antecedent is obvious, since one-quarter and one-third are greater than one-half; similarly from one-fifth to one-eighth is greater than one-half, and from one-ninth to one-sixteenth, and so on into infinity . . .

Appendix B

(From Descartes 1954, pp. 51–5)

Suppose the curve EC to be described by the intersection of the ruler GL and the rectilinear plane figure CNKL, whose side KN is produced indefinitely in the direction of C, and which, being moved in the same plane in such a way that its side KL always coincides with some part of the line BA (produced in both directions), imparts to the ruler a rotary motion about G (the ruler being hinged to the figure CNKL at L).

[See Descartes's picture (Fig. 4.). The triangle CNKL, more properly NKL, moves up and down AB; the ruler (as the picture more or less shows) is fixed to the triangle at L and passes through a loop or curtain-ring G which is fixed to the line AG. The curve is traced by the intersection C of the ruler and the (produced) side KN of the triangle.]

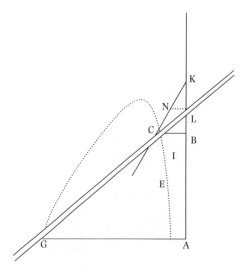

Fig. 4 Descartes's curve drawing machine.

If I wish to find out to what class this curve belongs, I choose a straight line, as AB, to which I refer all its points, and in AB I choose a point A at which to begin the investigation. I say 'choose this and that', because we are free to choose what we will, for, while it is necessary to use care in the choice in order to make the equation as short and simple as possible, yet no matter what line I should choose instead of AB the curve would always prove to be of the same class, a fact easily demonstrated.

Then I take on the curve an arbitrary point, as C, at which we will suppose the instrument applied to describe the curve. Then I draw through C the line CB parallel to GA [and meeting BA in B]. Since CB and BA are unknown quantities, I shall call one of them y and the other x. To the relation between these quantities I must consider also the known quantities which determine the description of the curve, as GA, which I shall call a; KL, which I shall call b; and NL parallel to GA, which I shall call c. Then I say that as NL is to LK, or as c is to b, so CB, or y, is to BK, which is therefore $\frac{b}{c}y$. Then BL is equal to $\frac{b}{c}y - b$, and AL is equal to $x + \frac{b}{c}y - b$. Moreover, as CB is to LB, that is, as y is to $\frac{b}{c}y - b$, so AG or a is to LA or $x + \frac{b}{c}y - b$. Multiplying the second by the third, we get $\frac{ab}{c}y - ab$ equal to $xy + \frac{b}{c}yy - by$, which is obtained by multiplying the first by the last. [This is the usual 'multiplying out' of an equation between ratios.] Therefore the required equation is

$$yy = cy - \frac{cx}{b}y + ay - ac$$

From this equation we see that the curve EC belongs to the first class, it being in fact a hyperbola.

[Without setting it as an exercise, you are encouraged to follow through this calculation to see how Descartes has derived his equation.]

Appendix C

(From Kepler, *Nova stereometria doliorum* (New measurement of wine-barrels), 1615 (in 1999).

Since, the wine-barrels are made up of the circle, the cone and the cylinder which are regular figures, they are suitable for geometrical proofs; and I shall gather these together in the first part of this investigation. Since they were investigated by Archimedes, to read a part of his work is enough to delight a lover of geometry. For absolute proofs which are exact in every number can be sought in these same books of Archimedes, if anyone is not frightened by the thorny reading of them.

However, we can pass the time in certain regions which Archimedes did not reach; and even the wiser readers can find things to please them there.

Theorem I. *First we need to know the relation of the circumference to the diameter. And Archimedes taught:*

The ratio of the circumference to the diameter is very near to the ratio which the number 22 has to 7.

[Note that this is not really what Archimedes said, and the use of terms like 'very near' would have been quite unacceptable in Greek mathematics. However, if this is bad enough, Kepler's 'proof' is even worse. I shall omit it, since it is not calculus so much as crude approximation using inscribed and circumscribed hexagons. (Remember Archimedes used 96-sided figures!)].

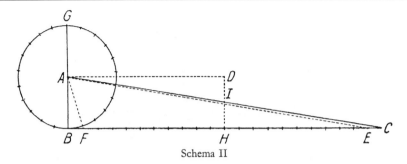

Fig. 5 Kepler's picture of the circle, the straight line, and the infinitely small subdivisions.

Theorem II. *The area of a circle compared to the square of its diameter has the ratio approximately 11 to 14.*

Archimedes uses an indirect demonstration, which leads to a contradiction, about which many authors have written much. But I think the meaning of it is this: [see Fig. 5]

The circumference of the circle BG has as many parts as it has points, say an infinite number; and of these any two may be regarded as the base of an isosceles triangle of side AB—so that inside the area of the circle there are to be found infinitely many triangles all coming together at a common vertex A, the centre. Let the circumference then be extended to a straight line, and let the length BC be equal to it [to the circumference], and AB perpendicular to BC. Then the bases of these infinitely many triangles (or sectors) are imagined as lying on one straight line BC, arranged next to each other. Let one little base [of the part of the circle] be BF, and let it be equal to CE [i.e. correspond to CE on the straight line]; and join the points F, E, C to A.

Since the number of triangles ABF, AEC are the same on the straight line as in the circle, and the bases BF, EC are equal, and they all have the same height BA, which is also the height of the sectors; the triangles EAC, BAF will have the same area, and any one will equal a sector of the circle. And since they all have their bases on the straight line BC, the triangle BAC, which is made up of all of them, will be equal to all the sectors of the circle, that is to the area of the circle consisting of all of them. This is what Archimedes wishes to prove by contradiction.

[The point is this. For Archimedes, you prove (say) that the area a of BG is not *greater than* $\frac{1}{2} \times$ (radius) $\times$ (circumference), by supposing it *is* greater. Since the perimeters of circumscribed polygons are as near as you like to the circumference of the circle, this means that for some such polygon P, a is greater than $\frac{1}{2} \times$ (radius) $\times$ (perimeter of P). But this is the area of P, which must be greater than the area a (see Fig. 6).]

Kepler is saying: if you choose the polygon to have an infinite number of infinitely small sides, then it 'is' already equal to the circle, its perimeter is the circumference, and you can avoid the tedious proof by contradiction. What could go wrong in this procedure?

The proof then ends by showing that the triangle BAC has area 'nearly' 11/14 times the square on the diameter, using theorem I.]

Appendix D

(From Galileo 1954, pp. 31–3)

SALV. I take it for granted that you know which numbers are squares and which are not.

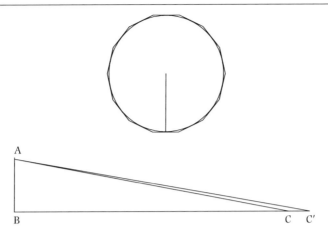

Fig. 6 Archimedes' idea: ABC is the triangle whose base is the circumference and whose height is the radius. BC' is the perimeter of the polygon, and its area equal that of triangle ABC'. If the area of the circle is greater than ABC, then it is greater than that of some ABC', since they approach ABC as near as you like; but then it is greater than the area of some polygon, which is absurd.

SIMP. I am quite aware that a squared number is one which results from the multiplication of another number by itself; thus 4, 9, etc., are squared numbers which come from multiplying 2, 3, etc. by themselves.

SALV. Very well; and you also know that just as the products are called squares so the factors are called sides or roots; while on the other hand those numbers which do not consist of two equal factors are not squares. Therefore if I assert that all numbers, including both squares and non-squares, are more than the squares alone, I shall speak the truth, shall I not?

SIMP. Precisely so.

SALV. But if I inquire how many roots there are, it cannot be denied that there are as many as there are numbers, because every number is a root of some square. This being granted we must say that there are as many squares as there are numbers because they are just as numerous as their roots, and all the numbers are roots ... [Salviati develops his point, and shows that as the proportion of squares gets smaller the more numbers we consider.]

SAGR. What then must one conclude under these circumstances?

SALV. So far as I can see we can only infer that the totality of all numbers is infinite, that the number of squares is infinite, and that the number of their roots is infinite; neither is the number of squares less than the totality of all numbers, nor the latter greater than the former; and finally the attributes "equal", "greater", and "less", are not applicable to infinite, but only to finite, quantities.

Solutions to exercises

1. The natural way to proceed would be as follows. Given a circle C, let a be its diameter. Construct a line B whose length is $3\frac{1}{7}$ times A—this is straightforward, Euclid quite early gives a method for dividing a line into (for example) seven equal parts. This we take to be equal to the circumference of C, and now the area of C is equal to that of a right angled triangle whose height is A and whose base is B. Finally, Euclid II.14 gives a method of constructing a square whose area is equal to the triangle (you construct a 'mean proportional' between A and B/2, a line whose length is $\sqrt{\frac{1}{2}AB}$).

2. Let us translate Euclid's statement into algebra. You start with a quantity a. You subtract from this $k_1 a$ where $k_1 > \frac{1}{2}$. Then you subtract again $k_2(a - k_1 a)$, where $k_2 > \frac{1}{2}$, and so on. Euclid claims that by repeating the subtractions $a - k_1 a - k_2(a - k_1 a) - \cdots$ you will arrive at a quantity less than any assigned quantity b.

In fact, after n subtractions, you have $a(1 - k_1)(\cdots)(1 - k_n)$. If each k_i equals $\frac{1}{2}$, you get $(1/2^n)a$, which tends to 0; even more so if each $k_i > \frac{1}{2}$.

However, the condition is actually too restrictive. By a 'classical' theorem, the product $a(1 - k_1)(\cdots)(1 - k_n)$ tends to 0 if and only if the sum $k_1 + k_2 + \cdots$ is divergent.

3. Guess 12 (a multiple of 3 and 4) for the height; then underground it is $4 + 3$, so above ground it is 5, which is wrong, as we need 30. So multiply the guess by $30/5 = 6$, giving 72. This is right ($24 + 18 = 42$ underground, leaving 30 above).

Guess 4; then $4 + 4 + 2 + 1 = 11$. We want it to be 99 (one denaro short of 100). So multiply the guess (4) by 9, giving 36.

You would need to convert the 2 years 7 months and 15 days to years (assuming the interest is annual—10 per cent per month seems extortionate, but such things did happen). Suppose a month is 30 days, and that there are 12 in a year, to simplify. This gives $2\frac{15}{24} \times \frac{10}{100} = \frac{21}{80}$ for the multiplier; you then have to multiply this by the principal. The calculation looks appalling; I do not know how a medieval student was supposed to do it.

(b) In a year—12 months—the interest is 12 pence in 1 pound, or $\frac{12}{240} \times 100$ per cent. This is clearly 5 per cent.

4. We have $u - v = 4$ and $uv = 1$ (divide 3 by 3 and cube). This gives $(u - v)^2 = 16$ and so $(u + v)^2 = 20$. Hence, $u + v = 2\sqrt{5}$ and $u = 2 + \sqrt{5}, v = -2 + \sqrt{5}$.

We now take cube roots and subtract: $x = \sqrt[3]{2 + \sqrt{5}} - \sqrt[3]{2 - \sqrt{5}}$.

(b) The point is of course that the difference *is* 1; you can check it first of all on a calculator. Both u and v are (if you think in terms of algebraic number theory!) exact cubes of expressions $a + b\sqrt{5}$; take $a = \pm\frac{1}{2}, b = \frac{1}{2}$. Hence, the difference of the cube roots is $(1 + \sqrt{5})/2 - (-1 + \sqrt{5})/2 = 1$.

(c) (1) Suppose the equation is $x^3 + cx = d$. Let $x = \sqrt[3]{u} - \sqrt[3]{v}$. Then $x^3 = u - v - 3(\sqrt[3]{u} \cdot \sqrt[3]{v})(\sqrt[3]{u} - \sqrt[3]{v})$, using the formula for $(a - b)^3$ and rearranging. If $uv = (c/3)^3$, then $\sqrt[3]{u} \cdot \sqrt[3]{v} = c/3$. So $x^3 = u - v - cx$. But $u - v = d$, so x does satisfy the equation.

(2) It is enough to construct a geometrical decomposition which verifies the formula $a^3 + 3ab^2 = 3a^2 b + b^3 + (a - b)^3$. From there you can 'geometrize' the argument in part (1). I leave the details to you.

5. This is not so pleasant, if no harder. We have $u - v = 500$, $uv = \frac{1}{27}$. So $(u+v)^2 = 250{,}000\frac{4}{27}$, and $u = \frac{1}{2}(\sqrt{250{,}000\frac{4}{27}}) + 500), v = \frac{1}{2}(\sqrt{250{,}000\frac{4}{27}}) - 500)$. Now write $x = \sqrt[3]{u} - \sqrt[3]{v}$ again. (This could be worked out with a calculator, if you have the patience; answer about 7.895.)

6. You are given $b = x - y$ and $d = x^3 - y^3$. Viète points out that if $u = x + y$, then $u + b = 2x$ and $u - b = 2y$ (obvious). Now $(2x)^3 - (2y)^3 = (u + b)^3 - (u - b)^3 = 6u^2 b + 2b^3$ by expanding the brackets and subtracting. But we know that $(2x)^3 - (2y)^3 = 8(x^3 - y^3) = 8d$.

We can therefore deduce u, since (by the above), $6u^2 b + 2b^3 = 8d$, so $u^2 = (4d - b^3)/3b$ (the displayed equation in the extract). Now we know $x - y = b$ and $x + y = u$, and so x and y.

7 The calculus

1 Introduction

By the help of this new Analysis Mr Newton found out most of the propositions in the *Principia Philosophiae*. (Newton 1967–1981 **8**, 598–9)

The 'new Analysis' was what we now call the calculus; a point in learning mathematics where many students give up, and which many others never reach. Intended to make everything easy, it is still found a stumbling-block. What, then, is it? A simple web encyclopaedia describes it as follows:

calculus, branch of mathematics that studies continuously changing quantities. The calculus is characterized by the use of infinite processes, involving passage to a limit: the notion of tending toward, or approaching, an ultimate value. (http://reference.allrefer.com/encyclopedia/C/calcul.html)

While not very clear unless you know what it is already, this definition would already have been acceptable at the end of the seventeenth century. As our opening quote suggests, the calculus begins with Newton; and in the extract he was, typically, discussing his own work (and somewhat bending the truth) anonymously in the third person many years after the event. He had already made himself into a monument, as his one-time colleague and later rival Leibniz never managed to do, and even today the work continues. The British Library, whose courtyard is incongruously dominated by a massive statue of the man,[1] lists in its catalogue 10 new books on Newton for the year 2001 alone; and one must suppose that many lesser books have appeared, together with articles learned and otherwise, student dissertations and entries on the many Newton websites. Experience shows that 'Newton-and-Leibniz' is easily the most popular subject for student essays in the history of mathematics—despite the difficulty of many of the early calculus texts. Is it not time to call a moratorium, a temporary halt to all this industry? Given that most scholars have arrived at a reasonable conclusion about the once burning question of priority in the discovery (to which this chapter will return later), must they still be concerned with the date when Newton discovered the inverse square law of gravitation, the reasons for his conversion from Cartesian geometry to the methods of 'the Ancients', or the extent to which his religion, the pursuit of alchemy, or political beliefs influenced his mathematics?

The answer is that at least some of the current research is important and necessary, even if much of it seems to be lacking in the 'social element' which we have had occasion to single out for praise in earlier chapters. This is partly because the story is still subject to myths and misunderstandings (e.g. of when the calculus became widely known and available, and to whom); and partly because, like the First World War (which has also had much too much written about its origins), the calculus is an important founding event in European history. And while Newton and Leibniz get more

---

1. Modelled, ironically one must suppose, on the drawing by Blake, for whom Newton represented the blindness and alienation of rationalism.

than their share of attention in history of mathematics courses, they are either ignored or grossly misrepresented even in 'historically' minded calculus textbooks where one or both feature with portrait as icons and founders, as Shelley Costa has pointed out:

> When writing of Newton and Leibniz, 20th-century authors of calculus textbooks tend to reduce their history to method and notation while exalting them as insightful, majestic intellectual forebears, perpetuating a mathematical mystique that rewards genius and ignores context. (Costa, n.d.)

While what seems important from a modern point of view is the easy access to powerful results, equally significant at the time was the questionable legality of the procedure. At least since the time of the Greeks, mathematics had rested its claims to certainty on rules of precision in reasoning. The new methods treated these rules with a degree of indifference from which they have never fully recovered. The major problem was the use (which was essential) of infinitely small quantities or 'infinitesimals' which were either zero or not zero, depending on where you were in the argument. It had, reasonably, been assumed that the Law of Contradiction ('if something is X it is not also not-X') operated in mathematics as elsewhere. The following quote from the first (and most important) early calculus textbook shows that its power was slipping:

> *Postulate 1.* Grant that two quantities, whose difference is an infinitely small quantity, may be taken (or used) indifferently for each other: or (which is the same thing) that a quantity, which is increased or decreased only by an infinitely small quantity, may be considered as remaining the same. (L'Hôpital 1696, cited Fauvel and Gray, extract 13.B.6)

The confusion underlying this extract was fundamental to the calculus, and was essential for its development. To make it explicit, as Bishop Berkeley was to do 40 years later, if two quantities which differ by an infinitesimal are the same, then what is the infinitesimal there for at all? More simply, if dx (as Leibniz called it) is infinitesimal, it differs by an infinitesimal from 0, and so is equivalent to 0. Is it then 0 or not?

The point is, of course, that any mediocre person can break the laws of logic, and many do. What Newton and Leibniz did was to formalize the breakage as a workable system of calculation which both of them quickly came to see was immensely powerful, even if they were not entirely clear about what they meant. The new methods built on Descartes's geometry; that geometry had raised a number of important questions about how to find tangents to curves, their lengths, and their areas, and the calculus was to provide the means of finding rapid solutions. Mathematics became, in a sudden transition, both easy and difficult; easy for the circle of initiates who learned how to use the method, and difficult for the outsiders who could understand neither what was being done nor how it was justified. 'There goes a man hath writt a book which neither he nor any body else understands' remarked a sceptical Cambridge undergraduate of Newton[2]; and the contradictory dogmas of the early calculus perhaps mark the origin of a widening split between the world of the 'serious' mathematician and the amateur.

Already it may be clear that this chapter differs from the preceding ones in covering a much shorter period; instead of hundreds of years, we are dealing with the relatively short time which separates the 1660s (when the calculus did not exist) from the 1720s (when it was on its way to becoming the dominant method for answering a wide range of mathematical questions). Indeed, appropriately for the science of the infinitely small, the historians often seem to be focusing

---

2. Cited Iliffe (1995, p. 174).

on extremely short intervals of time, so that the question of whether Leibniz received three communications about Newton's work in June 1676 or two in July (see Hofmann 1974, p. 231) comes to have capital importance.

Such infinitesimals are not properly the concern of this chapter, which will rather try to focus on the larger questions. How did the change take place? Why was such a dubious way of proceeding so quickly accepted? What end did it serve, and who profited? And what were the positive and negative effects of all the quarrels? This chapter will try to give a broad outline of the history, with reference to such questions.

Note. The material which is dealt with in this chapter is necessarily mathematically harder than that of the previous chapters. The reader who is unfamiliar with the ideas of the calculus (and they do still form a major part of mathematical culture) will therefore have to skip some of the detail; it is not possible to omit it from the text without losing what the history is about.

2 Literature

As has already been indicated, the literature is large and many-sided. To begin with the primary sources, the whole of Newton's mathematical work is available in an excellent modern edition, edited by D. T. Whiteside (Newton 1967–81). Substantial extracts of the *Principia* (from books I and III) are available on the Internet at www.members.tripod.com/gravitee—it is not clear whether it is intended to extend this selection. Fauvel and Gray's sourcebook is good on the period also, giving some essential Newton material and the bulk of Leibniz's key 1684 paper, plus the opening of L'Hôpital's book. The other works of Leibniz and the Bernoulli brothers are less accessible, and usually in Latin; and while L'Hôpital's book was quite well translated in the eighteenth century (see quote above), there is no modern edition.

With the secondary sources, the question is really that of differentiating between ways of thinking about history. There are, for example, numerous biographies of Newton, of which the definitive one is Richard Westfall's massive (1980), supplemented by Gjertsen's handbook (1986); and there is a comparable, if less comprehensive biography of Leibniz by Aiton (1985). Close to the biographical are studies of the small community of late seventeenth-century scientists who were able to appreciate and develop the calculus; for example, Hall (1980) and Hofmann (1974). And third, there are more technical, one could say 'internal', studies of what was involved in the early calculus, its techniques, and how practitioners saw what they were doing: these include particularly Guicciardini (1999), various essays of Henk Bos, with summaries and further thoughts in (1991), and Dupont and Roero (1991).[3]

However, all of these works, even the most 'external', are dealing with a tiny community, in comparison with the wider society in which the calculus was born and flourished. The obvious reason for this is that, at the outset and for quite a time afterwards, it was found incomprehensible outside a small circle. It is significant that Bishop Berkeley's damning critique in *The Analyst* dates from 1734, but still gives the impression that to a well-informed bishop the methods of the calculus were relatively new. Although L'Hôpital's 1696 text was an attempt at a popularization, intended to have the same impact as Descartes's *Géométrie*, written in French in a very similar style and

---

3. This book is often referenced. Apart from being in Italian, it seems almost impossible to find, but as a detailed study of Leibniz's paper, it is very interesting. Perhaps one can hope for a translation.

acknowledging the debt, its circulation was certainly more restricted. And yet the circle of *savants* with whom the principals corresponded, and who were interested in the same questions was much wider; such practical questions as, why did the planets move as they did? What was the reason for the tides, and could they be predicted? What was the shape of a freely hanging chain, of a loaded beam, or of a sail? It is rare to find this practical background included in modern discussions of a body of work which, however clearly it constituted a 'new mathematics', was embedded in a whole family of other practices which have now been forgotten.

Of these, more will follow later; but for the time being, we should refer to two key texts from the 1930s. The first, the Soviet historian Boris Hessen's *The Social Origins of Newton's Principia* in Hessen (1971), is occasionally referred to as an example of the crude Marxist approach; however, it has been recently reprinted, and is still worth reading. The second, the American sociologist Robert Merton's (1970), is a founding text for the sociology of science; but, as Merton acknowledges in his introduction, subsequent readers have focused more on the early chapters which (following Weber and Tawney) relate the pursuit of science to Puritanism and the 'Protestant Ethic' than on the later ones which, qualifying and extending Hessen's analysis, relate the problems studied by scientists to the demands of expanding capitalism. Merton's strength is his awareness of the wider (mainly English) scientific community, so that figures like Halley and Wren who hold a minor place in the mathematical history are seen as much more important in terms of ideology, patronage and influence. The specific nature of the calculus is not addressed by either writer; accordingly, as often in the interface of mathematics and physics, the kinds of source material available for study are not easy to harmonize. [And, it should be added, a problem for this chapter: how do we separate the history of the calculus from Newton's *Principia*? The latter is both broader (after all, Newton's concern was with physics, and the deduction of the system of the world from the basic laws of motion) and narrower (because of the peculiar way in which the calculus was used, or not used in the *Principia*—see later).]

Note on the use of texts. Few periods in the history of mathematics have been as badly served in translation as the prehistory and early history of the calculus. At a time when the exact notation which was being used by the participants (some of whom used traditional geometrical language, some cartesian equations, and many a confused mixture), many histories 'translate' the work of Wallis, James Gregory, and the early Newton and Leibniz indiscriminately into language which the modern reader can recognize—even when, like Gregory, they were hostile to Cartesian symbolism. Both Hofmann's (1974) and Westfall's (1980) are given to a free use of translation—even if it is usually acknowledged as such; so that for example, Collins is said in his 1675 report on English work for Leibniz to have shown how the arc [of a circle]

$$s = r\tan^{-1}(t/r) = \int_0^t r^2 \cdot dx/(r^2 + x^2)$$

may be found (Hofmann 1974, p. 135). The reader whose suspicions are aroused by the use of the integral sign which Leibniz was to invent in October of that year (and not publish for another ten years), will find on looking up the source that what Collins said was very different, but the fact that most of the original sources are hard to come by and mainly in Latin makes the question of accurate transcription the more important. Fortunately this situation is changing, and the more recent works of Guicciardini and Bos are textually faithful.

3 The priority dispute

We shall not discuss the shamefully bitter controversy as to the priority and independence of the inventions by Newton and Leibniz. (Boyer 1949, p. 188)

In one word he told me the secret of success in mathematics: Plagiarize! (Tom Lehrer, song, 'Lobachevsky')

Boyer represents a gentlemanly school of thought (now rather old-fashioned) according to which mathematicians should behave courteously towards one another. In this view, the controversy which opposed the supporters of Newton and Leibniz between 1700 and 1720 (with an acute period in the 1710s) was an unfortunate aberration. The most superficial look at the long history of mathematics and mathematicians shows that this is not in fact the case. We have already seen the quarrels of Tartaglia and Cardano in the sixteenth century, and there have been many worse ones since: one could cite Pascal and Torricelli on the cycloid, (1650s), Legendre and Gauss on the method of least squares (1790s), and countless others (an internet search for 'priority dispute' is instructive throwing up an instance in topology from 1996 in particular). What gave the argument about the calculus its peculiar bad taste was the involvement of the Royal Society, and of its president Newton, in adjudicating on the dispute.

Regrettable or not—and history can rarely afford regrets—the dispute provides a useful way into the history of the calculus and its diffusion. In fact, it is clear from the way in which the arguments developed that in 1690 'the invention of the calculus' was not a subject for discussion, while 20 years later it was generally agreed that something of great importance had been invented, and the question of who had copied whose prior invention was vital. British chauvinism (which one might suppose to be a major factor) played its part, but was a secondary issue. Indeed, Newton in his youth was notably more open to Continental ideas, those of Descartes in particular, than his seniors; and he continued to show repect for those foreign scientists, such as Huygens, who were not in direct competition. Equally, he had frequently engaged in more or less acrimonious disputes with British colleagues such as Hooke and Flamsteed. The scientific milieu of the time was generally suspicious and paranoid; discoverers withheld publication for fear of being copied, and then, when their unpublished discoveries were found by another, accused them of plagiarism. (And, one must suppose, they were sometimes right.) The practice of anonymous publication (in which authors often referred to themselves in the third person—'the distinguished Mr Leibniz has proved') was unhelpful, both in establishing who had done what, and in subsequent controversies often conducted via anonymous or pseudonymous attacks.

The question of priority (as distinct from plagiarism) was in fact settled quite early. As we shall see, Newton's version of the calculus was very similar to Leibniz's, both in its qualities and its defects, and dated from about 10 years earlier (1665 as against 1675). The question of plagiarism is more unpleasant and complex, since Newton's version was not published in his lifetime, but was seen in manuscript by a number of people. Leibniz in the early 1670s knew of some of Newton's work, chiefly by accounts in letters from London; and most famously, by two devious and obscure letters from Newton himself, which contained references to his most important work in the form of anagrams, which Leibniz was hardly in a position to decipher. This may seem absurd as a way of communicating scientific results but was again not uncommon at the time, and this may give some indication of what was supposed to constitute 'publication'. As Gjertsen says (in his entertaining article 'Anagrams', 1986, p. 16):

The advantages of the ploy were obvious. Priority was established yet nothing was given away to potential rivals... Invariably in Latin, clueless, and of immense length, they were virtually insoluble.

Modern scholars (see particularly Hall 1980) are agreed that Newton's communication was both too late and too obscure to exercise any serious influence on what was effectively an *independent* discovery by Leibniz, with his own ideas and notation. He too delayed publication, and was only forced into a brief announcement of his results in 1684 by fears that he in turn was losing priority.

It took about 20 years, and the rapid success of the Leibnizian calculus in the 1690s, for any question of priority/plagiarism to arise; and it entered its acute phase about 1710. Both players had by then acquired schools of students, defenders and partisans. Newton's attitude to those who disagreed with him, whether scientifically or politically, is often described as hostile to the point of paranoia; and while Leibniz, committed to a rational belief that this is the best of all possible worlds, was in many ways a contrast, he did not like suggestions that his own ideas were in any way unoriginal.[4] Following accusations and counter-accusations, the Royal Society in March 1712 appointed a committee composed almost entirely of Newton's supporters to investigate whether Leibniz had been unjustly accused by the quarrelsome Newtonian John Keill. Newton supplied the committee with documents proving his priority (his unpublished manuscripts, and letters from various of the parties involved), and apparently also drafted the final report (pompously titled *Commercium epistolicum*, the exchange of letters), which concluded:

That the Differential Method [Leibniz] is One and the same with the Method of Fluxions [Newton] Excepting the name and Mode of Notation . . . and therefore wee take the Proper Question to be not who Invented this or that Method but who was the first Inventor of the Method . . .

For which Reasons we Reckon Mr Newton the first Inventor; and are of Opinion that Mr Keill in Asserting the same has been noways Injurious to Mr Leibniz. (Newton 1959–77, **5**, p. xxvi)

Those who are looking for examples of Anglo-Saxon hypocrisy can find plenty in the *Commercium epistolicum*, and still more in the 'Report' on it which Newton wrote for the Royal Society's Transactions—an anonymous review of a document which he had largely written. (It is from this review that the quotation at the head of the chapter is taken.) All the same, the document did make clear, as Newton had never previously done, the nature and extent of his early work, and thus, however partisan, it cleared up the question of priority. The charge of plagiarism against Leibniz, who died soon afterwards, was never serious enough to stick; indeed one could ask what is the nature of intellectual property in mathematics, and how far plagiarism is to be condemned. Pierre de Montmort, one of several who heroically tried to reconcile the parties, made the key point: one should look at who used it, how, and with what results.

On the invention of the calculus he [Montmort] would not comment, he said, but Leibniz and the Bernoullis had been its true and almost sole promoters.

It is they and they alone who taught us the rules of differentiation and integration, the way to use the calculus to find tangents to curves, their points of inflection and reversal, . . . & who finally, by many and beautiful applications of the calculus to the most difficult problems of mechanics, such as the catenary, the sail, the spring, the quickest descent, and the paracentric, have set us and our descendants on the path of the most profound discoveries. (Westfall 1980, pp. 784–5, citing Montmort's letter to Brook Taylor of 18 December 1718, *Corr.* 7, 21–2)

It was Leibniz's calculus which was most successfully used, and which still dominates our notation; and when in the twentieth century the record was finally set straight with the publication of his manuscripts from the 1670s, no one was left to care.

---

4. His most devoted student Jakob Bernoulli found this to his cost early on when he misguidedly suggested that Leibniz's methods were similar to the earlier ones used by Isaac Barrow. He had to make a humble apology for what was clearly a mistake, but understandable when the calculus was still not fully understood, and Leibniz himself was not helping to make it clearer.

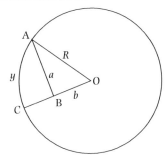

Fig. 1 Indian calculation of the arc. If y is the arc AC, $a = $ AB is the 'Sine' of the angle AOC, BO the 'Cosine', and R the radius. a/b is then the tangent, and the series gives y in terms of a and b (or a/b) and R.

4 The Kerala connection

We may consider Madhava to have been the founder of modern analysis... (Joseph 1992, p. 293)

Joseph's downright and still controversial statement calls attention to the 'other priority question'— whether the calculus was discovered (to put the claim at its strongest) in Kerala, south-west India, in the late Middle Ages.[5] Here we have a contest which is not among mathematicians (there is no record of an Indian seventeenth-century mathematician staking a claim for the calculus), but among contemporary historians, whose interests are different. Specifically, Joseph and others aim to attack the 'Eurocentric' story of mathematical discovery by drawing attention to parallel discoveries in non-European contexts.

The problem is here, that the material is both unfamiliar and very inaccessible, although it has been, in a very weak sense, known about since Charles Whish wrote about it in the 1830s. What seems undeniable is that a number of astronomical texts from Kerala, whose dating is probably between 1400 and 1600, give very sophisticated *infinite series* formulae for what we now call sin x, cos x, and the inverse tangent of x (the arc of the angle whose tangent is x, see Fig. 1). If the radius of the circle is R, when the angle is $45°$ the arc is $\pi R/4$ and the tangent is 1; and this gives in particular a formula for what we know as π. Series such as these were an important building block in the calculus as Newton developed it, in fact his early papers give equal importance to the calculus and the use of infinite series as 'new' methods. Similarly, the simplest of the formulae was found by Leibniz early in his career (but, as he was disappointed to hear, had been found by others before): it is commonly known as Gregory's series. In Jyesthadeva's sixteenth-century version it is usually quoted as follows (see for example, Joseph 1992, p. 290):

The first term is the product of the given Sine and radius of the desired arc divided by the Cosine of the arc. The succeeding terms are obtained by a process of iteration when the first term is repeatedly multiplied by the square of the Sine and divided by the square of the Cosine. All the terms are then divided by the odd numbers 1, 3, 5, ... The arc is obtained by adding and subtracting respectively the terms of odd rank and those of even rank. It is laid down that the Sine of the arc or that of its complement whichever is the smaller should be taken here as the given Sine. Otherwise the terms obtained by this above iteration will not tend to the vanishing magnitude.

---

5. The subject of medieval Keralan mathematics is a very promising one for the enterprising researcher. The field is small, and the source material is not vast, even if much of it may be hard to locate (the Royal Asiatic Society library, near Paddington, may be helpful). The aspiring researcher needs to be prepared to learn Sanskrit and probably Malāyalām, but the rewards could be substantial.

Here 'Sine' and 'Cosine' mean the lines AB, BO in the diagram (Fig. 1.)—sine and cosine multiplied by the radius. In some respects this is like Leibniz's series:

$$\text{arc} = Rx - R\frac{x^3}{3} + R\frac{x^5}{5} - R\frac{x^7}{7} + \cdots \tag{1}$$

which becomes our 'modern' series for $\tan^{-1}(x)$ by setting $R = 1$. Like the Leibnizian series, it is forthright about the use of an infinite number of terms, a novelty in India it would seem as it was to be in Europe. The equation (1) is, though, rather different from the form in which it was given by the Keralans, since (as the quotation above indicates) they wrote the formula in verse, and in words, without any use of symbols at all. One could add that since there is on the whole no explanation of how the series were arrived at, we can only guess at the methods; but we shall see that the seventeenth-century European mathematicians were often silent about how they had found their series and integrals, for their own reasons. In Keralan sources the original discovery is ascribed to Madhava, a famous fourteenth-century writer whose mathematical works are mostly lost. On the credit side, the practical Keralans realized that the series (1) is useless for computation; you can add 50 terms and you will still be making mistakes in the second figure of $\pi/4$, because your next term will be $\frac{1}{101}$, roughly 0.01. They accordingly refined the series to give a number of others, more useful, but still without explanation. In all this, they can rightly claim 100 years' priority, at least.

We have here a prime example of two traditions whose aims were completely different. The Euclidean ideology of proof which was so influential in the Islamic world had no apparent influence in India (as al-Bīrūnī had complained long before), even if there is a possibility that the Greek tables of 'trigonometric functions' had been transmitted and refined. To suppose that some version of 'calculus' underlay the derivation of the series must be a matter of conjecture.

The single exception to this generalization is a long work, much admired in Kerala, which was known as *Yukti-bhasa* by Jyesthadeva; this contains something more like proofs—but again, given the different paradigm, we should be cautious about assuming that they are meant to serve the same function. Both the authorship and date of this work are hard to establish exactly, (the date usually claimed is the sixteenth century), but it does give explanations of how the formulae are arrived at which *could* be taken as a version of the calculus.

As I have stressed before, in dealing with al-Samaw'al's algebra (for example), or the use of series by Oresme, this 'anticipation' as such (who was first?) is not a sensible object of history. Even if we could establish the existence of a 'transmission line' from sixteenth-century Kerala to seventeenth-century Europe (see Donald F. Lach 1965 for some evidence), we must recognize the existence of what Kuhn called incommensurability; different research efforts and a different language of science directed to different ends. What we know of Keralan society in the period is rather unlike the Europe of the Scientific Revolution, in particular there is no evidence of interest in the use of mathematics for warfare, mining, book-keeping, and so on.[6] The series of Madhava and his followers are—when translated into algebraic notation—the same as those found in Western Europe in the seventeenth century, but the translation has the effect of changing the aim and context of the work. It seems of more value historically to study the Kerala tradition in itself, with reference to its own society (how and where it was used, for example) than to argue about its role as a precursor. We can use the Keralan material to attack ideas of the uniqueness of Western discoveries, and for that matter to point out alternatives to the Western way of doing mathematics; but these projects ask us to see

---

6. This is not to say that evidence may not come to light in future.

the Indian work as, precisely, *not* a version of the (later) European 'method of infinite series', let alone that broader field which we call the calculus.

Exercise 1. *Derive equation (1) from Jyesthadeva's formulation.*

5 Newton, an unknown work

Observing that the majority of geometers, with an almost complete neglect of the ancients' synthetical method, now for the most part apply themselves to the cultivation of analysis and with it have overcome so many formidable difficulties that they seem to have exhausted virtually everything apart from the squaring of curves and certain topics of like nature not yet fully elucidated: I found not amiss, for the satisfaction of learners, to draw up the following short tract in which I might at once widen the boundaries of the field of analysis and advance the doctrine of series. (Newton 1967–81, 3, p. 33)

So what was it that Newton, and later Leibniz, invented (assuming with the Royal Society that they were essentially the same)? To clarify, consider the 'fundamental problem' of finding tangents. A variety of ad hoc methods were around by the 1660s, following on from the rather complicated one which Descartes had proposed (see Fauvel and Gray 11.A.9). For us today, (and this is how Newton and Leibniz saw it), the question can be posed:

Problem. Given a curve specified by an equation between x and y, to find the tangent at the point whose coordinates are (x, y) (see Fig. 2).

It is easy to see that we know the tangent if we know its gradient (its inclination to the x-axis), and this is now usually given—once you know calculus—by the expression 'dy/dx'. You will probably have been told that this is *not* a fraction, and that dy and dx do not mean anything on their own. Many students naturally find this idea confusing, if few question it, and indeed this is not how it started out. Instead, a number of writers in the mid-seventeenth century had arrived at the idea that the tangent was the line which joined two infinitely near points on the curve; and the challenge was to find a simple way of working out what that line was.

In one respect, Newton was more inventive. His fundamental idea was that the curve was described by the motion of a point in time, and that the tangent was the direction of its velocity at an instant. In itself, the idea could be thought of as a geometrical one; but if the curve was described in Cartesian form by coordinates x, y, then one could also think of the x-velocity and y-velocity

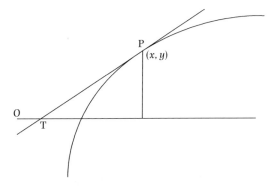

Fig. 2 PT is the tangent at the point P on the curve.

separately. This is how Newton's first private draft of his ideas begins, if with no explanation of where he is heading:

As if the body A with the velocity p describe the infinitely little line $(cd =)po$ in one moment, in that moment the body B with the velocity q will describe the line $(gh =)qo$. For $p : q :: po : qo$. So that if the described lines bee $(ac =)x$, & $(bg =)y$ in one moment, they will bee $(ad =)x + po$ and $(bh =)y + qo$ in the next. (Newton 1967–81, 1, p. 414)

There is no record that this draft was shown to anyone; and if it had been, it might well have been found strange. Newton was confidently using Descartes's notation of xs and ys for varying quantities, but he had added, as a new feature, that these quantities varied in time. He had, it appears, read Galileo (at least the *Dialogues*), but this bold introduction shows how far he was prepared to go beyond Galileo and Kepler in thinking of infinitely short distances and times. 'It is possible that Dr Barrows Lectures might put me on considering the generation of curves by motion, tho I not now remember it', he reflected many years afterwards (Westfall 1980, p. 131); but as Westfall points out, the idea was quite widespread (it went back to the Greeks, like so much else, in the case of Archimedes's spirals), and the cycloid (path of a fly glued to a rotating wheel) which was currently being bitterly argued about, was a prime example. In any case, if the idea of generating curves by motion was old, the idea of motion in an infinitely small instant was relatively new, and the use of coordinates to describe it even newer. Where Galileo, as we saw, had to go through a substantial argument to explain how one could make sense of 'velocity at an instant', and where Kepler avoided defining it at all, Newton took it for granted; and, moreover, by using an infinitely small time o, he worked the argument backwards, to deduce from a (supposed known) velocity p an infinitely small change po in the x-coordinate. The infinitely small quantities are not introduced or defended—they are simply there. [The idea of a 'moment of time' o presupposes some measurement of time which remains unclear, as Westfall points out (1980, p. 134), but the advance is still substantial.]

What Newton did next in the 1665 tract was naturally to bring in a curve, in its Cartesian form. He supposed that x and y were related by an equation:

$$x^3 - abx + a^3 - dyy = 0 \qquad (2)$$

where a, b, d are constants, and he showed how to find its tangent using his idea of change in time.

Rather than Newton's complicated equation (whose curve he did not draw), let us look at a simpler one, say (in seventeenth-century notation) $y + xx = ax$, which is the curve represented in Fig. 3. If the velocities—or, as Newton was later to call them, 'fluxions'—of x, y are p, q, then in the moment o, x becomes $x + po$ and y becomes $y + qo$. The point A′ whose coordinates are $x + po$ and $y + qo$ is (a) still on the curve and (b) infinitely near to the original one A.

Since the new point is on the curve, we have $y + qo + (x + po)(x + po) = a(x + po)$. *Because* the two points are infinitely near, the line AA′ which joins them is the tangent. Subtracting the original equation, and expanding the brackets, we get:

$$qo + 2xpo + ppoo = apo$$

We now divide this by o to get $q + 2xp + ppo = ap$. We want a relation between p and q (see Exercise 2). Discard the term ppo, which is infinitely small (this is the important part), and you end

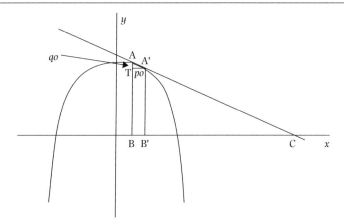

Fig. 3 A and A′ are infinitely close on the curve (a parabola, represented as a polygon with infinitely small sides). AA′C is the tangent at A; it meets the x-axis at C and BC is the 'subtangent'. In Leibnizian notation, BB′ = TA′ is dx, and AT (negative in the picture) is dy.

up with:

$$\frac{q}{p} = a - 2x$$

(If you are familiar with the calculus, you will notice that the right-hand side is obtained by differentiating $ax - xx$, according to the usual rules; and Newton began to formulate such rules for dealing with the velocities of curves.) In the seventeenth century, they found the tangent by finding the 'subtangent'—the length BC in Fig. 3.—which is equal to yp/q; in our case, $y/(a - 2x)$. [We can also say that y is a maximum when $x = a/2$, since then $q = 0$ so the tangent is horizontal.]

As Newton summed up his method:

Hence may bee observed: First, that those termes ever vanish in which o is not because they are the propounded equation [in our example, $y + xx = ax$]. Secondly the remaining Equation being divided by o those termes also vanish in which o still remaines because they are infinitely little. Thirdly that the still remaining termes will ever have that forme which by the first preceding rule [the rule for differentiating] they should have. (Newton 1967–81, p. 387)

Newton realized that he had made a major discovery; there are references in his texts on calculus from the 1660s to the possibility of solving with ease problems which had been difficult or impossible before. As we have seen, he took no serious steps to make it public. Under pressure (he said, from Dr Barrow), he finally in 1670 produced a serious Latin exposition, untitled, unfinished, generally known as the *Method of Fluxions and Infinite Series*. Where the 1665 notes had been essentially for his own use, this was (in intention) addressed to a circle of practitioners—'for the satisfaction of learners', as he puts it in the quotation which opens this section. Unfortunately, it did not reach even that narrow circle, although it was passed around among friends to the point of becoming dog-eared[7]; and its existence was only generally made known 30 years later in the priority dispute. However, it remains the clearest guide we have to what Newton's early calculus was like at the time of its discovery. The variable quantities x and y are in this text called 'fluents'; and their rates of change have the name 'fluxions' (which was to become a fixture in English mathematical language for a century). The relationship of differentiation—finding the fluxion—and

---

7. See Gjertsen (1986), p. 157. The exact date when the *Method of Fluxions* was observed to be dog-eared is a typical point of contention in Newton studies.

integration—finding the fluent—is established as the central problem; so is the fact that you can find areas by finding fluents. In Appendix A, I give his method of finding tangents—the easiest application—with an example. It is worth comparing with Leibniz's text for its greater clarity and (fairly) good explanations. The work, as far as it went, was well set out, and would have been more than a 'useful guide for beginners' if they had been allowed to see it.

There has been naturally considerable speculation (a) on why Newton did not publish it and (b) on what would have been the effect if he had. It might well, to begin with, have mystified its readers as Leibniz's later publication did, but the fact of publication would inevitably have brought clarification and improvements. Unfortunately, by 1670, his interests had already turned away from mathematics, and in Westfall's opinion '[n]early all of Newton's burst of mathematical activity in the period 1669–71 can be traced to external stimuli' (1980, p. 232). As we have seen, reasons for publication in the late seventeenth century were varied, and correspond badly with the image which is often presented of a new open scientific society. While some were free with their ideas, anxiety about theft and arguments about priority were widespread, and it was common for publication, whether by book or in a letter, to take place to forestall a potential rival and stake a claim for a discovery, rather than to reach an interested audience. 'Huygens, for example, had no intention of revealing his great discovery at once, but he did want to safeguard his priority by allusions in letters to his friends', is a typical comment (Hofmann 1974, p. 107); and Huygens was one of the more open publishers. The nascent 'community of savants' of the mid-seventeenth century had created a climate in which reputation and national rivalry rather than actual financial reward often encouraged scientists to be secretive; and Newton, always isolated and increasingly suspicious, needed little encouraging. For the next 15 years his main attention was focused on the pursuits of alchemy and biblical study, whose importance in his own estimation of his work equalled that of mathematics.

Exercise 2. *Suppose given infinitely near points $A = (x, y)$ and $A' = (x + po, y + qo)$ as above. Show that the line AA' has gradient q/p.*

Exercise 3. *Follow through the argument which leads to the gradient of the tangent to $y + xx = ax$, above. What are its strengths and weaknesses?*

Exercise 4. *Try to do the analogous calculation for the curve given by equation (2). What problems arise in finding the ratio q/p?*

6 Leibniz, a confusing publication

For what I love most about my calculus is that it gives us the same advantages over the Ancients in the geometry of Archimedes, that Viète and Descartes have given us in the geometry of Euclid or Apollonius, in freeing us from having to work with the imagination. (Leibniz letter to Huygens, 29th Dec. 1691, in Gerhardt 1962, 2, p. 123)

If no one knew what was going on in 1670, by 1690 it is fair to say that the few who did have an idea were either confused or misinformed. It would be quite inaccurate to suppose that there was a recognized object called 'the calculus' which was clearly destined to be *the* way ahead for mathematics. Leibniz's discovery, as is always pointed out, was different in many repects from Newton's; but they shared a common language to the extent that they could communicate, even if (as in the case of Newton's letters) partly for the purpose of concealment. Also like Newton, Leibniz was in no hurry to publish—it took nine years from his discovery of the method in 1675 to his first paper, which revealed very little; and he was only forced into publication by an apparent claim to priority by his friend Tschirnhaus. Rushing into print, he produced what must surely be one of the worst of all path-breaking papers, with no proofs, few and contradictory explanations and a long list of misprints.

About the 1684 paper a great deal has been said. It had almost no impact at the time, because it was so obscure; probably only Leibniz himself believed that he had revealed a revolutionary discovery to the world. And yet, today, in spite of the quite marked differences between seventeenth-century calculus and our own, a reader who knows what calculus is about can form a fair idea of his aims and method; as with Descartes's *Géométrie*, his language, which must have seemed so strange, has passed into common currency, and his rules for differentiation which his readers had to take on trust are (roughly) the ones we learn in school. A historical take on the paper, then, might properly contain two components:

1. What was Leibniz trying to communicate?
2. How might his communication have been received by a reader?

It would have been easier for readers who had been softened up by Newton's version of the calculus, but there were none of them, and the new notation (dx, dy for infinitesimals or 'differentials' and $\int$ for integrals) was quite unconnected with anything which had gone before. There are two extracts from the 1684 paper (Fauvel and Gray 13.A.3., pp. 428–34) in Appendix B to illustrate the difficulties.

What is noteworthy about Leibniz's exposition (and is often noted) is that at the outset the differentials dx, dy, and so on are not infinitely small. They are 'quantities' whose only property is that (for example) dy/dx is the gradient of the tangent to the curve specified by some equation between x and y. The *rules* which Leibniz then gives for working out differentials are introduced, to say the least, abruptly: 'Now, addition...'. However, they do make it possible to work out the relations between dx and dy. So, for example, using $d(xv) = x\,dv + v\,dx$, we can easily deduce that

$$d(x^2) = 2x\,dx$$

Equally, if $y = x^2$, $x = \sqrt{y}$; and, using the above equation,

$$d(\sqrt{y}) = dx = \frac{1}{2x}d(x^2) = \frac{1}{2\sqrt{y}}dy$$

And many other formulae can be obtained by more or less ingenious applications of the 'Leibniz rule' for multiplication. The problem is that the rule is stated without proof; and its proof depends on some sort of limiting argument—in the language of 1684, you need to use the fact that dx and dy are infinitesimal. Nor is the proof to be found later in the paper; the simple version due to L'Hôpital is given in Section 8.

What Leibniz does next must have seemed still more perplexing; as shown in the second extract, he uses the d procedure a second time to arrive at something called, for example, $d\,dv$. Since we have a rather vague idea of how dv has been defined, we may well ask how the preceding methods could possibly be applied to it to get a second differential. We shall have to wait, although the descriptions of its properties (in distinguishing maxima from minima, for example) show that it is useful.

However, it is only after three pages of this exposition of 'rules' for the calculus, which tells you how to get results with no idea of why they should be true, that Leibniz is prepared to assert that dx, dy, and so on can be taken infinitely small. In this sense, his paper is less open than Newton's early writings; among the reasons usually claimed are:

1. that the fact that the calculus worked was more important for him than the meaning of the symbols which it used;
2. that he was not entirely happy with the infinitely small himself, and was prepared to deny that he was using it if he could find another defence.

A particular viewpoint drawing on Leibniz's philosophical work holds that as a rationalist he was committed to a strong belief in the power of written signs to achieve results in the world. One of his projects in the late 1670s was for a 'Universal Characteristic', which would allow all meanings in all languages to be unambiguously expressed, and so finally put an end to human strife (since there would be nothing to argue about).

It was now clear to Leibniz that in order to discover the alphabet of human thought and realise the universal characteristic, it would be necessary to analyse all concepts and reduce them to simple elements by means of definitions, then to represent the simple concepts by appropriate symbols and invent symbols for their combinations, and finally . . . to demonstrate all known truths by reducing them to simple, evident principles. (Aiton 1985, p. 78)

Compared with this ambitious programme, which of course was never undertaken, the calculus seems a minor achievement. It might be thought that his calculus works in opposition to the rational aim of the characteristic; its status as 'marks on paper which perform a function' is quite divorced from its doubtful meaning. Still, in some way for Leibniz, that was its beauty. If the scientific community could accept that it worked, they would have agreed on a common project for the betterment of mankind. In fact, the heart of the paper—and this is perhaps the answer to the question about his aim—lies in his use of the word 'algorithm'. The reader is being told how to follow a set of mechanical rules (indeed, Leibniz had also invented a calculating machine) which will make it possible to solve with ease a vast number of previously unsolved problems. The justification of the procedure, which was present in Leibniz's unpublished notes from his time 10 years before in Paris, is secondary to its exposition as method.

Here there is perhaps another parallel between Newton and Leibniz which helps to explain their delays in publishing and the confusion of what they wrote: that both were still unsure about how much they needed the infinitely small, or whether given time and application, which neither of them had, they could dispense with it. Leibniz's attempts to explain infinitesimals in later years are many, and often contradictory—sometimes they exist, sometimes they are convenient fictions. A new generation of less sophisticated mathematicians had to adopt and promote the methods, often without much encouragement from their supposed patrons, before they became general currency.

Once we arrive at the infinitesimal 'arguments' in the 1684 paper, they are the ones—already in use before Leibniz' time—which were to become standard, however strange they may seem to us. A curve 'is' a polygon with an infinite number of angles, and its tangent 'is' simply one of the sides of this polygon, produced (see Fig. 3 again). The first of these ideas, as we have seen, goes back to Nicholas of Cusa (see Chapter 6) if not earlier. They could be used, as they were later, to justify the rules for differentiation, and much else.

Why (apart from the hurried publication already mentioned) did Leibniz produce his invention, which he had spent some time developing, in such an unsatisfactory form? The statements which we have claiming pride in the discovery, such as the one which opens this section, come in the main from some time later when others had taken the trouble to understand and explain it. And we can see the prestige of the calculus growing with the priority quarrel; that something had been plagiarized meant that it was worth plagiarizing, and vice versa. While the paper was clearly found more than difficult at the time, some of its faults cannot be judged by later standards. Since Descartes at least, the importance of proof of results had been declining. As Guicciardini (citing Bos as authority) says:

> [I]t was common in the early [and one might add, late] seventeenth century to give priority and publicity to the geometrical construction which solves the problem, rather that to the analysis necessary to achieve such a construction (an analysis which was often kept hidden by mathematicians) . . . (Guicciardini 1999, p. 98)

This is not to say that construction and proof were not often argued about, and unproved results or inadequate proofs challenged. However, Leibniz's paper was too murky even to warrant such a challenge, and only a series of later explainers (and a stream of errata issued in the journal) finally made it readable to later generations.

Interestingly, almost as an afterthought to the confusions of the paper itself, there appears a hint of what the new methods can do:

> It is required to find a curve WW such that, its tangent WC being drawn to the axis, XC is always equal to a given constant line a. [Fig. 4.] Then XW or w is to XC or a as dw is to dx. If dx (which can be chosen arbitrarily) is taken constant, hence always equal to, say, b, that is, x or AX increases uniformly, then $w = \frac{a}{b} dw$. Those ordinates w are therefore proportional to their dw, their increments or differences, and this means that if the x form an arithmetic progression, then the w form a geometric progression. In other words, if the w are numbers, the x will be logarithms, so that the curve is logarithmic.

Leibniz claims this is a problem which Debeaune proposed to Descartes, and which Descartes could not solve. Typically, this is not quite true; the problem is a simplified reformulation of Debeaune's problem, and Descartes did give a solution (see Fauvel and Gray 13.A.2 for the details, including

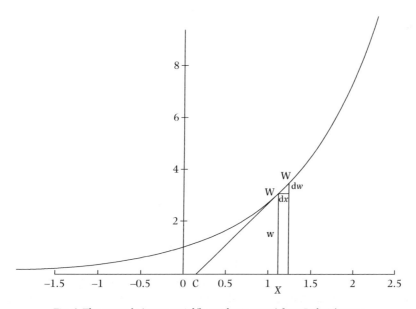

Fig. 4 The example (exponential/logarithmic curve) from Leibniz' paper.

Leibniz's manuscript notes). However, none of this matters in terms of the impact such a solution would have had on any supposed audience. Rapidly solving a problem which, in contemporary terms, was interesting and difficult, it must have been, of all the sections of the paper, the one which most invited the reader to make an effort, and it is unfortunate that it came at the end. In our terms, we would say that XC (the subtangent) is $w/(dw/dx)$, and we require this to be a constant a. We derive the equation

$$\frac{dw}{dx} = a.w \qquad (3)$$

which has the solution $w = Ae^{ax}$, or (as Leibniz says) x is proportional to the logarithm of w. Leibniz' solution is, in our terms, 'from first principles'. First, he has to assume that dx is constant. In modern notation this is *always* an unspoken assumption, based on the choice of x as the 'independent variable' in terms of which w is expressed. (Bos 1991, chapter 5 is an interesting essay on the nature of the 'early calculus' which brings this point out.) Next, he notices that as you add dx to x, you change w to $w + dw = w + (b/a)w = w(1 + (b/a))$. (Note that if dx is infinitely small, which it should be, then so is b, but this is not stated.) Hence, as Leibniz states, arithmetic progression in the xs corresponds to geometric progression in the ws. We would then say that w was a power function (e.g. e^x) of x, but in the 1680s they were not used to such functions. Hence Leibniz's statement goes the other way round: x is a logarithmic function of w.

Exercise 5. *Deduce the formula for $d(x^2)$ from that for $d(xv)$. Next, see if you can generalize to $d(x^n)$, where $n = 3, 4, \ldots$*

Exercise 6. *Solve the differential equation (3). What properties of differentials and integrals have you had to use?*

7 The *Principia* and its problems

After they had been some time together, the Dr asked him what he thought the Curve would be that would be described by the Planets supposing the force of attraction towards the Sun to be reciprocal to the square of their distance from it. Sr Isaac replied immediately that it would be an ellipsis, the Doctor struck with joy and amazement asked him how he knew it, Why saith he I have calculated it, Whereupon Dr Halley asked him for his calculations without any further delay, Sr Isaac looked among his papers but could not find it, but he promised to renew it, & then to send it him. (De Moivre quoted in I. B. Cohen 1971, pp. 297–8)

'The Dr' is Newton's friend Halley, and the above breathless quote (all those commas) is the classic account of how Halley's question on the paths of planets plunged Newton into three years' intense mathematical work which issued in the *Principia* (1687). If its three massive volumes may have been as strange and new as Leibniz's work, they were certainly more impressive, and backed up by a formidable apparatus of proof. As the quote above indicates, the work deduces (using Newton's three laws as starting point) that the observed paths of the planets are compatible with an inverse-square law of attraction; and—more uncertainly—that such a law is the only one which can account for the observations. Whatever he may have said—as in the opening quote of the chapter—about his calculus having been used to deduce his results, the book presents itself as new physics demonstrated by means of old (i.e. of course, Greek) mathematics. The reasons for this are complicated. As has been pointed out for example, by Hall, the work would have been doubly unfamiliar to its readers if it had used both a new physical and a new mathematical language:

> To the few skilled mathematicians in Europe around 1685 who were capable of understanding Newton's mathematical arguments at all, however expressed, the form he actually adopted in *Principia* was far more convenient and familiar than either the method of fluxions—known to no one but Newton—or the Leibnizian differential calculus... (Hall 1980, p. 30)

And there was already enough that they could and did disbelieve in the physics—notably the idea of a force of gravitation acting at a distance through a vacuum, which seemed pure mystification in contrast to Descartes's ideas. The *Principia* is intended to follow the model of Archimedes' physical texts—the *Statics* and *On Floating Bodies*, although it is far longer; with a set of first principles and rigorous deductions from them. Furthermore, Newton, who had as we have seen been an enthusiastic 'modernist' and follower of Descartes in the 1660s, had changed his position radically, for reasons which were mostly, it seems, concerned with his interests outside physics and mathematics. He now (whatever his private practice) took every occasion to attack the modern algebraic school of geometry to which he had once belonged, and scribbled 'Not Geometry' in the margin of his copy of Descartes. During the 1670s, he had immersed himself in his studies on alchemy and on the meaning and chronology of the Old Testament, on which his views were very unorthodox. Rather like Stevin 100 years earlier, he had come to believe in a golden age of 'first knowledge', which the Greeks had corrupted; that, for example, the rotation of the earth round the sun was known to the Egyptians and to Pythagoras. He planned additions to the *Principia* which would have explained this ancient learning (he had worked hard on reconstructing lost Greek texts); had they been published, the work would have been seen as definitely eccentric, and would probably have attracted much less admiration. For example, he drafted an explanation of the 'occult' force of gravitation which would have convinced none of the sceptics:

> Thus far I have explained the properties of gravity. But by no means do I consider its cause. However I will say in what sense the Ancients theorized about it. Thales held that all bodies were animate, inferring this from magnetic and electrical attractions... He taught that everything was full of Gods, and by Gods he meant animate bodies.[8]

We need to consider one example to clarify how the *Principia* might have appeared—and indeed how it appears to us, when we can read it. I have reproduced as Appendix C the crucial deduction of Kepler's area law, which is often cited as an example. The statement is Newton's version of the area law; as he had found, the law (equal areas are swept out in equal times) follows simply from the supposition that the body—always supposed to be a 'particle', concentrated at a point—moves under a force which is directed to the immovable centre S from which the areas are calculated. The proof is in two parts. The first part is good Greek geometry, if physically untrue. We think of the body as being moved in a succession of jerks ('by a great impulse') at equal time intervals, rather than moving smoothly, the impulses being directed to the centre S. We find (as in Fig. 5) that it moves in a polygon, and that the successive triangles have equal areas. We now suppose the number of triangles increased *'in infinitum'*, which is no longer Greek geometry at all, but the now familiar argument that an infinite-sided polygon is a curve. Newton asserts that the force now acts continually, and areas remain proportional to times.

---

8. Gregory MS fo. 13, cited Iliffe (1995, p. 172). Leibniz's ideas on the nature of atoms or 'monads' and the souls which animated them were equally strange; but they were not linked to an ideology which tried to validate ideas by their antiquity.

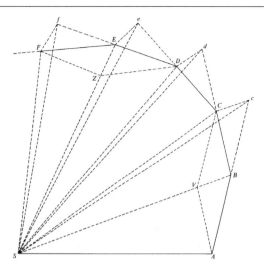

Fig. 5 Newton's picture for *Principia* I, proposition 1.

This argument does not use Newton's version of the calculus; it is a more careful version of the infinitesimal arguments which were being used by many of his predecessors (Pascal, Huygens, Wallis, and others). This does not make it any better, however. It is true that I have omitted the introductory material, in particular Cor. IV, Lem. III, which states that the limit of a polygon is a curve; and that this contains the theory justifying the passage to a limit. Such infinitesimal geometry gives the arguments in the *Principia* a superficial robustness. What Leibniz's (and Newton's) calculus has which the *Principia* does not is the security, and the ease of calculation supplied by algebra. In this respect, Newton's decision to turn away from Descartes's algebraic methods made things harder for him, and for his readers.[9] And this choice of a method of exposition made it quite unclear to what extent the two obscure new works, Leibniz's calculus and Newton's physics, were related.

8 The arrival of the calculus

By 1700 the calculus, or the method of fluxions, as it was now being called in England, had become a success story. This could hardly have been guessed from the beginnings as they have been sketched above. Both Newton and Leibniz had acquired a circle of interested students who attempted to find out what they could about the new methods, and to publicize them. In mathematical terms, Leibniz had some unexpected good fortune. Two Swiss brothers, Jakob and Johann Bernoulli, understood very early that something was to be gained from understanding his theory. Later, they were to claim that they toiled away to master it in a period of weeks; in fact, it took more like three years during which they wrote abjectly to Leibniz requesting some clarification:

Of this method of yours, if you could deign to impart some ray of light (which I earnestly beseech) to me—as much as you can spare given your very weighty affairs—by doing so you would make me not a mere admirer of your inventions but also a worthy esteemer and a publicist. (Letter of Jakob Bernoulli to Leibniz, quoted in Roero 1989)

9. Many stories circulate about the perplexity of the *Principia*'s immediate audience. John Locke (who was not a mediocre mathematician, in terms of the culture of the time) had to be given an outline by Huygens; 'he confessed that there were "very few that [had] Mathematicks enough to understand his Demonstrations" '. (Iliffe, 1995, p. 173.)

Leibniz was in Italy and did not reply to this crawling letter (there is more in the same vein). However, the brothers' persistence bore fruit; in the end, as Jakob had prophesied, they clarified, adapted, and extended Leibniz's method so that it could solve a vast range of problems; and they proved effective and aggressive propagandists. As Roero characterizes it, the Bernoullis can be credited with the 'rediscovery' of the Leibnizian algorithm (Roero 1989, p. 142).

Instructed by Johann, in 1696 the Marquis de l'Hôpital produced what might have seemed impossible 10 years earlier, an 'elementary' introduction to the theory, the *Analyse des infiniment petits, pour l'intelligence des lignes courbes*.[10] In this admirable book, the student can with relative ease learn both how the calculus works and why it works. A certain amount of what, in terms of all previous mathematical practice, would be considered nonsense must be accepted along the way (the problems are identical to those I have already indicated in Newton's theory); but within its own boundaries the theory works, and you will not make mistakes. The serious limitation of the *Analyse* is that it deals only with differential problems and does not touch integration. (Information on that was to come from various sources, notably Newton's *Quadratura curvarum*, in 1704.) However, you can for the first time find a published proof of the 'Leibniz rule' for the differential of a product.

Proposition II. To find the differentials of the products of several quantities multiplied, or drawn into each other. (Fauvel and Gray 11.B.6.)

The differential of xy is $y\,dx + x\,dy$: for y becomes $y + dy$, when x becomes $x + dx$; and therefore xy then becomes $xy + y\,dx + x\,dy + dx\,dy$. Which is the product of $x + dx$ into $y + dy$, and the differential thereof will be $y\,dx + x\,dy + dx\,dy$, that is, $y\,dx + x\,dy$: because $dx\,dy$ is a quantity infinitely small in respect of the other terms $y\,dx$ and $x\,dy$. For if, for example, you divide $y\,dx$ and $dx\,dy$ by dx, we shall have the quotients y and dy, the latter of which is infinitely less than the former.

Whence it follows, that the differential of the product of two quantities is equal to the product of the differential of the first of these quantities into the second plus the product of the differential of the second into the first.

The core 'problem' of early calculus is neatly set out here; and you cannot really derive the product rule for differentials in any other way. The differentials dx, dy are serious quantities which cannot be neglected, and which enter into the formula for $d(xy)$. And yet their product $dx\,dy$ can be neglected. You can get used to doing things this way, but its justification is still shaky. However, at least what was obscure in Leibniz has become transparently clear. If you suspend your worries about the infinitely small, it is easy to follow the instructions, and (for example) to find tangents to any curve.

From the moment that the new methods became at all understood, the Leibnizians had to defend them against those who held (with some justice) that they broke the rules of correct practice in mathematics. In the 1690s, Leibniz was already defending his 'new analysis' against Nieuwentijt (a Dutch philosopher), and the continued success of the calculus against such attacks is a good demonstration of how much importance its defenders attached to its value in producing results, as opposed to mere logical coherence. The most damning and serious attack was to come, some years after the founders were dead, from Bishop George Berkeley. Berkeley's *The Analyst* (1734) is logically (and in terms of classical mathematics) hard to fault, as well as being fine rhetoric; which perhaps goes to show that logic is not always what determines the progress of mathematics. He made the well-founded point that it is bad logic to claim that a theory is correct because it leads to correct conclusions (it is often possible to deduce the truth from false assumptions). And by a careful analysis of the version of the 'product rule' for differentiation given in the *Principia* he

---

10. One feels that L'Hôpital deserves some credit, although it is now recognized that his text is essentially due to Johann Bernoulli and his introduction to Fontenelle. Such is the destructive power of scholarship.

showed that it worked because two errors in calculation happened to cancel. (It just happened that one of the advantages of the early calculus was that this cancellation of errors had the status of a general rule.) Attacking the new scientists of what was becoming called the 'Enlightenment' for transferring their belief from theology to fictions in mathematics (he was particularly incensed against Halley, who was believed to have died an atheist), he asked for example:

> Whether mathematicians, who are so delicate in religious points, are strictly scrupulous in their own science? Whether they do not submit to authority, take things upon trust, and believe points inconceivable? Whether they have not their mysteries, and what is more, their repugnancies and contradictions? (Berkeley, in Fauvel and Gray 18.A.1, pp. 557–8)

The point was a good one, but it was largely ignored. Eighteenth-century mathematicians divided into those who genuinely believed that infinitesimals existed, and those who justified the methods of the calculus by their results in the belief that a sound foundation would arrive sooner or later. D'Alembert, the key Enlightenment figure and editor of the *Encyclopédie*, advised the student in theological terms: 'Allez en avant, et la foi vous viendra' (Go forward, faith will come to you). Faith and science, indeed, had neatly exchanged places.

9 The calculus in practice

So far, the story seems to be a purely 'internal' one, with problems within mathematics being argued over and solved in competing ways by professors (or diplomats, or men of independent means) in universities or coffee-houses. To go a little further, we need to consider the kinds of problems which were of interest, and what constituted a 'solution'; as well as the ways in which mathematicians—who are always trying to justify their bizarre activity to themselves and to the outside world—thought of the possible practical outcomes of their work. Some light is thrown on this by the often provocative essays of Henk Bos (1991); from whom I draw one particularly interesting example.

One of the classical challenge problems which the mathematicians of the 1690s set one another—in the interests both of propaganda for the calculus and of internal competition—was the description of the curve which a heavy chain follows hanging under gravity (Fig. 6). History books will in general tell you that the problem was solved by Leibniz, l'Hôpital and the Bernoullis using the calculus; but reference to their papers (which are neither easy to access nor to read, being in Latin in Gerhardt (1962, vol. V) shows that the story is somewhat more complicated. The naive reader would assume that the 'solution' was an equation for the catenary in the form $y = f(x)$, where x is horizontal and y vertical; and reference to a 'modern' textbook, if they still deal with the topic, will show that this is how it is now expressed:

$$y = a \cosh\left(\frac{x}{a}\right) = a \left(\frac{1}{2}(e^{x/a} + e^{-x/a})\right)$$

However, as Bos shows, despite the allegiance of all parties to Descartes's geometry, the idea that a curve was specified by its equation was not the norm. You could describe a curve 'by quadrature' (give the y-coordinate, say, as the area under another curve—obviously difficult, since you have to measure a curved area); or 'by rectification' (give it as the arc-length of a curve—much better, since you only have to stretch a string along the curve and measure it).

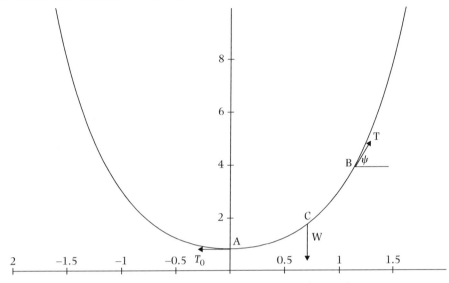

Fig. 6 The catenary $y = a\cosh(x/a) = (a/2)(e^{x/a} + e^{-x/a})$.

[Jakob Bernoulli] wrote that one should *at least* give a construction by quadrature of an algebraic curve. It was *better* to give a construction by rectification of an algebraic curve, or a 'pointwise construction' ... The *best* way to represent a curve, however, was a construction by curves 'given in nature' (as the *Elastica* or, e.g., the *Catenary*. (Bos 1991, p. 34).

We seem to have a vicious circle here; if the catenary is so natural that it is best to represent other curves by means of it, then what point is there in trying to find another way of describing it? The point is illustrated, in a slightly surreal way, in Leibniz's paper which 'solves' the problem. It should be noted that the solution (given in Gerhardt 1962, vol. V, pp. 258–63, in French for easy reading) is a purely geometrical construction of points on the curve, with no proof, either by old-fashioned geometry or by calculus; and that the same is true of the slightly different versions of the other competitors.

Having shown that the catenary was related to the logarithm (it involves the function e^x), Leibniz proposed that ships at sea should have a chain suitably suspended among their instruments, so that, if they lost their invaluable tables of logarithms, they could work them out from measurements on the curve.

Question. Given the equation of the catenary, how would you use measurements on a chain to work out $\log z$?

Question. Is this an entirely stupid and impractical idea, or is it on the contrary ingenious and practical?

The catenary was only one of a host of practical and pseudo-practical problems for which the calculus proved to be uniquely well-adapted, partly because it dealt with rates of change. So we find it applied to the flight of cannon-balls in a resisting medium; to the vibrations of stretched strings; to the shape of sails, and so on. These mundane problems took their place with the grander questions of the movement of the planets and the shape of the earth which Newton had discussed in the *Principia*.

And it is now time to return to a version—if possible updated—of Boris Hessen's thesis: that the whole concern of the *Principia* was with precisely those questions whose solutions were urgently required by the British bourgeoisie.

> At that time the most progressive class, it demanded the most progressive science ... The necessity arose of not merely empirically resolving isolated problems, but of synthetically surveying and laying a stable theoretical basis for the solution by general methods of all the aggregate of physical problems set for immediate solution by the development of the new technique.
>
> And since (as we have already demonstrated) the basic complex of problems was that of mechanics this encyclopaedic survey of the physical problems was equivalent to the creation of a harmonious structure of theoretical mechanics which would supply general methods of resolving the tasks of the mechanics of earth and sky. (Hessen 1971, pp. 170–1)

If for '*Principia*' we read 'calculus' and for 'British' we read 'European', is there any way of sustaining this thesis? The problem is worth the attention of the next generation of historians.

Exercise 7. *Using the equilibrium of a section of the chain, AB (where, say, A is the lowest point), derive the equation of the catenary given above.*

10 Afterword

Boyer's history of the calculus (1949) ends tidily with the nineteenth-century reformulation of analysis—usually ascribed to Weierstrass in the 1860s—which banished infinitesimals and made everything rigorous (in principle, after the model of Greek geometry) again. Whether this vindicated Berkeley or the hopeful analysts who expected that their methods would sooner or later (in this case much later) be justified, is unclear. However, it is worth mentioning that any impression that mathematicians stopped using such ill-founded methods once they had been shown the correct ones is very oversimplified, since mathematics is a large subject—by 1860 it was already out of hand—and not all areas are going to come under the same régime. In particular, the rigorous justification of the calculus is, on the whole, too hard to be taught outside universities, while its results are so useful that they need to be available much earlier. Hence, the language of infinitesimals lingered on. Here are two examples from personal experience.

1. At school, at the age of 17, we were taught to find areas of curves, in particular those given by 'polar equations' involving the coordinates r, θ as in Fig. 7. The heart-shaped curve shown is appropriately called a 'cardioid', and its equation is:

$$r = a(1 + \cos\theta)$$

To find the area, we used the 'element of area'; the infinitely small triangle of height r and angle $d\theta$. As a triangle, it has area $\frac{1}{2}r^2 \cdot d\theta$—this may not be immediately obvious, but it follows by neglecting second-order infinitesimals, and we were told that it was true. Since θ runs from 0 to 2π, the area inside the curve is

$$\int_0^{2\pi} \frac{1}{2}a^2(1 + \cos\theta)^2 \cdot d\theta$$

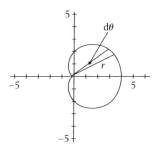

Fig. 7 A cardioid, $r = 2(1 + \cos\theta)$; and an element of area.

which (you can either believe this or work it out) is a relatively simple integral, leading to the answer $3\pi a^2/2$.

2. At university, calculus was taught rigorously (see above), and one had to forget all one thought one had learned about differentiation. However, when one came to study differential geometry—the geometry of curves and surfaces—it was another matter altogether. The textbooks (from the 1950s) which were used still picturesquely define the 'line element' ds on a surface as the distance between two neighbouring points, as though a point had a neighbour; and similarly we learned about infinitely small areas and the shape of an infinitely small ellipse near a point. (This language was indeed what Einstein used to formulate his general theory of relativity (1919), which was differential geometry par excellence.) Such formulations disappeared from universities, at least in England, in the 1960s with the arrival of a serious modernization drive, which introduced the ideas of Elie Cartan and Georges de Rham. The term dx was still allowed (and much the same things could be done with it), but it meant something finite, more abstract, sounder in logic but harder to grasp.

3. And today? Infinitesimals are still certainly used by working mathematicians in areas which have not been modernized; and physics, being more results-driven, is full of them. But to discuss the methods which physicists allow themselves would require much more space—see Chapter 10 for some thoughts on the subject.

Exercise 8. *Sketch some points on the curve $r = a(1 + \cos\theta)$, with a suitable choice of a, and verify that it looks as I have drawn it.*

Appendix A. Newton

(From *On the method of fluxions and infinite series* (in Newton 1967–81, **3**, pp. 121–7).

PROBLEM 4

TO DRAW TANGENTS TO CURVES

MODE 1.

Tangents are drawn to curves according to the various relationships of curves to straight lines [i.e. according to the coordinate system]. And in the first place let the straight line *BD* be ordinate to another straight line *AB* as base and terminate at the curve *ED*. Let this ordinate move through an indefinitely small space to the position *bd* so that it increases by the moment *cd* while *AB* increases

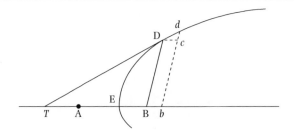

Fig. 8 Newton's picture of the tangent to a curve.

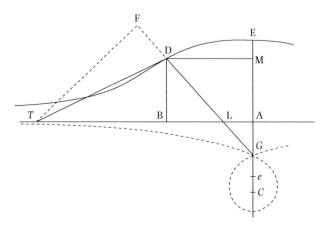

Fig. 9 Newton's 'cissoid'.

by the moment Bb, equal to Dc [Fig. 8]. Now let Dd be extended till it meets AB in T: this will then touch the curve in D or d and the triangles dcD, DBT will be similar, so that

$$TB : BD = Dc : cd$$

When therefore the relationship of BD to AB is exhibited in any equation by which the curve is to be determined, seek the relation of the fluxions by Problem 1 and take TB to BD in the ratio of the fluxion of AB to that of BD; then will TD touch the curve at D.

EXAMPLE 3. Let ED be a conchoid of Nicomedes described with pole G, asymptote AT and distance LD, and let $GA = b$, $LD = c$, $AB = x$, and $BD = y$. [Fig. 9]. Because of the similar triangles DBL and DMG there will be

$$LB(\sqrt{cc - yy}) : BD(y) :: DM(x) : MG(b + y)$$

and so $b + y$ times $\sqrt{cc - yy} = yx$. Having gained this equation I suppose $\sqrt{cc - yy} = z$ and thus I have two equations $bz + yz = yx$ and $zz = cc - yy$. With their help I seek the fluxions of the quantities x, y, and z (by Problem 1), [these are called m, n, r, respectively] and from the first there comes $br + yr + nz = nx + my$, and from the second $2rz = -2ny$, or $rz + ny = 0$. On eliminating r, there arises $-bny/z - nyy/z + nz = nx + my$. By resolving these there comes

$$y : z - \frac{by}{z} - \frac{yy}{z} - x(:: n : m) :: BD : BT$$

The Calculus

In consequence, since BD is equal to y, BT will be $z - x - (by + yy)/z$. This is $-BT = AL + (BD \times GM/BL)$. Here the sign $-$ prefixed to BT indicates that the point T must be taken on the side away from A.

Note. Newton's exposition of how to find tangents is very clear. Of his various examples, I think that the 'conchoid', whose equation and picture he gives, is the clearest; as Whiteside says in his note (1967–81) it is taken from an example of Descartes, who did not bother to write the equation. It is given here more as an example of style than as an encouragement to follow through the calculation One notational point: in Newton's ms, the fluxions of x, y, z are called m, n, r as above. This makes it unclear which fluxion belongs to which variable, and once Leibniz's d-notation became common, Newtonians adopted the clearer practice of writing 'pricked' (dotted) letters $\dot{x}, \dot{y}, \dot{z}$. Whiteside, perhaps again for clarity, uses these in his translation—it is one of the few points where he changes Newton's text—and I have changed them back, since this is how they appeared in the original.

Appendix B. Leibniz

From 'A new method for maxima and minima as well as tangents, which is neither impeded by fractional nor irrational quantities, and a remarkable type of calculus for them' (1684), reproduced from Fauvel and Gray 13.A.3.

1. Let an axis AX [Fig. 10][11] and several curves such as VV, WW, YY, ZZ be given, of which the ordinates VX, WX, YX, ZX, perpendicular to the axis are called v, w, y, z respectively. The segment

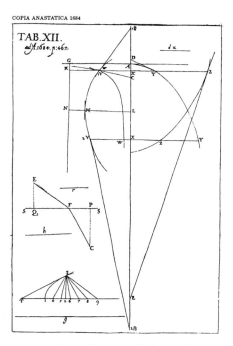

Fig. 10 Leibniz's illustration for his 1684 paper.

11. Dupont and Roero point out (1991) that the original picture given by Leibniz in the paper is usually wrongly reproduced in copies (in particular in Fauvel and Gray). Figure 10 is from Leibniz's works, and hopefully correct.

AX, cut off from the axis is called x. Let the tangents be VB, WC, YD, ZE, intersecting the axis respectively at B, C, D, E. Now some straight line selected arbitrarily is called dx, and the line which is to dx as v (or w, or y, or z) is to XB (or XC, or XD, or XE) is called dv (or dw, or dy, or dz), or the difference of these v (or w, or y, or z). Under these assumptions we have the following rules of the calculus.

If a is a given constant, then $da = 0$, and $d(ax) = a\,dx$... Now *addition* and *subtraction*: if $z - y + w + x = v$, then $d(z - y + w + x) = dv = dz - dy + dw + dx$. Multiplication: $d(xv) = x\,dv + v\,dx$, or, setting $y = xv$, $dy = x\,dv + v\,dx$. It is indifferent whether we take a formula such as xv or its replacing letter such as y. It is to be noted that x and dx are treated in this calculus in the same way as y and dy, or any other indeterminate letter with its difference. It is also to be noted that we cannot always move backward from a differential equation without some caution, something which we shall discuss elsewhere.

2. When with increasing ordinates v its increments or differences also increase (that is, when dv is positive, $d\,dv$, the difference of the differences, is also positive, and when dv is negative, $d\,dv$ is also negative), then the curve turns toward the axis its *concavity*, in the other case its *convexity*.

3. Knowing thus the *Algorithm* (as I may say) of this calculus, which I call *differential calculus*, all other differential equations can be solved by a common method. We can find maxima and minima as well as tangents without the necessity of removing fractions, irrationals, and other restrictions, as had to be done according to the methods that have been published hitherto. The demonstration will be easy to one who is experienced in these matters and who considers the fact, until now not sufficiently explored, that dx, dy, dv, dw, dz can be taken proportional to the momentary differences, that is, increments or decrements, of the corresponding x, y, v, w, z... We have only to keep in mind that to find a *tangent* means to connect two points of the curve at an infinitely small distance, or the continued side of a polygon with an infinite number of angles, which for us takes the place of the *curve*. This infinitely small distance can always be expressed by a known differential like dv, or by a relation to it, that is, by some known tangent.

Appendix C. From the *Principia*

Book I, Proposition 1, Theorem 1. (Reproduced from Fauvel and Gray 12 B.5.)

The areas which revolving bodies describe by radii drawn to an immovable centre of force do lie in the same immovable planes, and are proportional to the times in which they are described.

(See Fig. 5.)

For suppose the time to be divided into equal parts, and in the first part of that time let the body by its innate force describe the right line AB. In the second part of that time, the same would (by Law I.), if not hindered, proceed directly to c, along the line Bc equal to AB; so that by the radii AS, BS, cS, drawn to the centre, the equal areas ASB, BSc, would be described. But when the body is arrived at B, suppose that a centripetal force acts at once with a great impulse, and, turning aside the body from the right line Bc, compels it afterwards to continue its motion along the right line BC. Draw cC parallel to BS meeting BC in C; and at the end of the second part of the time, the body (by Cor. I. of the Laws) will be found in C, in the same plane with the triangle ASB. Join SC, and, because SB and Cc are parallel, the triangle SBC will be equal to the triangle SBc, and therefore also to the triangle SAB. By the like argument, if the centripetal force acts successively in C, D, E, &c., and makes the body, in each single particle of time, to describe the right lines CD, DE, EF, &c., they will all lie in the same plane; and the triangle SCD will be equal to the triangle SBC, and SDE to SCD, and SEF to SDE. And therefore, in equal times, equal areas are described in one immovable plane: and, by composition any sums SADS, SAFS, of those areas, are one to the other as the times in which they are described. Now let the number of those triangles be augmented, and their breadth diminished *in infinitum*; and (by Cor. 4, Lem. III.) their ultimate perimeter ADF will be a curve line: and therefore the centripetal force, by which the body is perpetually drawn back from the tangent of this curve, will act continually; and any described areas SADS, SAFS, which are always proportional to the times of description, will, in this case also, be proportional to those times. Q.E.D.

Solutions to exercises

1. If the 'Sine' is equivalent to $R \sin \theta$ (where θ is the angle, expressed in radians), then the 'Cosine' is $R \cos \theta$, and the arc is $R\theta$. Now, x in the series (1) is the tangent, $\tan \theta = \sin \theta / \cos \theta$; and this clearly is the same as the quotient of the Sine by the Cosine, since the Rs cancel. Now the first term is the product of the Sine and the radius divided by the Cosine (i.e. Rx). To get from each term to the next, we multiply by the square of the Sine and divide by the square of the Cosine; that is, we multiply by the square of the tangent, or x^2. We then divide successively by 1, 3, 5, ... and alternate the signs; the result is:

$$R\theta = Rx - \frac{(Rx)x^2}{3} + \frac{(Rx)x^4}{5} - \frac{(Rx)x^6}{7} + \cdots$$

which is the series (1).

2. The gradient of the tangent is (increase in y)/(increase in x) = AT/TA' = qo/po = q/p.

3. The strength is the (relative) simplicity of calculation. Once you have grasped the sequence (subtract the expression for A from the expression for A'; divide by o; cross out any terms left which still have o in them), it becomes automatic. The weakness is the fuzzy logic. Why is it all right to set $o = 0$ at the end of the argument but not at the beginning?

4. We have $(x + po)^3 - ab(x + po) + a^3 - d(y + qo)^2$ for the equation of A'. Subtracting the equation of A, we get

$$3x^2(po) + 3x(p^2o^2) + p^3o^3 - abpo - d(2yqo + q^2o^2) = 0$$

Divide by o and discard the terms which still have an o or o^2 in them; you get $3x^2p - abp - 2dyq = 0$ which gives the ratio of q to p, or, as we would say, dy/dx:

$$q/p = (3x^2 - ab)/2dy.$$

5. By the rule: $d(x^2) = x\, dx + dx\, x = 2x\, dx$. Generally, $d(x^n) = d(x^{n-1} \cdot x) = x^{n-1}dx + d(x^{n-1})x$. From this we deduce by induction (which was not very commonly used or well formalized in the 1680s, but still ...) the usual formula:

$$d(x^n) = nx^{n-1}dx$$

6. The simplest method of solution is to divide by w and multiply by x (so all like terms are together):
    $$\frac{dw}{w} = a \cdot dx.$$
    (We have had to assume that we can do this, but this is quite allowable on the assumption that differentials are 'quantities' of some kind. We now suppose that we know that $dw/w = d(\ln(w))$, that is, effectively we integrate both sides: $\ln(w) = ax + k$, or equivalently $w = Ae^{ax}$.
7. (Not easy.) Let the arc length from A to B be s, then the weight W is $\rho g s$ (ρ = density/unit length). The horizontal components give $T_0 = T \cos \psi$, and the vertical give $T \sin \psi = \rho g s$. Hence,
    $$\rho g s = T_0 \tan \psi$$
    This is called the 'intrinsic equation'. Note that $\tan \psi = dy/dx$. On the other hand, $ds^2 = dx^2 + dy^2$ (from an infinitesimal triangle), so $ds/dx = \sqrt{1 + (dy/dx)^2}$. Write u for dy/dx, and differentiate the intrinsic equation. You get:
    $$k \cdot \frac{ds}{dx} = k\sqrt{1 + u^2} = \frac{du}{dx}$$
    where $k = \rho g / T_0$. Now, if you know that the integral of $1/(\sqrt{1 + u^2})$ is $\sinh^{-1}(u)$ (look it up), the rest follows:
    $$kx = \sinh^{-1}(u); \quad u = \sinh(kx); \quad y = \frac{1}{k} \cosh(kx)$$
    using $u = dy/dx$, and noting that $u = 0$ when $x = 0$.
8. You have: $\theta = 0, r = 2a$, which has cartesian coordinates $(2, 0)$; $\theta = \pm \pi/2, r = a$, with coordinates $(0, \pm 1)$; and $\theta = -\pi, r = 0$, giving $(0, 0)$. These provide a basis, and you can add others.

8 Geometries and space

1 Introduction

Most people are unaware that around a century and a half ago a revolution took place in the field of geometry that was as scientifically profound as the Copernican revolution in astronomy and, in its impact, as philosophically important as the Darwinian theory of evolution. (Greenberg 1974, p. ix)

It is a general conviction that geometry, with all its truths, is valid with unconditioned generality for all men, all times, all peoples, and not merely for all historically factual ones but all conceivable ones. (Husserl 1989, p. 179)

The aim of this chapter is to consider one of the classic 'stories' in the history of mathematics: the origin of non-Euclidean geometry. Although in some ways a part of the arrival of modern, abstract mathematics (which is generally thought to be about itself) as a replacement for traditional mathematics (which is, again in our usual version, about things and the world), the story has traditionally been told as one of a particular process of discovery; the problem of Euclid's parallel postulate, and the invention of the non-Euclidean geometries by Lobachevsky and Bolyai in the 1820s. It has the virtues of good stories: a connected thread, even a hero/heroes, gropings for a solution followed by an unexpected twist. Its defects, as historians are sometimes anxious to point out, are that history is more complex, and to construct such a story some important details must be left out or simplified. The 'classical' history (Bonola 1955) which is full, careful, and scholarly, is nearly 100 years old, and not surprisingly for some time there have appeared criticisms, attempts to tell the story differently, or to tell a different story altogether. The questions raised are typical ones in the history of scientific revolution, which were already discussed by Bachelard in the 1930s: when mathematicians discover a completely different way of doing mathematics (in this case, geometry), are they adding to the old mathematics, replacing it, or giving us a new perspective from which the old (Euclidean) is a special case of the new? To what extent is the previous pursuit of Euclidean geometry made invalid or irrelevant? And so on.

Before we start, we face various problems; one particular one concerns geometry itself. Partly (but not only) as a result of the 'revolution' to which the quote refers, the study of geometry has gradually become something of a second-class subject, at least in universities. True, Pythagoras's theorem and the criteria for congruent triangles are still part of 'general culture', but their epistemological status tends to be hazy. So the reader might take a moment to reflect on two questions.

1. What is geometry about—what is its subject-matter?
2. How do we know that its results are true?

The answers to these will of course be influenced by your education as well as by personal opinion, but to have thought about them may help. In previous chapters, a too definite knowledge of

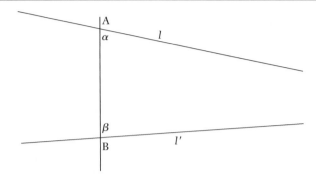

Fig. 1 The figure for postulate 5.

modern mathematics was perhaps a handicap to evaluating the mathematics of the past. But it is an equal handicap to start with no view at all, and the questions above are meant to elicit one.

To return to the story, the 'simplest' version, which still has wide currency and has the merit of simplicity, runs roughly as follows:

1. From the beginning of Euclid's geometry, and possibly even earlier, dissatisfaction with his apparently perfect system was centred on the so-called 'parallel postulate'. This says (in one version) that if the angles α, β in Fig. 1 add up to less than two right angles, then the lines l, l' meet. Another version, which is perhaps easier to understand, is 'Playfair's axiom': there is a *unique* straight line through A which is parallel to l', (does not meet it); and this line makes the angles add up to two right angles as stated. It was felt that this was not intuitively obvious, and should be provable using the other axioms, or from 'first principles'.

[It was, on the other hand, reasoned that if the angles added up to two right angles exactly, then A, B and l, l' would not meet (they would be parallel). A quick way of 'seeing' this is as follows. If the angles on one side are two right angles, so are the angles on the other. If the lines meet on one side, then by symmetry they must meet on the other side too. But this implies that there is not a unique straight line joining two points (the two points where they meet), which is unreasonable.]

2. For roughly two millennia there were attempts to prove the postulate. Recorded efforts were made by Proclus (fifth century), Thābit ibn Qurra (ninth century), ibn al-Haytham (tenth century), Khayyam (eleventh century), Naṣir al-Dīn al-Ṭūsī (thirteenth century); and, in 'modern' times, by a number of writers some well-known, others obscure. It is worth noticing that the 'parallels problem' was never regarded as a key question in mathematics. Obviously it was more important to those (like the Arabs) who valued the Greek classics, but even so, it was often seen as rather a blind alley, pursued by eccentrics.

3. The last major serious 'proof' within the context of classical geometry was due to an Italian priest, Gerolamo Saccheri, published in 1733. This refined a framework for the problem (division into three cases) which is perhaps originally due to al-Ṭūsī. We start by constructing a quadrilateral ABCD, (Fig. 2) with the angles at B and C both right angles, and the sides AB and CD equal. It is then easy to show that the angles at A and D are equal. If we have the parallels postulate, we can deduce that they are right angles (try to see how); but without it, we do not know this. Saccheri distinguished cases according to whether these two angles are right angles, acute, or obtuse; and describes these as the 'hypothesis of the right (acute, obtuse) angle'—HRA, HAA, HOA for short.

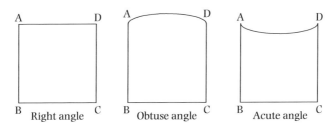

Fig. 2 Saccheri's three hypotheses.

HRA corresponds to Euclid's geometry with postulate 5 included; it is what we normally take to be true. His idea was to get a contradiction by carefully spelling out the consequences of the HOA and HAA, so that the HRA would be left as the only true geometry. The proof which he thought he had was in fact wrong, but the idea of the three hypotheses was to be very useful; and in developing his 'proof' he deduced a great many consequences which must follow if we assume the HAA; this, as we will see, is the difficult case, which amounts to denying Postulate 5.

Saccheri proved that in fact these are three mutually exclusive choices: if, say, the HAA is true for one quadrilateral then it is true for all. There are various other ways of looking at this distinction. For example, with the HOA there are no parallels (we shall consider how this can happen later), while with the HAA there are an infinite number of lines through a point P which do not meet a given line l. Again, with the HOA (the HAA), all triangles have angle-sum greater than (less than) two right angles.

4. After Saccheri, attempts at proof continued, but gradually new elements involving explicit measurement (such as trigonometry) were brought in—and at the same time we see an increasing tendency to doubt the possibility of effectively proving the postulate. Gauss[1] in particular became convinced (some time around 1800) that a consistent geometry in which the postulate was untrue could be constructed; but he confined his thoughts to private correspondence.

5. In the 1820s, two independent researchers, N. I. Lobachevsky and Janos Bolyai, both of whom had been trying to prove the postulate, chose a different aim: to construct a consistent geometry based on the 'acute angle' hypothesis. Note the similarity, though, to Saccheri's programme. In each case the idea was to assume such a geometry possible, but Saccheri hoped to deduce a contradiction, while Lobachevsky and Bolyai did not. Both published their results in obscure places (in Russia and Hungary) in the 1820s[2]; and both works were more or less forgotten. However, each of them proved some important and unexpected properties of the alternative 'non-Euclidean' geometry, and went a long way towards making it an interesting study in its own right. This is the 'Copernican revolution' referred to in our opening quote.

6. Although Lobachevsky and Bolyai had *constructed* their non-Euclidean geometries, they had not proved them consistent. This is not a merely pedantic point; it would theoretically still be possible to prove non-Euclidean geometry inconsistent, and so deduce postulate 5 after all. A wider variety of geometries (more or fewer dimensions, varying rules of measurement) were outlined by Riemann in his groundbreaking paper of 1854, and publicized in the years which followed by

---

1. Who should have much more than a passing reference; he was the dominant mathematician in almost all fields in the years from 1800 to 1840.
2. To be precise: Lobachevsky's first memoir, 'Theory of Parallels' in Russian, was in the *Kazan Messenger* in 1829; Bolyai's 'Science Absolute of Space', in Latin, was published in 1831 as an appendix to his father's *Tentamen*.

Helmholtz; and in particular, the meaning of 'non-Euclidean' was clarified. Proof of consistency was achieved in stages through the later nineteenth century by Beltrami, Klein, and Poincaré among others, by the characteristically modern method of defining 'models' for the new geometry. Of this again, more will follow later.

7. As a result of these developments, it was realized that there was no *unique* geometry, and the way was open to the modern understanding of a 'geometry' as the study of an axiom-system which asserts certain properties of objects called (for example) 'points', 'lines', etc., without reference to what these names may mean. The unique geometry of Euclid has been replaced by a multiplicity of geometries, which are equally valid as objects of mathematical study. Because they were the first to suggest an alternative to Euclid, Lobachevsky and Bolyai can be seen as the founding fathers of this revolution.

Note. The reader who has no idea of what non-Euclidean geometry is, let alone what a model of it might be, should consider the well-known picture 'Circle Limit III' by Moritz Escher (Fig. 3). In this, which is a picture of non-Euclidean geometry's version of a plane,

1. the curved lines are to be thought of as straight;
2. all the triangles (and all the fish) are to be considered as having the same size;
3. the bounding circle is 'at infinity', and lines which meet there are parallel. The calculations underlying the picture are set out by H. S. M. Coxeter at www.ams.org/new-in-math/circle_limit_iii.pdf.

Fig. 3 'Circle Limit III' by Moritz Escher.

We have given the traditional outline of the story, and it is easy to criticize. An up-to-date, serious history of mathematics such as this one claims to be obviously ought to be cautious of a narrative which (a) supposes that a single project has occupied researchers for over 2000 years (from Euclid's time to the nineteenth century) and (b) points to a single discovery, at the end of this time-span, as a founding event or revolution. The problem of the story of non-Euclidean geometry is the problem of stories in general in history. How far has a generally confused situation been simplified to produce a neat narrative? Has the meaning of the terms changed over the period? What other issues, of philosophy, or the varying meaning of the word geometry need to be taken into account?

In presenting the traditional history first, the intention is not to expose it to ridicule, but to raise some genuine problems. In a thoughtful discussion (cited in Fauvel and Gray 16.C.5), Gray raises the main problems of what he terms the 'standard narrative' for the major revolutionary period, that is, roughly from 1730 to 1860; but before dealing with these, a similar assessment needs to be attempted for the much longer earlier period. There seems to be an essential continuity in the history from 300 BCE to the mid-eighteenth century, and a discontinuity for some time after that, whether it is termed a 'revolution' or not. Is the continuity genuine, where does the discontinuity come from, and what do either of them have to do with wider questions about how we conceive of space and the world? The fact that Kant, whose famously influential ideas on space were founded on Euclidean geometry, wrote just before Lobachevsky and Bolyai is often remarked on; but is it just chance?

Instead of the usual lengthy discussion of source-material, it is easy to give a relatively short reading-list here; and at the head of it will naturally stand Jeremy Gray's excellent study (1979). This book is not only about the history of parallels, and it is the better for that; and it is useful in covering both mathematical and philosophical questions, with a natural bias to the mathematical. The lengthy history of attempts to prove the postulate (particularly in the Islamic period) is dealt with rather briefly, but this can be justified by the greater interest of the eighteenth and nineteenth centuries. And Gray consistently pays attention to the context—what other kinds of geometry were of importance, and receiving attention—so that non-Euclidean geometry is given its proper place as one contender in an often quite diverse field.

As Gray remarks, the older work of Bonola (1955) is still the final authority; it is all the more important since it includes the main founding works of Lobachevsky and Bolyai as appendices. Probably because of Gray's particular interest, the source material reproduced in the chapter on the subject in Fauvel and Gray (chapter 16) is generous, with extracts from Greek and Islamic writers in earlier chapters in particular; as regards the standard narrative, it is a very useful complement to Gray's book.

Dating from the same period as Gray is Torrelli (1978), which is important in covering, one would think, much the same questions (how did geometry change in the nineteenth century, and why?), but with relatively little common ground. Much more attention is paid to rival methods of axiomatics, and to ideas of what the subject matter should be. As a result, the cast of characters is richer, including not only mathematicians but physicists and philosophers as well as those who, like Helmholtz, Mach, and Poincaré, tried to combine the various disciplines. Finally, Joan Richards' (1988) provides an enlightening antidote to a narrative centred on research mathematicians, showing the reception of the new ideas on geometry in the rather special case of England, where the teaching curriculum and humanistic values played a central part in what one might have expected to be purely mathematical debates.

2 First problem: the postulate

Let us return to Euclid's postulate 5.[3] It is worth noting that postulates 1–4 require the reader to accept what might, in reasonable terms, be considered 'obvious': for example, that a circle can be described with any given centre and any given radius. However, even then there is a disjunction between geometry and 'experience'. Would the statement still be so obvious, one might ask, if the radius were chosen to be a million miles? As Torrelli points out:

> In the closed Aristotelian world not every straight line can be produced continuously as required by Postulate 2, and not every point can be the centre of a circle of arbitrary radius as demanded by Postulate 3. (Torrelli 1978, pp. 8–9)

This is an example of how the non-specificity of the postulates is useful; one does not have to consider such questions.

What postulate 5 says is different in kind. However, it is essential in constructing that Euclidean geometry which most of us would consider sensible. Not only the standard result on the angle-sum of triangles (two right angles), but the very existence of rectangles and squares (figures with four right angles at the corners) depend on it. And so, as a consequence, does much of what for the Greeks' predecessors was known and used, for example, the Pythagoras theorem.

In its classical form, postulate 5 reads:

> If a straight line, falling on two other straight lines, makes the two interior angles on one side less than two right angles, then the two straight lines, produced indefinitely, will meet on the side on which are the two angles less than two right angles.

This at first simply appears hard to understand; but what it states is indicated by Fig. 1 above. If we have such a diagram, and $\alpha + \beta < 180°$, then the two lines will meet as stated. Note the two other points:

1. 'produced indefinitely'; that is, we are allowed to make the lines as long as is necessary;
2. 'on that side'; that is, they will meet on one side, and not on the other.

While now perhaps comprehensible, the postulate—if we think about it—is asking us to accept quite a strong statement. Once again, if we were allowed to think in terms of numbers, we would find that it contains an assumption that geometry continues to work at arbitrarily large distances. Trigonometry tells us, for example:

Claim: If (in radians) $\alpha = \beta = (\pi/2) - 10^{-10}$, and the transversal is of length 1 cm, then the distance to the meeting point is roughly $\frac{1}{2} \cdot 10^{10}$ cm.

In other words, for postulate 5 to be true of 'the world', we must again conclude that the world has infinite extent. For evidence that the Greeks in general did not think this, it is enough to consider Archimedes (a very sophisticated Greek) who discusses the size of the universe in his *Sand-Reckoner* (see Fauvel and Gray 4.A.2). What figure he came up with is unimportant for our purposes; the main point is that it was a finite one. Hence, the ideally produced lines of geometry might go beyond the boundaries of the universe—and *conceptually*, one can see how an idealized straight line might continue after the universe had stopped. This points to an interesting disjunction between geometry (an ideal study) and the study of the real world. In this respect, Plato's point (see Chapter 2) that

---

3. Numbering in Euclid is sometimes problematic, and some authors (Bolyai in particular) call it axiom 11.

geometers are, or should be studying 'forms' rather than things in the world makes more sense than appears at first sight.

And yet, of course, Euclidean geometry was, and still is vital for surveyors, architects, and town-planners who care nothing for how far a line can be produced but who very much want to use the basic results about triangles, rectangles, lengths, and areas. The 'ideal' geometry, which in Greek terms is a fiction, founds the geometry which people need. Worse, a naive reading, at least of Euclid's early books, might lead us to think that we were studying on a flat earth, particularly if we use the variant definition of a straight line as 'the shortest distance between two points'. The Greeks knew the earth was round—Eratosthenes measured its radius. Al-Bīrūnī (see Chapter 5) used sophisticated spherical geometry, in which the shortest distance between two points is a great circle, to find routes between cities and determine the qibla. But in this 'geometry', all lines eventually meet, as is shown in Fig. 4, and the standard results referred to above are not true. They are, however, so nearly true that in (say) town planning, as opposed to long sea or air journeys, any error in Pythagoras's theorem could not be detected by our measuring instruments.

These two ways in which Euclidean geometry failed to measure up to the real world are worth bearing in mind when we consider its problems. For both Greek and Islamic geometers we find that the question is not strictly: 'How do parallel lines behave in the world'? Rather, it is how they behave in *geometry*. Here, then, we need to pause and make a historicist evaluation of what, in proving postulate 5 (or any other Euclidean study) Euclid's successors were after. The title 'Ideas of Space' given to Gray's book is not, or has not always been a characterization of geometry.

Be that as it may, the influence of Euclidean thinking over subsequent geometers was naturally enormous, even when they misunderstood him. Consequently, it is not surprising that the terms in which Proclus stated the problem (in the sixth century CE—and they were already old by then) remained constant for so long:

This ought to be struck from the postulates altogether. For it is a theorem—one that raises many questions, which Ptolemy proposed to resolve in one of his books—and requires for its demonstration a number of definitions as well as theorems. (Proclus 1970, cited in Fauvel and Gray 3.B1)

In other words: (a) the postulate is *not* 'reasonably acceptable' in the sense that the others are, and (b) it should be proved to be a consequence of assumptions which *are* acceptable. This was the long-term programme at least up to 1700. As can be seen, it was not, in any sense, a problem about the coherence or otherwise of Euclidean geometry, which was by universal agreement *the* geometry. Rather, it was a problem about constructing a proof of the postulate. Notice also that,

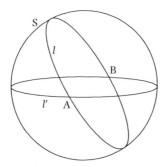

Fig. 4 Geometry on the surface of a sphere. Any two 'straight lines' *l* and *l'* meet (twice), at A and B.

while Proclus agreed that to prove the postulate one would need extra definitions and theorems, he did not say that one would need an alternative postulate. One needed, essentially, to provide a way of seeing why what did not appear self-evident could be made so. Such were the terms of reference, again until the seventeenth century. The recurring problem is that what is self-evident to one writer may be illegitimate to another; and that almost anyone who made a serious attempt agreed that the problem was difficult (necessarily, or it would not have preoccupied so many eminent predecessors) and accordingly a proof could not be simple.

In a detailed analysis reproduced as Fauvel and Gray 6.C.3, Youschkevitch describes the various proofs from the Islamic period and their flaws. As an example of one which seems quite adequate, we could consider ibn al-Haytham's proof, reproduced as Fauvel and Gray 6.C.1. Khayyam criticises it (loc.cit. 6.C.2) for using the idea of motion, but in fact this is not the major problem. Euclid used ideas of motion implicitly in places, and Thābit ibn Qurra in discussing the same question gave a reasoned defence. Here is how ibn al-Haytham begins:

Let us start with a premise for that, and that is: 'When two straight lines are produced from the extremities of a finite straight line, containing two right angles with the first line, then every perpendicular dropped from one of these two lines on the other is equal to the first line, which contained two right angles with these two lines.' [The meaning must be (see Fig. 5) that the angles actually are two right angles, not as sometimes in Euclidean-language that they add to two right angles—otherwise the statement is obviously untrue.] Thus, every perpendicular dropped from one of the afore-mentioned lines on the other contains a right angle with the line from which it was dropped. An example of this is as follows: there is extended from the two extremities of line *AB* two lines *AG*, *BD*, and the angles *GAB*, *DBA* are each right. Then point *G* is assumed on line *AG* and from it perpendicular *GD* is dropped on line *BD*. I say, then, that line *GD* is equal to line *AB*. The proof of that is that nothing else is possible. (Fauvel and Gray, p. 235)

Notice first, that ibn al-Haytham states clearly what he is using as a replacement for the parallel postulate. (It is again equivalent, since it essentially asserts the existence of rectangles.) Second, he does not consider it self-evident, despite the bold 'nothing else is possible'; because he goes on to prove his statement from what he *does* consider self-evident, or at least more basic geometry. The details are quite complex; you can follow them through in the source and find, if you can, where something 'equivalent' to postulate 5 is being used as an assumption.

Exercise 1. *(a) How does the 'angles of a triangle' theorem' follow from the parallel postulate in either of the forms cited above? (b) Let AB be a straight line; construct AC, BD perpendicular to AB and on the same side of it, so that AC = BD. What would you need to show (1) that the angles at C and D are right angles, (2) that AB = CD?*

Exercise 2. *Prove the 'Claim' above about the distance of the meeting-point.*

Fig. 5 Ibn al-Haytham's method—basically to construct a rectangle. Angles A, B, and G are right angles; the statement is then that AG must be equal to BD. (We shall find Lambert discussing quadrilaterals of this type much later.)

3 Space and infinity

As lines (so loves) oblique may well
Themselves in every corner greet:
But ours so truly parallel,
Though infinite, can never meet. (Andrew Marvell, *The Definition of Love*, in 1972, p. 50)

The point has already been made that the axioms of Euclid's geometry do not properly apply to what Greeks thought of as 'space'; and this distinction between the two objects of study seems to have remained constant until about 1600. That lines can indeed be infinite is, on the other hand, apparently assumed by Marvell in the quote above (about 1650), a neat poetic formulation of postulate 5. Something had changed in the way space was thought about which was to pose further problems. Geometry had not only to be 'self-evident' in terms of some idea of common sense, it was necessary that it should describe the world.

It would appear that Thomas Bradwardine (fourteenth century), in a typically scholastic approach, was among the first to consider the idea of infinite space—if not as where we live then as a property of God:

God must imagine the site of the world before creating it; and since it is absurd to imagine a limited empty space, what God imagines is the infinite space of geometry...'Indeed, He coexists fully with infinite magnitude and imaginary extension and with each part of it'. (Torelli 1978, p. 28, quoting Bradwardine, *De causa Dei*)

Of course, such speculations coexisted with the more extreme ones such as whether God could create a triangle whose angles did not add to two right angles; but in terms of a unification of geometric space with the actual universe, the idea gained ground through more radical early modern thinkers such as Giordano Bruno (sixteenth century), Descartes, and finally Newton. Was Descartes's need for infinite space ('the extension of the world is indefinite') related to his revolution in geometry? It would not appear so, since Descartes's plane is still an abstract one, a copy of Euclid's with numbers introduced. [And if you look back to the extracts from the *Geometry* in Chapter 6, you will see that he needs parallels to introduce the numbers.] Rather, it is a consequence of his *physical* law, equivalent to Newton's First Law, that

a freely moving body will always continue to move in a straight line—thereby perpetually performing the construction demanded by Euclid's second postulate—and this would be impossible if every distance in the world were less than or equal to a given magnitude. (Torelli 1978, p. 24, referring to Descartes Princ. Phil)

We noted in Chapter 7 the importance of Greek geometry for Newton's later work, specifically the *Principia*. Indeed, Newton went further than Descartes by constructing a vast scheme of how all matter in the universe behaved. Here the universe was explicitly identified with the space of Euclidean geometry, in which straight lines have indefinite extension. (Infinite, if you are freer in your language.) Again, it was impossible for the laws of physics to work without such an identification, but it is important to stress that this was relatively new. If Newton had to some extent borrowed the idea from Descartes he certainly made it much more explicit in the whole geometric and deductive structure of the *Principia*. With Descartes and Newton, a great step forward is achieved, in that geometry *can* apply to physical space. As for the application of postulate 5, it means that any inclined lines (in the same plane) will meet somewhere in the universe. The drawback of this is that questions about geometry may become identical with questions about the

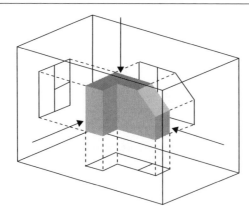

Fig. 6 Classical descriptive geometry as it is still practised today. The three projections are united to give a general view, using the algebra of vectors.

world, at least for some geometers.[4] Of course, in a sense this has always been so; but the idea that the study of geometry was derived from knowledge of the world, whether innate (part of our mental structure) or empirical (derived from observation) became a dominant one for the next two centuries. The contrast with Plato's view that geometrical ideas were in some way above practice is striking.

The main developments in geometry from Newton's time on—and this is important when we come to consider the relative importance of the 'new' geometries—concerned, naturally, the increasing introduction of coordinates and of calculus as tools. To study curves and surfaces meant to study their equations, even if diagrams were used as aids to understanding. In the late eighteenth century, the French geometer Gaspard Monge developed what was called 'descriptive geometry' a key subject in the immensely influential École Polytechnique. Very fashionable throughout the nineteenth century, and still surviving as an essential part of practical training although unknown in most mathematics departments, descriptive geometry was the study of three-dimensional figures via their projections—plans, elevations, and so on; the breaking down of a figure into its projections, and its reconstitution from them (Fig. 6); and it leaned heavily on calculus in its more sophisticated parts. Partly because the use of coordinates was central, partly because of the importance of practical application, it was not concerned with the question of the world. In 1837 we find Monge's follower Michel Chasles praising him precisely for avoiding those diagrams ('figures') which were an essential starting point in thinking, say, about parallels.

[A]lthough descriptive geometry...by its nature makes an essential use of figures, it is only in its practical and mechanical applications, where it plays an instrumental part, that it needs them: no one more than Monge thought of and practised geometry without figures. There is a tradition in the École Polytechnique that Monge knew to an amazing degree how to make his audience imagine the most complicated forms of extension in space, and their most hidden properties, without any other aid but his hands, whose movements followed his words admirably... (Chasles 1837, p. 209)

4. In an earlier draft, I used the phrase 'confused with' rather than 'identical with'; but the confusion is more that of the historian, who has to try to understand, from Newton on, whether a geometer is describing an abstract construction or the empirical universe. Sometimes, but not always, the geometer will explain.

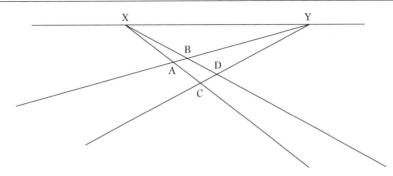

Fig. 7 Traditional perspective generates the 'ideal line' at infinity XY of projective geometry. The lines CA, DB are parallel and meet at infinity at X in the 'extended plane'; similarly AB, CD meet at Y.

Indeed, even the study of projective geometry, which *is* still taught as a very abstract subject, was initially an offshoot of descriptive geometry[5] and was regarded as the study of space enriched by those ideal points and lines 'at infinity' which we find in perspective drawings (Fig. 7). Parallel lines, in defiance of postulate 5, could meet provided that their meeting point was in that imagined exclusion zone which was termed the 'line at infinity'; the Euclidean structure of space was not challenged, however strange that may seem. Perhaps the strongest expression of the prevailing orthodoxy was given by Bolzano in 1817 when he attacked the use of geometry to prove results in analysis:

But it is clear that it is an intolerable offense against *correct method* to derive truths of *pure* (or general) mathematics (i.e. arithmetic, algebra, analysis) from considerations which belong to a merely *applied* (or special) part, namely, *geometry*. (Bolzano, in Fauvel and Gray 18.B.1, p. 564)

Geometry was an applied subject, since its truth was derived from our knowledge of the world. This point would have been almost unquestioned by Bolzano's readers, even if they did not share his conclusions. It was not contested by Lobachevsky and Bolyai, and it would not be for some sixty years. It is only with hindsight that we see non-Euclidean geometry as pointing towards a democracy of geometries in which all may have equal status and truth-claims are no longer the issue.

4 Spherical geometry

When God Almighty intended the creation of mankind, He purposely designed the creation of the earth at first, and gave it the consolidating force to evolve its natural shape, I mean that which is truly spherical. (Al-Bīrūnī 1967, p. 24)

We are about to confront the major problem of this chapter; that the mathematics which underpins non-Euclidean geometry is, at times, difficult both conceptually and in terms of sheer calculation. The ideas of Lobachevsky and Bolyai need their formulae to work, and the formulae are far from intuitive. Following a common precedent, we shall therefore consider first the 'geometry' which is defined on a sphere—think, as usual, of the Earth—by taking 'straight lines' to be lines of shortest distance, that is great circles. This corresponds to Saccheri's HOA. By what one might call

5. The founder, Poncelet, was one of Monge's students.

coincidence, it has always been dismissed as the HAA has not, as clearly contrary to what is obvious. In Section 1, I gave it a quick dismissal; a more interesting one rests on Euclid's proposition I.16.

In any triangle, if one of the sides is produced, then the exterior angle is greater than either of the interior and opposite angles.

Euclid's proof, together with a picture which illustrates how it fails to work on a sphere, is given as Appendix A. In his discussion of the 'standard account', Gray raises the question of why spherical geometry (which was, as he says, 'well known throughout the entire period') was not seen as an answer to the question (i.e. whether the fifth postulate could be deduced from other self-evident facts). It may be the mere fact that obvious Euclidean postulates, like the existence of circles of arbitrary radius, were untrue in spherical geometry; or it may be that other simple defects such as are brought to light in the failure of proposition I.16 were responsible. In any case, the aim of this section is not so much to discuss this 'what if' question (why did geometers not see the sphere as an answer?), but to look at what was known of that geometry and how it influenced later thinking.

Already in the time of the Greeks, as has been mentioned, it was recognized that a line of shortest distance on a sphere (let us call the sphere S) was an arc of a 'great circle'—the intersection of S with a plane through the centre (Fig. 4). Because of their importance in astronomy, the Greeks, in particular Ptolemy gave attention to understanding spherical triangles (triangles whose sides are shortest lines on a sphere), and the Islamic mathematicians who had (roughly) our trigonometric functions to help them were able to derive the key formulae for 'solving' them, which are given in Appendix B.

The formulae are essential if we are to find our way about on a sphere; they have been used by geographers ancient and modern. Of much less interest to geographers, but more to mathematicians, is something about the *angle-sum* of spherical triangles which was discovered by Albert Girard in the seventeenth century: that

1. the angle-sum $A + B + C$ is always greater than π (so much is obvious);
2. the 'excess' $A+B+C-\pi$ increases with area; in fact, it is precisely equal to $(1/R^2) \times$ area(ABC).

This is easy to see for the triangle all of whose angles are $\pi/2$, which makes up an eighth of the sphere (why?). To see that the excess is simply a multiple of the area is a subtler argument, but acceptable if you are prepared to take a little time thinking about pasting triangles together.

The way in which Girard's formula might lead to a better understanding of what it means to deny postulate 5 seems first to have occurred to Johann Heinrich Lambert, whose posthumous *Theory of parallels* appeared in 1786. (Indeed, Lambert is singled out as a key transitional figure by Gray (1979, ch. 5) for this and other reasons.) It was Lambert who, by reasoning with quadrilaterals again, arrived at a key consequence of the HAA:

[I]t is not only the case that in every triangle the angle sum is less than 180°, as we have already seen, but also that the difference from 180° increases directly with the area of the triangle. (Lambert, in Fauvel and Gray p. 518)

Second, he saw that in consequence an HAA geometry must, like a spherical geometry, have an absolute measure of length. This comes from reasoning with a quadrilateral ADGB (Fig. 8) in which $AB = AD$ and the three angles A,B,D are right angles, but the fourth (G) may not be:

Since the angle has a measure intelligible in itself [i.e. as a fraction of 360°], if one took e.g. *AB = AD* as a Paris foot and then the angle *G* was 80° this is only to say that if one should make the quadrilateral *ADGB* so big that the angle *G*

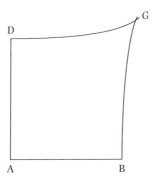

Fig. 8 Lambert's quadrilateral.

was 80° : then one would have the absolute measure of a Paris foot on $AB = AD$. (Lambert, in Fauvel and Gray, pp. 517–8)

Lastly, Lambert recognized that the area formula he had found was the 'negative' of the area formula for spherical triangles; the defect $\pi - (A + B + C)$ replaces the excess as the measure of area. As Gray remarks, he was nearly there. In another sense, he was not there at all; he could see very clearly what a non-Euclidean geometry must be like, but he went no further in claiming its existence than the unfortunate statement that it might hold on an imaginary sphere—which no amount of modern reinterpretation can make sense of. This perhaps is the key point at which one is justified in asking Gray's question (1979, p. 155), why did the development take so long—in this case, from the 1780s to the 1820s? Not excessively long, perhaps; and it should be remembered that the pursuit of parallels was, as already mentioned, outside the mainstream. It was notoriously a problem for masochists, eccentrics, or those with unrealistic ambition.

Exercise 3. *(a) What does Lambert's statement about absolute measurement mean? (b) How could it be justified?*

5 The new geometries

In fact, one sees not only that no contradiction is reached, but one soon feels oneself facing an *open* deduction. While a problem given a proof by contradiction should head fairly quickly for a conclusion where the contradiction is clear, the deductive work of Lobachevsky's dialectic settles itself more and more solidly in the mind of the reader. Psychologically speaking, there is no more reason to expect a contradiction from Lobachevsky than from Euclid. This equivalence will no doubt later be technically proved thanks to the work of Klein and Poincaré; but it is already present at the psychological level. (Bachelard 1934, p. 30)

At this point, rather than continue with the detail of the story,[6] the reader may reasonably want to know what is meant by saying that Lobachevsky and Bolyai 'constructed a geometry'. What is it to construct a geometry? This is the 'Copernican' aspect of the discovery—no one before had tried to do such a thing. Following on the Euclidean model, one would reasonably ask for a set of rules or axioms—maybe not this time self-evident—which are full enough for a substantial theory to be deduced from them. Let us suppose, as the innovators did, that you simply deny postulate 5.

---

6. The key stages between Lambert and Lobachevsky–Bolyai can be found in Gray (1979), chs. 6–9 or Bonola (1955, Chapter III).

This means that you are adopting what Saccheri called the hypothesis of the acute angle. The simple negation of postulate 5, though, is:

There exists a straight line, falling on two other straight lines, which makes the two interior angles on one side less than two right angles, such that the two straight lines, produced indefinitely, never meet.

This is impossibly vague, and cannot be made a basis for any serious deductions. To have a workable geometry, one would need—at least in the terms understood in the 1820s—to have rules for when triangles were congruent, rules for angle sum, rules for areas of figures, and so on. In other words, one would need measurement of a kind, and to construct a new geometry was in a certain way to define a new way of measuring. This was to be made explicit by Riemann in the 1850s, but it was not how Lobachevsky and Bolyai proceeded. Their expositions were surprisingly similar, each with its advantages; Bolyai's is perhaps the clearer, Lobachevsky's the more complete.

It is easiest to start with Lobachevsky's clarification of the vague statement above. Properly analysed, he claimed, it must go as follows:

All straight lines in a plane which go out from a point can, with reference to a given straight line in the same plane, be divided into two classes—into *cutting* and *not-cutting*.

The *boundary lines* of the one and the other class of those lines will be called *parallel* to the given line.

From the point A let fall upon the line BC the perpendicular AD, to which again draw the perpendicular AE (Fig. 9)...

In passing over from the cutting lines, as AF, to the not-cutting lines, as AG, we must come upon a line AH, parallel to DC, a boundary line, upon one side of which all lines AG are such as do not meet the line DC, while upon the other side every straight line AF cuts the line DC.

The angle HAD between the parallel HA and the perpendicular AD is called the parallel angle (angle of parallelism), which we will here designate by $\Pi(p)$ for $AD = p$. (Lobachevsky, in Fauvel and Gray 16.B.3, pp. 524–5)

The above definition is hardly more sophisticated than the work of Lambert. Its essential importance is that it changes the imprecise negation of postulate 5 into a precise statement about angles and their relation to lengths; with every p is associated a $\Pi(p)$. It is, of course, crucial in constructing a sensible geometry that $\Pi(p)$ depends only on p, but this follows from the 'elementary' results which Lobachevsky gives at the outset. These include the standard rules for when two triangles

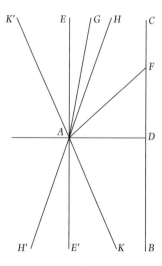

Fig. 9 Lobachevsky's diagram.

are congruent, of which more later. Staying for the moment with 'internal' factors, we can see an obvious reason for the relatively *late* development of the theory; that the actual function Π(p) was quite a sophisticated one, whose formulation would have been difficult for (say) Leibniz but was fairly accessible a century later. It is in fact given by

$$\sin(\Pi(p)) = \text{sech}(Kp)$$

or

$$\tan\frac{1}{2}(\Pi(p)) = e^{-Kp}$$

Here K is the constant of 'curvature' appropriate to the space. It is analogous to $1/R$ for a sphere of radius R—the bigger K is, the more curved the space. From this, a great deal of detailed understanding of 'non-Euclidean space' follows; in particular the triangle formulae (Appendix B), and Lambert's area-defect formula of the previous section. For the fuller development of Lobachevsky–Bolyai geometry, you are referred either to the expositions in Gray (1979), or the actual sources, which are quite readable, in Bonola (1955). For a modern explanation of what it is, and how it works, in what is called the 'Poincaré model', see Thurston (1997, Chapter 2).

Exercise 4. *How would you prove that two equal line-segments determine the same angle of parallelism?*

Exercise 5. *Check that the two formulae given for $\Pi(p)$ are equivalent; and that $\Pi(p)$ is a decreasing function of p, with $\Pi(0) = \pi/2$, $\Pi(p) \to 0$ as $p \to \infty$.*

6 The 'time-lag' question

Gray's question—why did it take so long?—actually divides into two parts, as the story is usually understood. The first is the delay from Saccheri (1733) to Lambert to Lobachevsky–Bolyai (1820s); this, it has been argued, can be accounted for. More serious is the delay from the invention to the wider reception of the new geometries, which was in about 1866, that is, roughly 40 years. One could ask (for example):

1. Why did it take so long for the new geometries to reach the 'public domain'?
2. If the conditions were favourable for two (three counting Gauss) independent discoveries in the 1820s, why were there no further such discoveries in the next 40 years?

If we say that the discoveries occurred in the 1820s because the problem, and its particular solution were 'in the air', we have to explain why the solutions were neither noticed nor duplicated in the years that followed. Conversely, if the historical conditions were not right for a solution we have more of a problem in explaining the occurrence of three. There is no particularly easy answer to this dilemma. The isolation of Lobachevsky and Bolyai and the caution of Gauss are usually invoked as sufficient reasons for the neglect of their work.

The development of non-Euclidean geometry in Central and Eastern Europe was half-hidden from the public owing to the obscurity of two of its creators and the shyness of the third. (Torelli 1978, p. 110)

However, as Torelli implicitly acknowledges, the fact is that, since geometry was now firmly believed to be about 'space', or the world of physics and of everyday life, the question which non-Euclidean geometry addressed—what sort of a world do we live in?—was not considered. In this sense, the new geometries were outside the mainstream which we have referred to above. One of Lobachevsky's expositions of his system, the 'Géométrie Imaginaire' was published in the highly respected *Journal für die reine und angewandte Mathematik* in 1837; it aroused no response. This was the year of Chasles's 'History' referred to above, in which all the important modern developments focused on were in projective geometry, then (as we have seen) considered as an extension of Euclidean geometry.

The question of the geometry of space, as a possible subject for doubt, was famously raised in a new and finally extremely influential form by Gauss's student Bernhard Riemann in his 1855 paper 'On the Hypotheses which lie at the basis of Geometry'. This also suffered a time-lag; it was not published until 1866, and although it is a key document of modern mathematics it is not an easy read. (W. K. Clifford's English translation is to be found on www.maths.tcd.ie/pub/HistMath/ People/Riemann/Geom/WKCGeom.html, and elsewhere.) Because it allows for a very large variability of structure, it is *both* a founding text of modern physics, specifically Einstein's General Theory of Relativity *and* an opening towards the modern mathematical view which divorces the study of geometries from any ideas of what the world may be like.

> It will follow from this that a multiply extended magnitude is capable of different measure-relations, and consequently that space is only a particular case of a triply extended magnitude. But hence flows as a necessary consequence that the propositions of geometry cannot be derived from general notions of magnitude, but that the properties which distinguish space from other conceivable triply extended magnitudes are only to be deduced from experience. Thus arises the problem, to determine the simplest matters of fact from which the measure-relations of space can be determined; a problem which from the nature of the case is not fully determinate, since there may be several systems of matters of fact which suffice to determine the measure-relations of space—the most important system for our present purpose being that which Euclid has laid down as a foundation. (Riemann 1873, p. 14)

Riemann's aims here deserve some closer attention. They are, in his words, to determine the 'simplest matters of fact' from which we can discover the geometry of space. Such matters of fact might include (for example) the rules for congruence of triangles; the possibility of prolonging lines indefinitely; even the axiom of parallels. Equally, they might not, in which case one would have to include something else in their place. While not questioning that geometry was the study of space, he wished to examine what presuppositions we bring to that study and how far—by experiment, intuition, or whatever—we can justify them, and use them for deducing what space must be. It had three dimensions, that much was certain; that is the meaning of the forbidding phrase 'triply extended magnitudes'; and one had rules for measuring the lengths of curves within it. Guided by the analogy of Gauss's work on surfaces, Riemann thought of 'straight lines' in space as curves of shortest length and gave rules, at least in outline, for how such lines could be found. He also considered the question of what geometry space would need in order to satisfy one reasonable presupposition: that rigid bodies could be moved around in it without changing shape. (This is to speak in mechanical terms. A more geometrical view is that the ordinary rules for congruence hold.) The answer is that what Riemann called the 'curvature' of space would have to be constant from point to point; and that this is satisfied in three cases:

1. Euclidean geometry;
2. Geometry on a sphere, or something like it (which Riemann considered)—this is Saccheri's HOA;

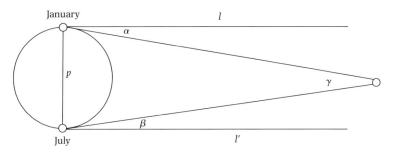

Fig. 10 The parallax of a star. p is the width of the Earth's orbit, and the lines l, l' are perpendicular to the diameter. In January (July) the line to the star makes an angle $\alpha(\beta)$ with the line in question. The parallax is the angle $\alpha + \beta$, which equals γ in Euclidean geometry, and so is very small for large distance. In non-Euclidean geometry it can never be smaller than $\pi/2 - \Pi(p)$.

3. Lobachevsky–Bolyai or 'hyperbolic' geometry. This was not considered by Riemann, but when his ideas came to be publicized, particularly by Helmholtz in the 1870s, it had become widely known and could be seen as another candidate.

Helmholtz wrote a number of articles setting out his view that alternative models for space should be considered (and tested). In particular, he wrote for the new English journals *Nature* and *Mind*; and an extract from one of his articles is included as Appendix C. By this time it had been established that hyperbolic geometry was free from contradiction (the model argument). However, this did not settle the question of whether it was worth considering, which hinged on whether space could conceivably have such a geometry. Lobachevsky had already considered the question of measurements to determine this, and Helmholtz clarified the point:

All systems of practical mensuration that have been used for the angles of large rectilinear triangles, especially all systems of astronomical measurement which make the parallax of the immeasurably distant fixed stars equal to zero (in pseudospherical space the parallax even of infinitely distant points would be positive), confirm empirically the axiom of parallels and show the measure of curvature of our space thus far to be indistinguishable from zero. It remains, however, a question, as Riemann observed, whether the result might not be different if we could use other than our limited base lines, the greatest of which is the major axis of the earth's orbit. (Helmholtz 1979, p. 258)

The 'parallax' of a star S (Fig. 10) is the angle $\alpha + \beta$ in the diagram, which in Euclidean geometry equals γ (and so is vanishingly small when the star's distance is much bigger than p, the diameter of the Earth's orbit). In non-Euclidean geometry, the smallest possible value of $\alpha + \beta$ is (roughly) $\pi/2 - \Pi(p)$. Interestingly, the fact that this is, for practical purposes, zero was used as an argument against the Copernican theory; if the Earth moved, it was argued, the stars would have a measurable parallax. By the nineteenth century it was accepted that the Earth did move, but the parallax was too small to measure.[7]

7 What revolution?

Let us not forget that no serious work toward constructing new axioms for Euclidean geometry had been done until the discovery of non-Euclidean geometry shocked mathematicians into reexamining the foundations of the former. We have the paradox of non-Euclidean geometry helping us to better understand Euclidean geometry! (Greenberg 1974, p. 57)

---

7. Sirius, the obvious candidate as it is both bright and close, has a parallax of 0.0377 seconds of arc.

I discovered [about 1890] that in addition to Euclidean geometry there were various non-Euclidean geometries, and that no one knew which was right. If mathematics was doubtful, how much more doubtful ethics must be! (Russell, cited in Richards 1988, pp. 204–5)

The term 'Whig history', which is only used by historians to denigrate the work of other historians, describes a way of constructing the narrative so that the process of history is seen as one of development towards the present; which itself is seen as a good state of affairs, if not the best. Mathematicians, who find it difficult enough to imagine that mathematics could be done in any way which is better than our present one, are particularly given to Whig history, and the quotation above (non-Euclidean geometries were good because they led to the construction of necessary axiom-systems) is a mild example. In its defence, it does at least provide (a) a structure for the bare record of events, and (b) a starting-off point for more sophisticated narratives, which can criticize it. In this section, we shall examine how far the story of the discovery and assimilation of non-Euclidean geometry, outlined above, can be made sense of as part of a wider development of geometry.

The natural endpoint of that development, as implied by Greenberg, is not the mere acceptance of non-Euclidean geometry, but the modernization of the subject as a whole. The problem is that the latter—the development of axiom-systems, and the increasing insistence that geometry was not about 'space' at all, but about *any* entities which might satisfy the axioms—occurred later still, during the years from 1890 to 1910. Peano (1889), was the first to produce an axiom-system which he declared to be 'free-standing', that is, independent of any possible meanings which one might give to terms like 'point', 'line', and so on.

We are given thus a category of objects called points. These objects are not defined. We consider a relation between three given points. This relation, noted $c \in ab$, is likewise undefined. The reader may understand by the sign 1 any category of objects whatsoever, and by $c \in ab$ any relation between three objects of that category[...] The axioms will be satisfied or not, depending on the meaning assigned to the undefined signs 1 and $c \in ab$. If a particular group of axioms is satisfied, all propositions deduced from them will be as well. (Peano 1889, quoted in Torelli 1978, p. 219)

More picturesquely, Hilbert, whose axiom system became the most influential put it:

If I conceive my points as any system of things, e.g. the system *love, law, chimney-sweep*, ... and I just assume all my axioms as relations between these things, my theorems, e.g. the theorem of Pythagoras, will also hold of these things. (Hilbert, cited in Torelli 1978, p. 251)

This *was* a radical change in how geometry was thought about. That the views of Hilbert and his followers were not generally accepted—and are not universally believed even today among research mathematicians—is less important than that they were voiced, and carried weight. They outlined a programme for a new view of geometry which (since Hilbert was not obscure, indeed he was the most respected of mathematicians) had to be taken seriously.

Non-Euclidean geometry was 'revolutionary' in its successful construction of a geometry which was not Euclidean, so much is obvious. How much in this subsequent development of geometry can be attributed to it is an altogether more problematical question. From a broader point of view, the axiomatization of geometry, while it clearly owed something to the problems which had arisen, should be seen as a part of the wider tendency to axiomatization in mathematics during the late nineteenth century. Peano, indeed, is more often remembered for his axioms for the natural numbers than for his geometric ones, and Hilbert and Russell similarly had a strong belief that the way to make mathematics 'safe' was through the construction of axiom systems.

In fact, in the whole period from the 'rediscovery' both of Lobachevsky–Bolyai and of Riemann in the late 1860s up to 1900, the main questions about geometry were not about 'foundations'. This is where Joan Richards (1988) provides a useful view of working mathematicians concerned not only with research but with what provided the best education for young men at Cambridge, what should be taught in schools, what was most uplifting, and many other questions which hardly seem now to be on the agenda.[8] Her restriction to England is not a serious one; although English mathematicians were certainly less research oriented and tended to be more conservative than their French, German, or Italian counterparts, they were in touch with the debates which were going on and contributed to them. Helmholtz's propaganda for non-Euclidean geometry, as we have seen, was published in England, and promoted by Clifford, while at the end of the century the generation of Russell and G. H. Hardy abruptly set out to force Cambridge mathematics into the continental mainstream.

Almost coincidentally, quite different events in physics separated geometry from empirical investigations of the world. What geometric form the universe might have was an interesting scientific question for Riemann, for Helmholtz, indeed until 1906. In Newton's theory it was a perfectly flat three-dimensional Euclidean space, in which one could (theoretically) determine the place and time of any event. The nineteenth-century revisions of geometry amounted to questioning the nature of the space component. Much more serious problems were raised, however, by Einstein's special theory of relativity which—by denying the idea of simultaneity— effectively killed off the concept of a unified three-dimensional space as a physical object of study. The geometers had been studying something which had no physical reality. The general theory of course reintroduced Riemannian geometry (which Einstein learned with considerable difficulty), but in a way so complex that questions about the shape of the universe were turned into questions about the nature of solutions to some difficult differential equations. Axioms, and the shape of triangles, in the Einsteinian universe were not the guide which they always had been.

In trying to present alternative versions of the simple story of 'Copernican revolution' with which this chapter opened, there is no need to belittle or downgrade the work of the founders of non-Euclidean geometry; the record speaks for itself. Rather, we hope that the reader may be encouraged to think about geometry itself, its changing nature at different times, and how far the work of Lobachevsky and Bolyai may be said to have influenced those changes.

Appendix A. Euclid's proposition I.16

In any triangle, if one of the sides is produced, then the exterior angle is greater than either of the interior and opposite angles. Let ABC be a triangle, and let one side of it BC be produced to D.

I say that the exterior angle ACD is greater than either of the interior and opposite angles CBA and BAC.

---

8. There was a typically 'Victorian' view that the certainty of geometry supported the certainty of theological arguments for the existence of God. It accordingly acquired a religious, even a political importance.

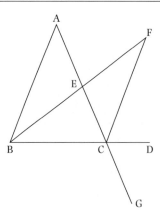

Fig. 11 The picture for Euclid I.16.

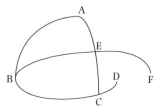

Fig. 12 A 'large' triangle on a sphere, and the way in which proposition I.16 fails.

Bisect AC at E. Join BE, and produce it in a straight line to F. Make EF equal to BE, join FC, and draw AC through to G (see Fig. 11)

Since AE equals EC, and BE equals EF, therefore the two sides AE and EB equal the two sides CE and EF, respectively, and the angle AEB equals the angle FEC, for they are vertical angles. Therefore, the base AB equals the base FC, the triangle ABE equals the triangle CFE, and the remaining angles, equal the remaining angles, respectively, namely those opposite the equal sides. Therefore, the angle BAE equals the angle ECF.

But the angle ECD is greater than the angle ECF, therefore the angle ACD is greater than the angle BAE. Similarly, if BC is bisected, then the angle BCG, that is, the angle ACD, can also be proved to be greater than the angle ABC. Therefore in any triangle, if one of the sides is produced, then the exterior angle is greater than either of the interior and opposite angles. Q.E.D.

This proposition *fails* in the 'geometry' of shortest lines on a sphere. In fact, it is not hard to construct triangles in which it is untrue; one such is shown in Fig. 12. Where does the proof fail? The point F constructed above no longer falls 'inside' the angle ACD, as seems obvious from the Euclidean picture; and so the angle ECD is no longer greater than the angle ECF. Hence, although the previous steps in the proof work, the crucial use of 'greater than' does not.

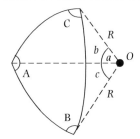

Fig. 13 Solving a spherical triangle.

Appendix B. The formulae of spherical and hyperbolic trigonometry

To state these, we need a convention, since the formulae will vary with the radius of the sphere. The ancient convention was to work with a sphere of radius 60, but let us simply call the radius $R = 1/K$, K being the 'curvature' of the sphere. So, the bigger K is, the more curved the sphere—the smaller its radius.

Now for any arc on S of length a, Ka is a number between 0 and 2π. And the two key formulae in 'solving' a spherical triangle ABC as in Fig. 13, already mentioned in connexion with al-Bīrūnī's work, are:

$$\frac{\sin(Ka)}{\sin A} = \frac{\sin(Kb)}{\sin B} = \frac{\sin(Kc)}{\sin C}$$

the analogue of the 'sine formula', and

$$\cos(A) = -\cos(B)\cos(C) + \sin(B)\sin(C)\cos(Ka)$$

one of two analogues of the 'cosine formula'. The first of these goes over into the ordinary sine formula when Ka, Kb, Kc are small (tend to zero).

The hyperbolic formulae are simply related to the spherical; one 'replaces' cos by cosh and sin by $i$sinh when dealing with lengths (but not with angles, since only the ordinary functions apply to angles). They are, then,

$$\frac{\sinh(Ka)}{\sin A} = \frac{\sinh(Kb)}{\sin B} = \frac{\sinh(Kc)}{\sin C}$$

and

$$\cos(A) = -\cos(B)\cos(C) + \sin(B)\sin(C)\cosh(Ka)$$

Exercise 6. *Prove the statement above, about the limit of the spherical sine formula for small values of Ka, Kb, Kc.*

Exercise 7. *Use the second formula in the case when the angle $A = 0$ to deduce the formula for $\Pi(p)$.*

Appendix C. From Helmholtz's 1876 paper

[Reproduced in (1979, pp. 249–50. Helmholtz, to simplify his argument, considers two-dimensional 'beings' constructing geometry from observation of the world in which they live.]

But intelligent beings... might also live on the surface of a sphere. Their shortest or straightest line between two points would then be an arc of the great circle passing through them. Every great circle passing through two points is divided by them into two parts. If the parts are unequal, the shorter is certainly the shortest line on the sphere between the two points, but the other, or larger, arc of the same great circle is also a geodesic, or straightest, line; that is, every smallest part of it is the shortest line between its ends. Thus the notion of the geodesic, or straightest, line is not quite identical with that of the shortest line...

Of parallel lines the sphere-dwellers would know nothing. They would declare that any two straightest lines, if sufficiently extended, must finally intersect not only in one but in two points. The sum of the angles of a triangle would be always greater than two right angles, increasing as the surface of the triangle grew greater. They could thus have no conception of geometric similarity between greater and smaller figures of the same kind, for with them a greater triangle must have greater angles than a smaller one. Their space would be unlimited, but would be found to be finite or at least represented as such.

It is clear, then, that such beings must set up a very different system of geometric axioms from that of the inhabitants of a plane or from ours, with our space of three dimensions, though the logical processes of all were the same; nor are more examples necessary to show that geometric axioms must vary according to the kind of space inhabited. But let us proceed still further.

Let us think of reasoning beings existing on the surface of an egg-shaped body. Shortest lines could be drawn between three points of such a surface and a triangle constructed. But if the attempt were made to construct congruent triangles at different points of the surface, it would be found that two triangles with three pairs of equal sides would not have equal angles. The sum of the angles of a triangle drawn at the sharper pole of the body would depart further from two right angles than if the body were drawn at the blunter pole or at the equator. Hence it appears that not even such a simple figure as a triangle could be moved on such a surface without change of form.

Solutions to exercises

1. (a) The 'Aristotle proof' (see Fauvel and Gray 3.B.4 (b) and (c)) is the nicest. Let ABC be a triangle; draw PAQ through A parallel to BC. (See Fig. 14.) Then ∠PAC + ∠ACB are two right angles; and ∠PAC + ∠CAQ are also two right angles (angles on a straight line). So ∠ACB = ∠CAQ. (This is just the alternate angle theorem.) Similarly, ∠ABC = ∠PAB. But the sum of the three angles ∠PAB, ∠BAC, ∠CAQ is two right angles (straight line again); so the same is true of the three angles in the triangle, which we have proved equal to them.

   (The key point is that the line PAQ has the 'alternate angles property' with respect to *both* the transversals AB,AC, which will only be true if the property is equivalent to being the unique parallel.)

   (b) (See Fig. 15.) This is best thought of as a result about so-called 'Saccheri quadrilaterals', see later. To stay strictly in the framework of what Euclid would have done (one imagines): the

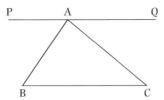

Fig. 14 Proof of the 'angles of a triangle' theorem.

Fig. 15 Figure for Exercise 1(b).

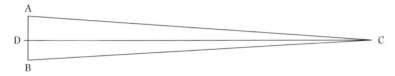

Fig. 16 The figure for Exercise 2; AB is 1 cm.

point is that CD must be parallel to AB. In fact, if the angles ∠BAC, ∠ACD are greater than two right angles, then CD must meet AB on the left, say at E, and by symmetry it must also meet on the right at F. But then there are two straight lines joining E,F.

Accordingly, CD does not meet AB, so ∠BAC, ∠ACD are together equal to two right angles; since ∠BAC is a right angle, this means that ∠ACD is one, and similarly for the fourth angle. Once we have this, it is easy to use congruent triangles to show that AB = CD.

2. See Fig. 16. We have: distance to meeting point is length of CD, or $0.5 \tan \alpha = 0.5/\tan(10^{-10})$ For such a small angle, we can approximate $\tan x$ by x, arriving at $CD = 0.5 \times 10^{10}$ as stated.

3. (a) (See Fig. 8) What is meant is that the angle G is a (decreasing) function of the length AB, and so to any angle G there corresponds a unique length AB of the side of the quadrilateral. As a result, you can define a measure of length not by arbitrary choice, as is done in Euclidean geometry, for example, by giving a fixed value to the length of a metre or 'Paris foot'; but by stating that your unit will be that length AB which corresponds, say, to 80° for the angle G.
(b) This follows from the angle-sum theorem. In fact, the area of ABGD is twice the area of triangle ABG. By the angle-sum theorem, this is $2C(\pi - \angle GAB - \angle ABG - \angle BGA) = 2C(\pi - \pi/4 - \pi/2 - \alpha/2) = C(\pi/2 - \alpha)$, where C is the relevant constant for the geometry and α is the angle at G. It follows that the angle G is a decreasing function of the area ABGD, and so of AB.

4. Let AD be perpendicular to BC, and A'D' to B'C', and suppose that AD = A'D'. Now let AF be a cutting line from A to BC, and construct F' on B'C' so that D'F' = DF. Then it is easy to see that ADF, A'D'F' are *congruent* triangles (equal sides and included right angle), and so

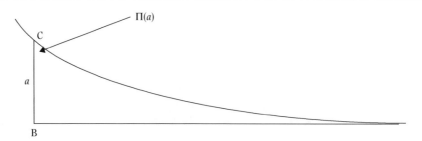

Fig. 17 Figure for Exercise 7.

∠D'A'F' = ∠DAF. (This is where it is essential that the rules for congruent triangles work, but Lobachevsky has assumed they do.) Hence angles at A' define cutting lines on B'C' if the same angles at A define cutting lines on BC—and obviously vice versa. This is enough to conclude that the angles of parallelism are the same for the two line-segments.

5. If $t = \tan(x/2)$,
$$\frac{2t}{1+t^2} = \frac{2\tan(x/2)}{\sec^2(x/2)} = 2\sin(x/2)\cos(x/2)$$
(writing tan as sin/cos, and sec as 1/cos), which is $\sin x$. Hence, the second formula is equivalent to:
$$\sin(\Pi(p)) = \frac{2e^{(-Kp)}}{1+e^{-2Kp}} = \left(\frac{1}{2}(e^{Kp} + e^{-Kp})\right)^{-1} = \text{sech}(Kp)$$
As p increases from 0 to ∞, $\cosh(p/K)$ increases from 1 to ∞, and its inverse sech decreases from 1 to 0. Since by definition $\Pi(p)$ is between 0 and $\pi/2$, the first equation implies that $\Pi(p)$ decreases from $\pi/2$ ($p=0$) to 0 ($p \to \infty$).

6. As $x \to 0$, $\sin x/x \to 1$. Hence, to first order in the lengths a,b,c, the 'sine formula' can be replaced by $a/\sin A = b/\sin B = c/\sin C$, the usual formula.

7. If $A = 0$, we are in the situation where the sides b,c are parallel in Lobachevky's sense. We consider a triangle (cf. Fig. 17) with $B = \pi/2$ and $C = \Pi(a)$. The formula gives $1 = 0 + \sin(\Pi(a))\cosh(Ka)$, using $\cos(\pi/2) = 0$, $\sin(\pi/2) = 1$. This can be transformed into the first version of the formula for $\Pi(a)$.

9 Modernity and its anxieties

1 Introduction

If in summing up a brief phrase is called for that characterizes the life center of mathematics, one might well say: mathematics is *the science of the infinite*. (Weyl 1949, p. 66)

Pure mathematics is the class of all propositions of the form 'p implies q', where p and q are propositions each containing at least one or more variables, the same in the two propositions, and neither p nor q contains any constants except logical constants. (Russell 1903, p. 3)

The 'long twentieth century', which should end our narrative, has seen more mathematics, as well as more changes to what mathematics means to those who do it or those who use it than the whole of preceding history. Those who had tried to define what mathematics was in its long past had certainly not come up with answers as extremist, or as 'unmathematical' in appearance as either Hermann Weyl or Bertrand Russell; and yet both answers now seem to belong to a bygone era. Even Alan Turing's paper on computable numbers, which more than ever stands as a founding document for 'where we are now' is hard for the modern reader to construe; not only because Turing was writing in a difficult field, but because the problems he was addressing belong to a time which, a mere 70 years later, has long disappeared. As a minimal strategy in managing the material, we have had to divide it in two, taking Gödel's 1931 paper as a useful cut-off point. This chapter, then, will deal with the central concerns which led up to the crisis of the years from 1900 to 1930; who was affected and how they dealt with it; and how, in some sense, it ended. At the same time, it seems essential to remember that the crisis was the concern only of a few mathematicians, although those were among the most important ones. Hence, in the interests of balance—and also because foundational questions provide painfully few opportunities for pictures—we shall consider parallel developments in algebra and topology, particularly knot theory. These are also part of the story, in that if there is a 'twentieth-century outlook' characterized by increasing abstraction and formalism, it can be seen spreading even to such apparently down-to-earth subjects as the classification of knots. Naturally, a very large part of the field has still been omitted, most particularly all that has to do with physics. There may be some compensation in the next chapter, but the reader must remember the arbitrariness of our selection.

As we shall see, the natural beginning of the story precedes the twentieth century by some 30 years. The world of mathematicians by that time was substantially professionalized around great institutions of teaching and research in Germany and France and lesser ones in many other countries. While this state of affairs remained constant, it should be borne in mind at each stage (a) that the number of people so employed was tiny in comparison with today and (b) that it was more or less constantly growing in response to the demands of society—not for mathematicians (who needs them?) but for engineers, accountants, statisticians, and the like. That this growing community chose to concern themselves chiefly with the definition of the numbers, or with how to tell two knots apart, was their business; one can speculate on how this related to what was going on in the economic sphere, and we shall try to raise some questions. However, their main function as employees of the state (which they usually were) was to teach, and to uphold the prestige of their institutions.

2 Literature

The mathematics of the early twentieth century has been patchily studied. Because the crisis of foundations provides such rich material for the historian, it is easily the best covered, with very full sourcebooks edited by van Heijenoort[1] (1967) and Mancosu (1998), as well as chapters in Grattan-Guinness (1980). Corry's recent work on the origins of abstraction (2004), if rather dry, is useful on ground which in the main will not be covered here. Added to these, and to a growing volume of articles in journals, the period has naturally been an attraction to biographers who have often had access to their subjects or to close friends; one could cite Cantor (Dauben 1990), Russell (Russell 1967; Monk 1997, 2001), Hilbert (Reid 1970), Brouwer (van Dalen 1999), Weyl (Wells 1988), Ramanujan (Kanigel 1991), Noether (Dick 1981), and so on. They are not properly 'histories', but they can be excellent sources. All this, as well as the texts themselves—which, it must be said, are usually extremely difficult as should be expected of mathematics today—are useful material. Two interesting early twentieth-century works of fiction have 'mathematicians' as their heroes—Musil's (1953) and Ford Madox Ford's (2002); however, the fact is fairly marginal to the lengthy unfolding of the two novels.

Because of their difficulty, the remarks made in Chapter 8 apply even more here. Almost no twentieth-century mathematical *discoveries* find their way into an undergraduate course, although any course on linear algebra, or group theory, or analysis will be taught in a way that was only settled around 1950. We shall face constant problems of presentation, and must hope for the reader's patience. Modern mathematics does not easily lend itself to being democratized; Hilbert (1900), introducing his very difficult list of problems for the next century, attributed to an unnamed 'old French mathematician' the saying: 'A mathematical theory is not to be considered complete until you have made it so clear that you can explain it to the first man whom you meet on the street', but little of Hilbert's own work, or of what has been done since, stands up to the test.

3 New objects in mathematics

Es steht schon bei Dedekind [That's already in Dedekind]. (Emmy Noether (frequently), quoted in Dick 1981, p. 68)

As a clutch of Victorian professors, avuncular, ascetic and a little disheveled, they [Dedekind and Cantor] were gathering unawares around the cradle of an infant Briar Rose that would one day be christened Modernism. (Everdell 1997, p. 31)

To arrive at real proofs of theorems (as e.g. $\sqrt{2}\sqrt{3} = \sqrt{6}$), which to the best of my knowledge have never been established before. (Dedekind 1948, p. 22)

---

1. It is usual to point out that Jean van Heijenoort was Trotsky's secretary during the 1930s, only later becoming a distinguished historian of mathematical logic. And it is indeed an interesting footnote.

Mathematicians can only feel flattered by William Everdell's breathless placing of them as the forerunners of Rimbaud, Freud, Joyce, Picasso et al., even if Dedekind's statement of what he did seems something of a let-down. He was aware of this himself:

[T]he majority of my readers will probably be disappointed in learning that by this commonplace remark the secret of continuity is to be revealed. (Dedekind 1872, in Fauvel and Gray 18.C.1, p. 575)

Nonetheless, the work of Richard Dedekind and his more adventurous friend Georg Cantor on numbers, the continuous and the infinite, did lead to a reshaping of mathematics if not the whole world-view. Indeed, after a relatively short time it brought about the 'crisis of foundations', which began some time around 1903, became acute in the 1920s, and was in some sense killed off, if in no way settled, by the work of Kurt Gödel in 1931. The problems which arose were about sets; and a first reasonable question is, how did mathematics, which as long as we have known it has been about numbers and geometry, come to concern itself with sets? It has to be understood that now even more than before, the world of mathematics was becoming fragmented, and these concerns were not those of the average university teacher, let alone the engineer or statistician. We are concerned for the moment with a relatively small research élite working mainly in France and Germany, and the crisis as it developed came out of their attempts to make some sense of the calculus which, as we have seen (Chapter 7) made very little sense as theory although it worked well in practice.

Dedekind's statement on what he could prove stands as an important pointer. To discuss what problems his definition was meant to solve would take us too far back but the fundamental idea was, in the words of one commentator:

to find definitions from which the basic theorems on limits could be proved. (R. Bunn, in Grattan-Guinness (ed.) 1980, p. 222)

Briefly, you needed limits to define both derivatives and integrals properly; and hence to deal with the problems of the calculus, and with numerous other problems, notably the behaviour of Fourier series, which had arisen since. Dedekind's definition of real numbers, as the necessary foundation for the calculus of limits, is reproduced in Appendix A. It is more popular and easier to understand than Cantor's—though still not much taught in calculus courses—and as such it is indeed a founding document of modernism in mathematics, if nowhere else. Given the set R of all rational numbers (i.e. fractions—$\frac{1}{3}$, $\frac{7}{5}$, and so on), which for the time being we consider unproblematic, Dedekind considers the problem of characterizing, say, $\sqrt{2}$. This is not rational—there is a 'hole' in the rational numbers where the square root of 2 should be. The idea is to *define* the real number to be the hole. Less mystically, we consider the 'cut' defined by the two sets:

$$L = \{x \in R : x < 0 \text{ or } x^2 < 2\}; \quad U = \{x \in R : x > 0 \text{ and } x^2 > 2\}$$

(See Fig. 1.) Everything in L is less than $\sqrt{2}$, everything in U is greater; $\sqrt{2}$ is the missing point between.

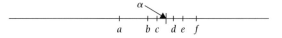

Fig. 1 Dedekind cut. The number α divides the left-hand class L (which contains rational numbers a, b, c) from the right-hand class (which contains d, e, f).

In Dedekind's definition, one takes $\sqrt{2}$ to be the cut. This might be considered slightly vague too, and later writers who subscribed to Cantor's set theory (such as Russell) defined it to be the lower set. Which you do is a detail. The two important points are:

1. Once the definition has been made, it is easy to do arithmetic (adding, multiplying, even taking roots etc.) with such numbers.
2. The further operation of taking limits (e.g. of an increasing sequence $x_1, x_2, \ldots$ which is bounded above) is *equally* easy.
3. On the other hand, however you look at it, you have 'defined' a number to be something which *is not* a number. For thousands of years, mathematics has been about numbers and geometrical figures. It now, suddenly, is about something else. Has it then changed?

Underlying all this were ideas which were to come much later: that the objects of mathematics were not actual things-in-themselves (as one thinks of a triangle, say, or the number '7'), but the rules which they obeyed. Any way of constructing objects which obeyed the rules endowed them with existence, and two different ways of construction, if the results obeyed the same rules, could be thought of as the same—we would now in suitable circumstances use the word 'isomorphic'. We have already seen this in Chapter 8, where the non-Euclidean plane was constructed as a surface with new rules about what constituted 'straight line' and 'angle'.

Even today, such metaphysics are considered beyond the scope of the high-school student or (often) the first-year college student. At the time they were new, just beginning to be explored, and only a strong feeling—backed up by examples—that intuitive ideas of number were not reliable enough drove the process forwards. Nothing makes clearer the fundamental change underlying the new outlook than the fact that it seemed immediately necessary to go back further and set the natural numbers $\{0, 1, 2, 3, \ldots\}$ on a secure foundation, although they had previously troubled no one. Gottlob Frege was in 1884 to define these as sets too. '3', for example, meant the set of all sets which could be put in a 1-1 correspondence with a (previously defined) set, say S_3, which had three elements—so that '3' meant 'the set of all sets with three elements'. [It is Frege's way of defining S_3 which stops the definition from being circular.]

Bit by bit, among these mathematicians—mainly German, but to include Peano (Italian), Russell (British), ...—more and more things which had seemed obvious were to need proof; when part of the edifice seemed sound, one started to worry about its underpinnings, so that by the 1920s we find Hilbert, probably the ablest mathematician of the time, taking time out to show how one could prove that $1 + 2 = 2 + 1$.[2] The drive for sound foundations was a strong one, and on the whole fruitful; it is interesting that it is an episode which can be considered closed, in that mathematicians have returned to a naive condition of assuming that what they do works (although the procedures of physicists may still worry them). The process of investigation, however, brought the worlds of mathematics and philosophy into a much closer relationship.

The relationship was by no means a new one; almost all philosophers since Plato had reflected on mathematics, and many mathematicians (Descartes, Pascal, Leibniz, Bolzano) were philosophers as well. But the dependence on set theory and logic introduced a new outlook into both mathematics

---

2. Hilbert, 'The New Grounding of Mathematics', (1922) in Mancosu (1998, p. 207). The implication of triviality is of course unfair; Hilbert was showing how a formal minimal axiom system for arithmetic could be used to establish all necessary results. All the same, the image is a striking one.

and philosophy, and in mathematics it meant a return to doubt, to the search for what was wrong, what Morris Kline many years later was to call the 'loss of certainty' [Kline 1980].

Exercise 1. *Check that the relation $\sqrt{2}.\sqrt{3} = \sqrt{6}$ does indeed follow from the cut definition by showing*

(a) *that there is a unique positive real number $\sqrt{n}$ whose square is n, for any integer $n > 0$;*
(b) *that if we define a.b for positive real numbers in the obvious way—you may need to take a little care in doing this—$\sqrt{2}.\sqrt{3} = \sqrt{6}$;*
(c) *(Not as difficult as it looks...) Let $x_1, x_2, x_3, \ldots$ be a sequence of real numbers (defined by cuts) which is bounded above, that is, there exists M such that $x_i < M$ for all i. Prove that there exists M_0 (least upper bound) such that:*
   (1) $x_i \leq M_0$ *for all i;*
   (2) *if $y < M_0$, then for some i, $x_i > y$.*
(d) *Define a real number to be an infinite decimal, that is, a series of type $x = a + .a_1 a_2 a_3 \ldots$, where a is an integer and the a_is are numbers between 0 and 9. In other words, x is the sum of the series*

$$a + \frac{a_1}{10} + \frac{a_2}{10^2} + \frac{a_3}{10^3} + \cdots$$

What problems arise in devising a rule for adding such numbers?

4 Crisis—what crisis?

It seemed unworthy of a grown man to spend his time on such trivialities, but what was I to do? There was something wrong, since such contradictions were unavoidable on ordinary premises. (Russell 1967, p. 147)

In [this] light, mathematics appears as a monstrous 'paper economy'. Real value, comparable to foods in economics, is only possessed by the singular, the quintessentially singular. Everything general, and all existential statements partake in it only indirectly. And yet we, as mathematicians, very seldom consider the redemption of this 'paper money'! It is not the existential theorem that is the treasure, but the construction carried out in the proof. (Weyl, 'On the New Foundational Crisis in Mathematics' (1921), in Mancosu 1998, p. 98)

The behaviour of many leading mathematicians in the years 1900–1930 is so uncharacteristic that the reader may feel more in tune with the mathematical aims of the Chinese than with those of Hilbert, Brouwer, Russell, and their contemporaries. Why was the mathematical enterprise suddenly seen as so insecure? How did it come about that mathematicians were bitterly divided into competing schools of thought, who went so far as to call each other 'Bolshevist' (Hilbert against Brouwer and Weyl) or 'non-Aryan' (Brouwer against his opponents); and to fight about presence at conferences and editorship of journals? The year 1900 saw Hilbert's calm summing-up of the progress of mathematics, and his famous list of problems awaiting solution in the new century. His mood was optimistic:

This conviction of the solvability of every mathematical problem is a powerful incentive to the worker. We hear within us the perpetual call: There is the problem. Seek its solution. You can find it by pure reason, for in mathematics there is no *ignorabimus*. (Hilbert 1900)

Still, it should be noted that the first two problems deal with foundations: Cantor's 'continuum hypothesis' and the consistency of the axioms for arithmetic (however defined). Already, it appeared, work needed to be done on the axioms. There were some disagreements about what should be done, but Cantor's set theory underpinning Frege and Dedekind's arithmetic seemed to provide a good programme.

Nonetheless, as Gray points out in a recent article (2004), there were signs of what could be called 'anxiety' about foundations. They could be traced back much earlier; in 1810 Bolzano had said:

However the greatest experts of this science [i.e. mathematics] have in fact always answered, not only, that the edifice of their science is still no completely finished and self contained building; but also, that the foundation wall of this otherwise splendid building is itself not yet completely firm and regular; or, to speak without pictures, that even in the elementary lessons of all mathematical disciplines many gaps and imperfections are to be found. (Bolzano, (1810))

However, it was only at the end of the century that a general optimism about the power of mathematics began to give way to doubt—and indeed it would be hard otherwise to account for the amount of work that was by then under way in an attempt to shore up the building. Any familiarity with critical work on the culture of the period will show that anxiety and modernism go together like a horse and carriage. How far mathematics caught a general infection, and how far it contributed to it are questions yet to be settled.

The crisis came naturally: it appeared that set theory as (following Cantor) it had been freely used (a) was inconsistent and (b) demanded extra articles of belief which were hard to accept. Inconsistency followed from the so-called 'Russell paradox' of 1903.[3] The problem—if you have not seen it before—is the following. Dedekind and Cantor had introduced sets to deal with numbers. However, if real numbers 'were' sets, one also wanted to deal with sets of sets, and so on (perhaps) indefinitely. In Cantor's general set theory, which he imagined worked, given any property P one had a set of all things with property P (the 'axiom of comprehension').

The problem was that, once the language became this general, the subject became the province of philosophers, who may choose to replace numbers by philosophical objects such as the golden mountain or black swans. The way was wide open for Russell, standing between mathematics and philosophy, to devise the simple example of the property:

P: x is a set and x is not an element of itself ($x \notin x$).

For example, the set of all sets is an element of itself (this already shows signs of an infinite regress). So is the set of all things which are non-human; or the set of all things which can be described in English sentences of less than 18 words. On the other hand, most sets (numbers, black swans, students in the classroom) are not elements of themselves. If S is the set determined by the property P, then one can derive a contradiction both from $S \in S$ and from $S \notin S$. Our first quote shows Russell himself, disturbed at the paradox, feeling both that grown men should not worry about such things (but then who should?) and yet that they had to be settled.

Various attempts were made to be more restrictive about how sets were used; they had become too much a part of how the leading mathematicians thought to be given up altogether. Zermelo around the same time tried to produce a system of axioms for set theory which would both avoid paradoxes and do what mathematicians needed. He came up with what was in a way as shocking as Russell's paradox: the 'Axiom of Choice' (1904). This could be thought of as a modern analogue

---

3. As so often, there is a priority question here; Zermelo had already described the paradox in a letter to Husserl.

to Euclid's postulate 5: no one liked it, but it became clear (and Zermelo pointed out) that a great many people[4] had, without acknowledgement, been using it. The usual statement is:

Given a set of sets $\{X_\alpha\}$ indexed by $\alpha \in A$, there exists a function f on the indexing set A such that for each α, $f(\alpha) \in X_\alpha$.

In other words, given *any* collection of sets, you can pick out—think of it as 'electing'—one representative from each of them. This is easy to agree if the set A is finite, although it might be far from practical if (say) it contained 10^{25} elements. It is when it is infinite that it begins to look dubious.

At this point, the problems probably disturbed some mathematicians intensely, but they did not seriously divide them. Again, Gray finds a point in O. Perron's 1911 inaugural lecture to demonstrate the existence of argument, indeed doubt about the adequacy of procedures:

Indeed, there is one branch of mathematics today over which opinion is divided, and some consider right what others reject. This is the so-called set theory, in which the certainty of mathematical deduction seems to be becoming completely lost. (Perron, cited in Gray 2004, p. 41)

It was the attack initiated by L. E. J. Brouwer on a much more fundamental principle, the 'Law of the Excluded Middle', which created a situation in which mathematicians became intemperate and, for a short period, made the world of mathematics more exciting than it had been since the time of Newton and Leibniz. This is not surprising, because the Law of the Excluded Middle underpins the kind of mathematics which derives from the Greeks. From the simple, and apparently harmless statement:

Either P is true, or P is false.

applied to a proposition P, the Greeks derived their peculiar method of 'proof by contradiction', which is still such a favourite. To prove that P is true, you suppose that it is not. By a chain of deduction, you derive a contradiction ('Which is absurd'). Therefore the assumption that P is not true must have been wrong, and hence it must be true.

This principle was used constantly by Euclid: to take a random example, as early as book I proposition 7 on isosceles triangles ('If in a triangle two angles equal one another, then the sides opposite the equal angles also equal one another'). You suppose one of the two sides greater, and derive an absurd conclusion.

Many students, to be sure, feel uncomfortable about this kind of proof, but they learn to consider it acceptable. What was considered doubtful, even wrong, what Weyl (under the shadow of the approaching German hyper-inflation) described as 'paper money' was the use of the law in existence proofs applied to infinite sets. Ironically, one of the neatest examples of such a proof is due to Brouwer himself, and is still an essential element in beginning topology courses. This is the *Brouwer Fixed Point Theorem*, which asserts:

Let D be the disk $\{(x,y) : x^2 + y^2 \leq 1\}$, and let f be a continuous mapping from D to D (i.e. for $(x,y) \in D$, $f(x,y)$ is also in D, and depends continuously on (x,y)). Then there exists a *fixed point*: for some (x_0, y_0), $f(x_0, y_0) = (x_0, y_0)$.

The proof of this proceeds by supposing that there is no fixed point (Fig. 2); we join each $f(A)$ to A, and continue to the boundary circle C, which it hits at $g(A)$. The mapping g, which fixes C, is shown to be 'impossible' by methods which had been developed a little earlier (see 'topology' below).

4. Particularly in France, where opposition was strongest.

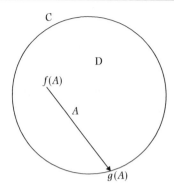

Fig. 2 Brouwer fixed point theorem.

We have therefore a contradiction from supposing that there are no fixed points. Are we allowed to deduce that there is one? Brouwer the prover of the theorem would have said yes, but Brouwer the philosopher was emphatically against the idea. The theorem (see Weyl above) gives you no idea of where the fixed point is.

Unlike the axiom of choice, the law of the excluded middle is venerable and would be hard to give up. Yet it was simple for the intuitionists to find examples where it seemed not to work, often relating to conjectures which were unsolved such as Fermat's Last Theorem; an example from Weyl is given in Appendix B(i). It is interesting since the conjecture which it draws on (whether any numbers of form $2^{2^n} + 1$ are prime for $n > 4$) is still unsettled, and is the object of searches taking years of computer time and superspeedy search programs. Hence one might say there is still a question of belief for mathematicians, at least for those who are interested: is it the case that *either* the statement 'no such numbers are prime' *or* 'one of the numbers is prime' is true? An intuitionist would say that you cannot assert such an either/or, since implicitly it means that you can search through all the integers and decide the question.

It is important to realize how radical the intuitionists were in their aims. Where the work of making the calculus rigorous left all the *results* in place, only changing the proofs so that they made sense without the infinitely small, Brouwer and Weyl were forced into a situation where they declared large parts of contemporary mathematics to be unacceptable and meaningless. Brouwer's concise formulation of his programme, to which he gave the name 'intuitionist' some time after 1907, is in Appendix B(ii). Note that he bans two things: the free use of sets and the law of the excluded middle.

The problem for the intuitionists, as time went on, was that their destructive programme was much easier to understand than their various attempts to be constructive. Weyl agreed that the 'honest' real numbers were the ones which one could calculate by a well-defined procedure; examples include $\sqrt{2}, \pi, \cos(0.5)$, or even $\Sigma_{i=0}^{\infty}(1/i^s)$ for s rational; roughly what Turing not long after was to call 'computable numbers'. Yet he also wanted to keep, in his words, 'the continuous 'spatial soup' that is poured between these [Euclidian] points' (In Mancosu 1998, p. 132). Bogged down in an attempt to preserve something of the 'continuum', even the infinite, intuitionist mathematics, which was seen as the way of the future by many in the 1920s, became a difficult and rather arcane specialist study by the 1930s—and survives as such.

There is a tempting analogy with the famous study of 1920s German physics by Paul Forman (1971). Forman argued that physics, in the Weimar republic, had to adapt to a milieu in which precision and abstraction were seen as impoverished forms of thought, associated with Germany's defeat; and that the stress on uncertainty and subjectivity in the new physics of Heisenberg and Schrödinger deflected such criticisms. Intuitionism—which had in fact started some time before—perhaps owed some of its popularity to a similar reaction.

Exercise 2.
(a) *A shop contains an infinite number of pairs of socks $S_1, S_2, \ldots$. I want to choose one sock from each pair.*
   (1) Why will I need the Axiom of Choice to do it?
   (2) Why will I not need the axiom if I am dealing with pairs of shoes?
(b) *Show using the Law of the Excluded Middle that there exist numbers x, y such that x, y are irrational but x^y is rational. [Hint: Start with $x = \sqrt{2}$.] What would be an intuitionist view of this argument?*
(c) *(Bolzano–Weierstrass theorem—hard!). Let $x_1, x_2, \ldots$ be an infinite sequence of numbers in the unit interval [0, 1]. (1) Show that there is a subsequence $x_{i_1}, x_{i_2}, \ldots$ which tends to a limit x. (2) What, from the intuitionist point of view, has gone wrong here?*[5]

5 Hilbert

I remember how enthralled I was by the first mathematics class I ever attended [at the University]...It was Hilbert's famous course on the transcendence of e and π. (Weyl, quoted in Reid 1970, p. 201)

In mathematics...we find two tendencies present. On the one hand, the tendency towards abstraction seeks to crystallise the logical relations inherent in the maze of materials...being studied, and to correlate the material in a systematic and orderly manner. On the other hand, the tendency towards intuitive understanding fosters a more immediate grasp of the objects one studies, a live rapport with them, so to speak, which stresses the concrete meaning of their relations. (Hilbert 1999)

We have delayed mentioning David Hilbert perhaps longer than we should, because although a central figure in the crisis, he was much more. His broad achievements and immense influence have made him something of a folk-hero, at least among mathematicians, although we are unlikely to see a film of his relatively uneventful life. For 30 years he dominated mathematics at Göttingen, and made Göttingen the centre of the world; and one would have to go a long way today to find a teacher who could transfix students on the subject of the transcendence of e and π,[6] if that particularly late nineteenth-century subject is still taught. He has been well served by Constance Reid's biography (1970), with an excellent mathematical section by his favourite student Weyl. Genial, productive, liberal, he remoulded the style of mathematics in algebra (particularly), and in the foundational disputes which were to be such a central preoccupation, where he stood in direct opposition to the intuitionists—and here Weyl's defection was to be a cause of distress, if not permanently so.

Rather than a film, Hilbert could make a good subject for a Greek tragedy, of downfall resulting from an excess of ambition. We have already seen his announcement in 1900 of his belief that any problem could be solved, to which Brouwer took such exception. The attacks of the intuitionists and the notable weaknesses in set theory forced him into constructing an ingenious position; but

---
5. This exercise and the preceding one have been borrowed, with their solutions, from Assaf J. Kfoury at BU.
6. That is, that neither of them satisfies an algebraic equation $a_n x^n + a_{n-1} x^{n-1} + \cdots + a_0$ where $a_n, \ldots, a_0$ are integers.

one which with hindsight is more ambitious than anyone before or since has thought necessary or possible. The proposal was as follows. One would define mathematics to be (a) a set of formulae constructed on simple lines and (b) a finite set of axioms which characterized some of those formulae as true; plus standard rules for deduction. If you take the axioms to be those which define basic properties of natural numbers (and Hilbert did, as did Russell and Whitehead in *Principia Mathematica* (1910–1913)), then you can of course deduce arithmetic of a simple kind; and if you add some infinite processes, you can define real numbers à la Dedekind and deduce calculus and geometry. To a certain extent, you *do not care* what the axioms, or formulae refer to. The question is not about what numbers are, but about how they behave; and this is why Hilbert was (by others) called a 'Formalist'. 'In the beginning was the sign', was a typical quote, which has been interpreted in any number of ways. (Part of) his statement of aims, which at the time he seems to have believed completed, is reproduced in Appendix C.

You now, in Hilbert's programme, shift to a different register called 'metamathematics'. (One could make a parallel with other twentieth-century studies where the activity, once pursued for its own sake, becomes the object of study.) Looking at the process of forming provable formulae, you ask two questions:

1. *Completeness.* Is it possible, given any formula P, to determine that it is true or false? An obvious example is the formula:

There exist natural numbers x, y, z and $p > 2$ such that $x^p + y^p = z^p$

necessarily capable of being proved true or false? (True, we know the answer now, but Hilbert did not.) If so, the system is called complete.

2. *Consistency.* Is it impossible, in the system, to deduce both P and 'not P'? If so, the system is called 'consistent'.

We have stressed the immense ambition of this programme, but the field was completely new, and Hilbert had an often justified confidence in the rapid progress of mathematics. Around 1929, his student von Neumann seemed to have a proof, which only needed a little patching to make it work.

The next part of the story is well known. In 1931 a young Austrian, Kurt Gödel, announced a proof that neither completeness nor consistency were provable; more strictly, that they could not be established by the finite methods which (under pressure from the intuitionists) were seen as necessary. It was a perfect piece of Hilbert-type mathematics—and Gödel was much more a formalist than an intuitionist—but it effectively destroyed the programme.

When Hilbert first learned about Gödel's work from Bernays, he was 'somewhat angry'... The boundless confidence in the power of human thought which had led him inexorably to this last great work of his career now made it almost impossible for him to accept Gödel's result emotionally. (Reid 1970, p. 98)

Retired, and with his last attempt to make mathematics secure defeated at least provisionally, Hilbert had to watch in disbelief as his best colleagues and students—Courant, Landau, Noether, Weyl, and von Neumann, were either forced out of Göttingen by the Nazis or left because they were unable to endure life under the Third Reich. He survived a few more years in Göttingen among the ruins:

Sitting next to the Nazis' newly appointed minister of education at a banquet, he was asked, 'And how is mathematics in Göttingen now that it has been freed from the Jewish influence?'
'Mathematics in Göttingen?' Hilbert replied. 'There is really none any more'. (Reid 1970, p. 205)

6 Topology

In an attempt to show some of what went on outside the world of foundational disputes, we consider the rise of topology. This, it has to be said without any personal *parti pris* is *the* success story of twentieth-century mathematics, barely existing at the beginning of the century and intruding into all other fields by the end. While there are obviously multiple reasons for this, we could give two: first, that any problem which requires the passage from a simple local statement to a more difficult global one (what can electromagnetic fields be like in the presence of currents? what shape can the space-time of Einstein's relativity have?) is a topological question; and second, that the machinery was in place, or could be developed to solve such problems. A great many problems in topology are hard, but not as hard as the continuum hypothesis, or Langlands's conjecture on automorphic forms; and a great many of the ablest mathematicians have devoted their time to them. It can therefore be seen as a domain of Hilbertian optimism, in which questions are successively raised and dealt with.

The time has now arrived when topologists consider their subject, however young, has a history; rightly so, and luckily there are two substantial contenders, Dieudonné (1989) and James (1999). These provide the groundwork which more professional historians are now sifting over and commenting. It is normal to consider the subject, properly considered, as just over a hundred years old. Aside from two famous excursions by Euler, it was originated and given its name by Möbius and Listing, classifying surfaces in the mid-nineteenth century; but serious methods of study became available at the end of the century with Poincaré, who extended the field to higher dimensions and gave it its first major theorem ('Poincaré duality') and its most enduring problem (the 'Poincaré conjecture').

At this point, the methods of argument were not such as would have been recognized elsewhere, for example, in algebra; and topology was perhaps fortunate in having as its originator Poincaré, who was more interested in finding results than in defining exactly what he was talking about. His object of study was:

(1) what Riemann would have called (and we now call) 'manifolds'—curves, surfaces,... up to any number of dimensions (think of the unit sphere in $\mathbf{R}^{n+1}$)[7]
(2) under the relation of 'homeomorphism' (same shape); continuity is preserved, but not distance.

To give standard examples: (a) manifolds of different dimensions are not homeomorphic (e.g. the circle C and the torus T in Fig. 3); (b) nor are the sphere S and T; (c) but T and the knotted torus T′ are (Fig. 4). Poincaré's method was to decompose a manifold into 'cells', rather like the faces of a polyhedron; and to derive numbers from the cell decomposition.

---

7. This field of acceptable objects was later to be substantially extended; no longer manifolds, no longer finite dimensional,...

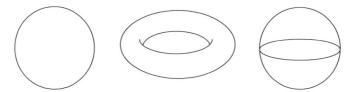

Fig. 3 Circle, torus, and sphere.

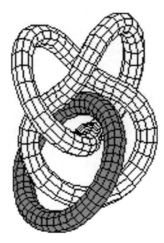

Fig. 4 A torus T and a knotted torus T' (which, incidentally are linked), are homeomorphic.

Again, this was very unlike Göttingen mathematics. It looked forwards (nothing like it had been done before) and backwards (it went against the growing current of abstraction and the need for certainty). How was one to be sure that two different cell decompositions gave the same numbers? It took some time for a satisfactory proof to emerge, but in the meanwhile the handful who were interested in topology were happy to make a start with Poincaré's ideas and methods, and his 'invariants'. And, interestingly and importantly, his main followers came not from France, but from America (Veblen, and then Alexander) and Germany (Dehn, Heegaard, and then Reidemeister). Under their influence, and that of the Russian Alexandrov, a close associate of Noether, topology in the 1920s and 1930s became 'algebraic topology'. In discussions with Noether, Alexandrov realized that Poincaré's invariants concealed a group.[8] What had been the intuitive subject par excellence had been forced to define itself. As we shall see, worse was to come. All the same, as we meet it in Seifert and Threlfall's classic *Lehrbuch der Topologie* of 1934 (reprinted as 1980), the language may seem strange, and almost all of today's methods are missing (homotopy groups, exact sequences, fibre spaces, . . .) but the basic objects are in place.[9] What perhaps most strikes today's reader is that the *Lehrbuch* is packed full of attractive and illuminating pictures (see Fig. 5). As an image of the kind of work that topologists do (cutting out knots and gluing them in with a twist, say), they are an invitation to read further, even though the text is not always easy. While topology was a great deal more abstract, it was still the most graphic of mathematical pursuits.

---

8. There is an account of this in Corry (2004, p. 245), with mention of a possible claim by Mayer.
9. As time went on higher-dimensional applications became more important, so that Seifert-Threlfall is still the 'easiest' reference for two and three dimensions.

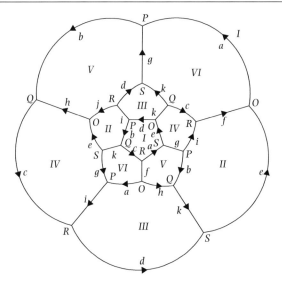

Fig. 5 The 'dodecahedral space' from Seifert and Threlfall's book. This is obtained from a solid dodecahedron by gluing opposite faces (marked I, II, III, IV, V) as shown in the picture.

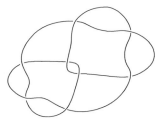

Fig. 6 A knot ('true lover's knot') with eight crossings.

The changes which took place can be well illustrated by the specific case of knots. Unlike the subject-matter of topology in general, these are easy to understand, and what they are, in basic terms, has remained the same since they were first systematically studied by Tait (in response to a failed theory of Lord Kelvin) in the 1870s.[10] For that reason, they provide a particularly interesting index of what does change. One thinks of a knot (Fig. 6) as a closed curve (a 'circle') in three dimensions, and defines two knots to be equivalent if one can be deformed into the other; so much is more or less obvious. Also immediately clear to Tait, and probably to the reader, is that one can represent a knot unambiguously by projecting it down into a plane (as in the figure) using broken lines to mark which strand goes under at each crossing. Of course, this 'diagram' is far from unique, and the very simplest question is how one tells whether two diagrams determine the same knot. Since that would require a language of diagrams, it is already perhaps too difficult. Tait determined all knots which had diagrams with up to eight crossings, and made some important and hard conjectures. This, it must be remembered, was *without* using any of the still-to-be invented tools of topology.

10. Tait referred to earlier work of Gauss and Listing, but he is usually considered the founder.

Around 1910, Dehn and Wirtinger were aware of the tables of knots (or knot-projections) compiled by Tait, and could see that beneath them a question of topology, treatable by Poincaré's new methods, might lie. The problem was that the knot K was just a circle, the ambient space just $\mathbf{R}^3$. The answer was to consider the 'difference', $\mathbf{R}^3 - K$. Not a 'manifold' in Poincaré's sense, since it was infinite, this still seemed a good subject for treatment. It is attractive in philosophical terms to note that the first step forward (rather like Dedekind's?) replaced studying the knot by studying the hole which was left when you removed it.

One of Poincaré's most interesting invariants was a *group*, which we now call the 'fundamental group' of X, $\pi_1(X)$; and he had given a means of computing it from a cell decomposition. Although the 'cells' in $\mathbf{R}^3 - K$ were not obvious, Dehn and Wirtinger did arrive at a description of generators and relations for the group $\pi_1(\mathbf{R}^3 - K)$.[11] An excellent first stage, this ran up against serious problems relating to how little was known about such presentations. When did two define the same group? Was the problem even (in intuitionist, or Gödelian terms) decidable? (It is not.) Information can be gathered about the group when you are lucky, but how can you enforce luck?

The next major advances were due to Alexander and Reidemeister in the 1920s. There may be various priority questions to disentangle here, on which Epple has commented (2004), but they need not concern us here. The first point is that a new definition of a knot was found useful. 'Simplexes' (triangles, tetrahedra, etc.) were seen, correctly, to be a more precise way of finding your way around than Poincaré's more general 'cells'; and so a knot K was defined to be a closed *polygon* in three dimensions. An 'elementary equivalence' was defined to be one of the type shown in Fig. 7 where you replaced the side AB of the triangle by the sides AC, CB, or vice versa—provided that K did not meet the interior of the triangle. And, finally, K and K' were equivalent if you could get from one to the other by a sequence of elementary equivalences.

This was a substantial change. Was it proved equivalent to the previous definition? I am not sure, and in a way it is not so important as the new language. While the new knots look pretty much like the old, and in actual drawing topologists often smooth out the corners for reasons of aesthetics or laziness, there is still a greater precision which is involved specifically in the 'elementary

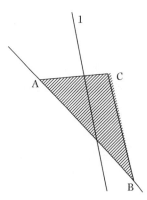

Fig. 7 Elementary equivalence. The shaded triangle ABC does not meet the other strands of the knot, so AB can be replaced by the two edges AC, CB.

---

11. That is, π_1 is the set of all expressions in (say) $x_1, x_1^{-1}, \ldots, x_k, x_k^{-1}$ (generators), subject only to the rules which follow from requiring certain expressions $R_1, \ldots, R_l$ in the xs (relations) to equal 1. See a book on group theory.

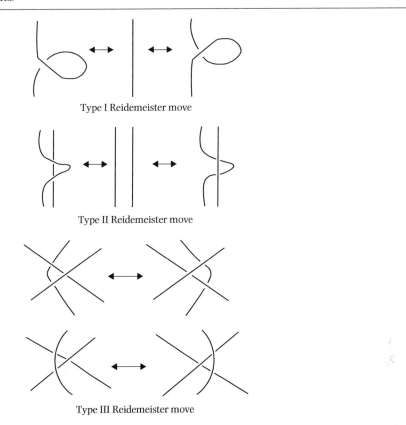

Fig. 8 The three Reidemeister moves. Type I flips a loop over; type II pulls one looped strand over another; type III takes a strand through a crossing.

equivalence'. Given this definition, Reidemeister was able to show that two knot *projections* were equivalent if and only if you could go from one to the other by a sequence of 'Reidemeister moves', as shown in Fig. 8.

On the face of it this is a rather unambitious result; and indeed it was a rather small basis for the construction of more and easier invariants by Reidemeister and Alexander. However, again it illustrates a major change in the way mathematicians treat their subject-matter. For Tait, a knot is obvious; it is represented by a picture, and we know what it means to say that two knots are the same. For Reidemeister, we have to construct meticulous (and finitistic!) definitions both of the object 'knot' and of the relation 'the same'. The payoff is a relation between diagrams which guarantees sameness. This too may not be easily checkable; however, if we construct a function of a knot diagram (as several authors were to do in the 1980s), we can show that we have a 'knot invariant' by showing that the function does not change under the three Reidemeister moves. The study of knots has become a kind of algebra.

Exercise 3.
(a) Show that the two knots shown (Fig. 9) are equivalent directly.
(b) Show that they are equivalent using Reidemeister moves. How many have you used?

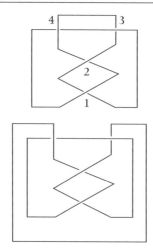

Fig. 9 Two knots, each with four crossings. The crossings on knot 1 are labelled 1, 2, 3, 4 (and obviously those on knot 2 could be labelled similarly).

7 Outsiders

I have not trodden through a conventional university course, but I am striking out a new path for myself. I have made a special investigation of divergent series in general and the results I get are termed by the local mathematicians as 'startling'. (Ramanujan to Hardy, January 1913)

If one proves the equality of two numbers a and b by showing first that $a \leq b$ and then that $a \geq b$ it is unfair; one should instead show that they are really equal by disclosing the inner ground of their equality. (Noether, quoted in Weyl 1935)

The community of mathematicians in 1900 was restricted almost entirely to white males, as one might expect. Today, the restriction is less complete, although one would not want to make any very strong claims about the progress which has been achieved. At the beginning of the century stand two very different figures, Srinivasa Ramanujan and Emmy Noether, whose stories are often told as exemplary, and who certainly show the different ways in which unusual individuals could break into the closed world of mathematicians; what they could achieve, and what were the necessary limits of that achievement. They were both quite exceptional—and would be today—for completely different reasons.

When Ramanujan came from Madras to study with the already famous number theorist G. H. Hardy in Cambridge in 1914, he came from a situation where mathematical research was not far advanced, and Indians had little chance of making headway in it. (Despite this, he had had articles published in the *Journal of the Indian Mathematical Society*.) In the well-known story, he was forced to leave his wife, to make an unwelcome adjustment in Cambridge to a hostile climate and uneatable food, and finally to ruin his health, in order to study with the only mathematician who had taken the trouble to respond to his extraordinary letters. Rather than looking back on the outsider status which he had to endure as something belonging to a distant past, one might wonder how likely it is today that a clerk without a university degree, writing to a professor at Princeton (say) in such terms would be fortunate enough to get a similar response.

It is a cliché in writing about Ramanujan to describe the difficulty of assessing what he contributed to modern mathematics—quite aside from the difficulty of the mathematics itself.

To take a 'proper' historical viewpoint one needs to have some understanding of what his work meant to him, what it meant to number theorists in Cambridge at the time, and why it has continued to be important.[12] In the quotation above, he describes his interest in 'divergent series'. This at least restricts the field a little, and it helps to fix ideas (if it gives no idea of the scope of his thinking) to remember that at the age of 17 he was investigating the simplest such series, $\Sigma 1/n$ (see Oresme, Chapter 6), and had calculated Euler's constant

$$\gamma = \lim_{n \to \infty} \left(\Sigma_{k=1}^{n} \frac{1}{k} - \log n \right)$$

to 15 decimal places. One sees an interest in infinity, and in its control; in regularity as infinity is approached. And Hardy's attempts to understand his friend's thought—as between Cambridge public school atheist and pious Brahmin—convey images of an infinity which is capable of generating all primes.

Whatever ideas or traditions may have underlain Ramanujan's thought, his practice was solidly in a successful nineteenth-century tradition of hard number theory (modular functions) which went back to Dirichlet. Indeed, his weakness in supplying proofs would not have been as suspect in Dirichlet's time as it was a hundred years later. However, he had the special advantages both of working with Hardy (undoubtedly among the best in the field) and of his own 'intuition'. Both of these enabled him to go further than his contemporaries. It also helped that his directions were not, in general, ones which anyone else had thought of or imagined would be worth pursuing. A famous example, whose exact history still seems uncertain, is the formula for the partition number; and since this (unlike some of his other work) is easy to describe, it is worth a mention.

For any natural number n, the partition function $p(n)$ is defined to be the number of ways in which n can be written as a sum of natural numbers (unordered): easily, $p(1) = 1$ ((1)), $p(2) = 2$ ((2), (1, 1)), $p(3) = 3$ ((3), (2, 1), (1, 1, 1)), and $p(4) = 5$ ((4), (3,1), (2,2), (2,1,1), (1,1,1,1) in obvious notation). Work had been done since the time of Euler to find formulae for $p(n)$; and Ramanujan claimed that he had such an exact formula. Hardy was unable to believe this—the usual best hope would be for an 'asymptotic formula', which describes limiting behaviour. Naturally, too, Ramanujan was unable to explain, if he knew, why his formula was right. As a result, what we now have is known as the rigorously proved 'Hardy–Ramanujan asymptotic formula', found in 1916–17:

$$p(n) \sim \frac{1}{4n\sqrt{3}} e^{\pi \sqrt{2n/3}}$$

Here '$a_n \sim b_n$' means that $a_n/b_n \to 1$ as $n \to \infty$. The formula is at least relatively simple, and gives an idea of how fast the partition function grows. The much more complicated exact function, which may or may not be what Ramanujan found, was proved by Rademacher in 1937.

To say that Ramanujan's mathematics was tangential, let alone marginal to twentieth-century mathematics would be absurd given the influence of his published work, his unproved conjectures, let alone his unedited notebooks. Perhaps his work should stand rather as an image of modern mathematics' capacity for absorbing *anything* however 'different' and setting it to work in its vast theorem factory.

---

12. To the extent that (for example) Deligne's solution of the Ramanujan conjecture in the 1970s with the full apparatus of late twentieth-century algebraic geometry (I forbear to give details) is for many mathematicians an achievement on a level with Wiles's proof of the Taniyama–Shimura conjecture.

While Ramanujan had to receive an accelerated schooling in what then counted as up-to-date number theory, and much else, from Littlewood and Hardy, Emmy Noether was in a much more usual European situation; trained in a rather old-fashioned approach to a central topic—crudely, the relation of algebra to geometry—she came into contact with Hilbert and others who were in the process of transforming it, abandoned her earlier lines of work and took the ideas which she had received so much further as to have a crucial influence on the next generation. It would be easy to point out how far she was from receiving the kind of recognition which a man would have received in her career, and being a Jew and a Communist did not help in 1920s Germany. Forced out of her untenured position at Göttingen by the Nazis, she found refuge at the women's college of Bryn Mawr, and by the time of her death had already made a decisive impact on the course of mathematical history. Again, the central ideas are not easy to explain; however, one should try since they are so much at the centre of what happened in twentieth-century mathematics (and at least they are easier than those of Ramanujan). Partly they came from editing Dedekind's posthumous papers, partly from current work on polynomials, but she unified the two into what, as a result of her work has become known as the general theory of 'rings' and 'ideals', the latter defined by Dedekind. [A *ring* R is a set in which addition and multiplication can be defined, satisfying (to simplify) the same sensible rules as they do for integers; a subset $I \subset R$ is called an *ideal* if (a) $a + b \in I$ when a and $b \in I$, and (b) $ab \in I$ when $a \in I$ (not necessarily b). Example: the set of all multiples of 6 in $\mathbf{Z}$ is an ideal, usually written (6).] These ideas entered into two apparently disparate areas of mathematics.

1. In algebraic geometry, it had become common to consider not so much the curve C (for example) defined by an equation like $a^2 x^2 - b^2 y^2 - 1 = 0$ (Fig. 10), but the *ring* of all polynomial functions in x and y; the functions which vanished on C were then an ideal. Similarly for algebraic manifolds (varieties) of higher dimensions. Instead of studying the geometric object, (it came to be realized) one could equally well study the ideal.
2. In number theory, one routinely considered 'number rings', such as the set of all $m + n\sqrt{-5}$, where m and n are integers. It was to study these, their sometimes strange division and factorization properties, and so to solve arithmetical problems that Kummer and Dedekind had introduced ideals, or 'ideal numbers' in the first place.

What Noether observed is clear from the above description—but only because the description has been framed in the terms which she devised: that these two families of questions were both concerned with the structure of ideals in a particular type of ring (it is now called 'Noetherian').

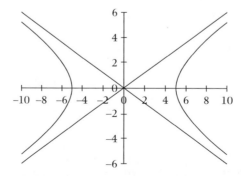

Fig. 10 Curve (hyperbola) with equation as in text; $a = 1/5, b = 1/3$.

The idea of *structure* unified two apparently disparate areas of mathematics. It was to become more and more important through those who drew inspiration from her writings.

As we have suggested earlier, Noether is too unusual a figure to be in any sense 'typical' in the short list of women in mathematics. Women emerge to take a place as major figures in this history almost at its last moment. And while one can and should find space for the skilled amateur Ada Byron, Countess of Lovelace as a 'joint forerunner' of the computer, or for Sofia Kovalevskaya as a high-quality contributor to nineteenth-century analysis, the place of Noether, like that of Ramanujan, goes well beyond the category of prizes awarded in a special outsiders' category. Without her, the drive to abstraction which we shall chart later might well have developed unstoppably, but her work of the 1920s certainly set a particular shape on it.

Appendix A. The cut definition

(From 'Continuity and irrational Numbers' reproduced in Fauvel and Gray pp. 575–6.)

From the last remarks it is sufficiently obvious how the discontinuous domain R of rational numbers may be rendered complete so as to form a continuous domain. [Earlier] it was pointed out that every rational number a effects a separation of the system R into two classes such that every number a_1 of the first class A_1 is less than every number a_2 of the second class A_2; the number a is either the greatest number of the class A_1 or the least number of the class A_2. If now any separation of the system R into two classes A_1, A_2, is given which possesses only *this* characteristic property that every number a_1 in A_1 is less than every number a_2 in A_2, then for brevity we shall call such a separation a *cut* and designate it by (A_1, A_2). We can then say that every rational number produces one cut or, strictly speaking, two cuts, which, however, we shall not look upon as essentially different; this cut possesses, *besides*, the property that either among the members of the first class there exists a greatest or among the numbers of the second class a least number. And conversely, if a cut possesses this property, then it is produced by this greatest or least rational number.

But it is easy to see that there are infinitely many cuts not produced by rational numbers.

Appendix B. Intuitionism

(Weyl, in Mancosu 1998, p. 97)

(i) [Weyl] Let us, for example, assume that 'n has the property E' means that $2^{2^{n+4}} + 1$ is a prime number, and that property $\bar{E}$ means the opposite ($2^{2^{n+4}} + 1$ is a composite number). Now consider the following. The view that it is in itself determined whether there is a number with property E, or not, is surely based on the following idea: The numbers $1, 2, 3, \ldots$ may be tested, one by one, for the property E. If such a number with property E is found, the answer is *yes*. But if such a termination does not occur, that is to say, after a *completed run* through the infinite number sequence, no number of kind E is found, then the answer is *no*. Yet this point of view of a completed run through an infinite sequence is nonsensical.

(ii) [Brouwer] 1. The Axiom of Comprehension, on the basis of which all things with a certain property are joined into a set ... is not acceptable and cannot be used as a foundation of set theory. A reliable foundation is only to be found in a *constructive* definition of a set. 2. The axiom of the *solvability of all problems* as formulated by Hilbert in 1900 is equivalent to the logical Principle of the Excluded Middle; therefore, since there are no sufficient grounds for this axiom and since logic is based on mathematics—and not vice versa—the use of the Principle of the Excluded Middle is *not permissible* as part of a mathematical proof. The Principle of the Excluded Middle has only scholastic and heuristic value, so that theorems that in their proof cannot avoid the use of this principle lack all mathematical content.

Appendix C. Hilbert's programme

(Hilbert, 'The New Grounding of Mathematics', in Mancosu 1998, p. 204)

But we can achieve an analogous point of view if we move to a higher level of contemplation, from which the axioms, formulae, and proofs of the mathematical theory are themselves the objects of a contentual investigation. But for this purpose the usual contentual ideas of the mathematical theory must be replaced by formulae and rules, and imitated by formalisms. In other words, we need to have a strict formalization of the entire mathematical theory, inclusive of its proofs, so that—following the example of the logical calculus—the mathematical inferences and definitions become a formal part of the edifice of mathematics. The axioms, formulae, and proofs that make up this formal edifice are precisely what the number-signs were in the construction of elementary number theory ... and with them alone, as with the number-signs in number-theory, contentual thought takes place—that is, only with them is actual thought practiced. In this way the contentual thoughts (which of course we can never wholly do without) are removed elsewhere—to a higher plane, as it were; and at the same time it becomes possible to draw a sharp and systematic distinction in mathematics between the formulae and formal proofs on the one hand, and the contentual ideas on the other.

In the present paper my task is to show how this basic task can be carried out in a rigorous and unobjectionable manner, and to show that our problem of proving the consistency of the axioms of arithmetic and analysis is thereby solved.

Solutions to exercises

1. (a) We have already done this—the definition of '$\sqrt{2}$' works for the square root of any number. (That is, except when the number is already a square; then of course one defines (for example) $\sqrt{4}$ to be 2, which is already rational, and not a problem.)

    (b) First we must define the product of two positive real numbers. We use the upper classes, otherwise we keep having to make special provision for the negative numbers. Let U and U' (sets of rational numbers) be the upper classes belonging to the positive real numbers a and a'. Then all numbers in each of these are positive. Define V to be the set of all z such that $z \geq uu'$ for some $u \in U, u' \in U'$. (This is complicated, but ensures that V obeys the rules for an upper class; that is, if $z \in V$ and $z' > z$, then $z' \in V$.) By definition, aa' is the real number whose upper class is V.

    [What happens when aa' is rational—for example, π and $2/\pi$?]

    Now let V be the upper class for $\sqrt{2}.\sqrt{3}$, and V' the upper class for $\sqrt{6}$; we have to show they are the same. If $z \in V$, $z \geq aa'$, where $a^2 > 2$ and $(a')^2 > 3$. So $z^2 > a^2(a')^2 > 6$, and clearly

$z \in V'$. The other way round is harder. Suppose $z^2 > 6$, we must decompose z as a product. For some n, $z^2 > 6 + (1/n)$. Choose a rational number a with $a^2 > 2$ and $a^2 < 2 + (1/3n)$. Then,

$$\frac{z^2}{a^2} > \frac{6 + (1/n)}{2 + (1/3n)} = 3$$

so $z = aa'$ as required.

(c) Here it is much easier to use the lower classes. Let L_i be the lower class belonging to x_i, and simply define L to be the *union* of the L_is: all rational numbers a which are in at least one of the L_is. Then L is *not* all of the rationals; in fact, if $a > M$, a is in none of the L_is, so not in L. And it is easy to check that L satisfies the requirements for a lower class. Hence, L defines a real number M_0.

Since each L_i is contained in L, M_0 is greater than each x_i. Now suppose y is a real number $< M_0$. Then there is a rational number a such that $y < a < M_0$. From the definition of M_0, a is in L_i for some i. Hence, $a < x_i$ and so $y < x_i$.

(d) I put this in because, naively, one might think that defining real numbers as infinite decimal fractions was good enough, and one needs to realize that adding them is complicated by the possibility of having to 'carry' indefinitely far back. Consider the difference between:

$$0.12345 + 0.87654 = 0.99999$$

and

$$0.12345 + 0.87655 = 1.00000$$

A change in the fifth decimal place changes all the results of the addition.

2. (a) Socks being identical, one needs a function which to each pair of socks S_i assigns just one sock s_i. The axiom of choice will ensure this (but you cannot do it by hand). On the other hand with shoes, there is a natural rule: simply choose the left shoe of each pair. [In set-theoretic terms, it is a difference between ordered and unordered pairs.]

(b) The number $\sqrt{2}^{\sqrt{2}}$ is either rational or irrational. If it is rational, we have the example we are looking for (since $\sqrt{2}$ is irrational anyway). If it is irrational, then take $x = \sqrt{2}^{\sqrt{2}}$, and $y = \sqrt{2}$. $x^y = \sqrt{2}^{\sqrt{2}^2} = 2$ is rational and we are done in this case too.

Intuitionist critique: The procedure is not constructive; since, if we do not know which of the two alternatives is true, we do not know which procedure to follow.

[Further complication; it can be proved that $\sqrt{2}^{\sqrt{2}}$ is irrational. If we use this, then of course we have a proof without the law of the excluded middle.]

(c) Let $a_0 = 0$ and $b_0 = 1$. We construct inductively an interval $[a_n, b_n]$ of length $1/2^n$. Suppose $[a_n, b_n]$ constructed. Then:

1. if $[a_n, (a_n + b_n)/2]$ contains infinitely many elements of the set, let $a_{n+1} = a_n$ and $b_{n+1} = (a_n + b_n)/2$;
2. if not (so the other half contains infinitely many elements), let $a_{n+1} = (a_n + b_n)/2$ and $b_{n+1} = b_n$.

The sequences a_n, b_n, clearly both converge, to the same limit x; and at least one of them must contain infinitely many distinct elements. This is the sequence we are looking for.

Intuitionist critique: we have no way of telling which of the two cases (1), (2) holds at each point; so we cannot apply the law of the excluded middle.

3. The simple way of seeing this is that the right-hand knot is obtained from the left-hand one by pulling the top loop out and lifting it downwards, so as to surround the whole diagram. If we want to use Reidemeister moves, begin with knot 2 and tilt the diagram slightly. The long edge at the bottom can then be taken up through the diagram to the top (say above it) using Reidemeister move 2 for the two bottom loops and move 3 for each of the four crossings in turn. When this is finished, you find that you have not quite knot 1, but one which differs by two loops which can be removed by Reidemeister move 1. The total number of moves (I think) is then $2 + 4 + 2 = 8$.

10 A chaotic end?

1 Introduction

I am not thinking of the 'practical' consequences of mathematics ... at present I will say only that if a chess problem is, in the crude sense, 'useless', then that is equally true of the best mathematics; that very little of mathematics is useful practically, and that that little is profoundly dull. (Hardy 1940, p. 29)

Mathematical formalism, however, whose medium is number, the most abstract form of the immediate, ... holds thinking firmly to mere immediacy. Factuality wins the day; cognition is restricted to its repetition; and thought becomes mere tautology. (Adorno and Horkheimer 1979, p. 27)

In 1936, the year when the Spanish Civil War broke out, Lancelot Hogben wrote a book called *Mathematics for the Million*—an unexpected best-seller, which still sells today. The aim was to educate the masses in mathematics, since:

The mathematician and the plain man each need one another. Maybe the Western world is about to be plunged irrevocably into barbarism. If it escapes this fate, the men and women of the leisure state which is now within our grasp will regard the democratization of mathematics as a decisive step in the advance of civilization. (Hogben 1936, p. 20)

Hogben's judgement stands in opposition to that of Hardy (what is good in mathematics is not useful); and equally to that of Adorno and Horkheimer, also writing in the 1940s (mathematization is everywhere, and to mathematize ideas is to deprive them of their creativity). The mathematics which he aimed to democratize was not easy—it included algebra, the calculus, and statistics—but both in content and in presentation it would have been found trivial by Hardy, and not surprisingly, it included no recent results. How has the democratization of mathematics fared during the last 70 years—and is anyone still convinced that it is either possible or desirable?

It seems difficult, in mathematics, to approach the history of the present—or even of the recent past. School history programmes and TV history channels thrive on the wars and oppressions of the last 100 years; but they have a clear narrative line to help them, and ample resources in the form of film and picture archives. With regard to the history of any science, and of mathematics in particular, there seems to be too much of it; it is too difficult, too diffuse, and it is hard to put the various narratives together to point a moral or adorn a tale. The historian begins by being grateful to Andrew Wiles, who, having proved the outstanding theorem of mathematics ('Fermat's Last Theorem') in the closing years of the twentieth century, has at least provided the story with a neat, if provisional ending. However, it is a rather special kind of ending, centred on the rarefied world of universities and extremely hard pure mathematics. It could serve as a model for what, in mathematics, is undemocratic. A contemporary Hogben could, at a pinch, include a chapter on the exposition of Gödel's Theorem (proved in 1931), but might find it pointless except as a lead-in to Turing's work, and so, as we shall see, to computers. He would have no time at all for Wiles's proof, which is perfectly in the Hardy mode: 100 pages long, understood only by a small circle, and, while possibly applicable in its results (see later), certainly not in the methods it takes to get there.

And yet mathematics, at all levels of subtlety and difficulty, is everywhere present in the world we live in. One may ridicule the failures of 'smart bombs' to hit their targets, or of satellites to find the weapons they are supposedly searching for; but the technology which they rely on, and that which guides robots on Mars, and makes it possible to find optimal routes for motorists, is all underpinned by mathematics of various kinds. The computer, which we shall have to consider, is an essential part of much of this technology, and is in part a spin-off of the early twentieth century's preoccupation with logic and the constructible. But there are many more diverse inputs. When an architectural design programme translates building specifications (breeze-blocks, windows, doors, joists) into three-dimensional views of the projected construction, it is *using* the ability of the computer to translate keystrokes or mouse movements into the eighteenth-century language of Monge's descriptive geometry (see Chapter 8). At the more advanced level of *Tomb Raider* or in the animations of *The Matrix*, the same trick is being worked for jumps and turns in three dimensions, using the classical dynamics developed by d'Alembert and Euler. When we buy, pay bills, or consult our bank balance on a 'secure website' our data are encrypted using (perhaps) the properties of large prime numbers, or even elliptic curves, which also play a part in the design of CDs; this most classical part of mathematics, (Wiles's preoccupation, one could say) is now intensively modernized and even—for obvious reasons—subject to intellectual property law. If, rather than historians, we were simply surveyors of the field of what mathematics is doing in 2004, it would be hard to avoid intoxication, an endless list, and, somewhere along the line, a word like 'awesome'. We could hardly any longer agree with Hardy's assertion that practical mathematics is by its nature trivial. On the other hand, the pessimists among us might, like the 1940s Marxists Horkheimer and Adorno, conclude that what is creative about human thought has been lost in its universal mathematization.

From the historian's viewpoint, then, what stands out about the development of mathematics in the later twentieth century? First, we have an extremely rapid *growth* in the uses of mathematics, the number of people employed, and the amount of work done; this really begins after the Second World War. Second, as far as pure mathematics is concerned, we have seen a tendency towards increasing *abstraction*, which has now perhaps peaked, but which was very influential in the mid-century. Third, we have the rise of *new forms of 'applied' mathematics*; most obviously computer science and statistics, but including a number of others (operational research, control theory, . . .). There have been other major and important developments, but these will be enough to be going on with. Even to connect such a brief list into some kind of coherent account will obviously involve leaving out a great deal, and the variety of what is left may still seem confusing. It's the mathematics which we have now, for better or worse, and for that reason alone it deserves our attention.

2 Literature

The historians of mathematics, who are so dedicated and scholarly on the Greeks and the Chinese, have not been as productive on the present. A vast amount of historical work has been produced on the Second World War; even the collapse of the Soviet Union a mere 16 years ago has its serious historians. Where are the comparable works on the development of chaos theory, category theory, and financial mathematics? Naturally there is a problem: many of the actors are still alive, and the writer must usually be cautious about describing them. As in the previous chapter, the best sources are often biographies of mathematicians who are either dead (Alan Turing, Hodges 1985, John von Neumann, Macrae 1992) or cooperative (John Nash, Nasar 1998, Smale, Batterson 2000). The best of these add valuable information on the wider scene—Hodges on Cambridge, Nasar on Princeton, at key periods. Mathematicians are also given, in their retirement, to producing autobiographies and reminiscences, of variable value. And individual spectacular developments are covered in more or less journalistic accounts which attempt to popularize and promote a view of what the writer finds exciting: Gleick (1987) on chaos, Singh (1997) on Fermat's Last Theorem, even perhaps Hofstadter's over-the-top mathematical rhapsody (1979).

Corry (2004) on abstraction and Dieudonné (1989) and James (1999) on topology, mentioned in the last chapter, continue to be useful; and we shall draw on Segal's interesting specialist account (2003) of mathematics under the Third Reich. However, the full history of the period is still to be told.

The previous chapter's remarks about difficulty apply again, of course; even if one can follow the popularizers of chaos theory with their coffee-table pictures, what can one do about Wiener measure, étale cohomology, or topological quantum field theory? Faced with the mathematical world of today, the popular consumer is naturally tempted to give up on the content and settle for an experience of awestruck wonder: who are the people who carry on this strange, remote, abstracted activity, and why do they do it? Accounts of how they operate always seem to miss something.

Professor Mazur sipped his cappuccino and listened to Ribet's idea. Then he stopped and stared at Ken in disbelief. 'But don't you see? You've already done it! All you have to do is add some gamma-zero of (M) structure and just run through your argument and it works'. (Singh 1997, p. 221)

Is it, then, just a question of getting access to Barry Mazur's brand of cappuccino? In Singh's portrait of Wiles, we have the mathematician as secretive solitary obsessive—set against the background of a research community which has the opposite values, exchanging ideas over coffee. In Sylvia Nasar's book on Nash, and still more in the 'spectacularly dumb' (Taylor 2001) Russell Crowe film version of it, we have the mathematician as paranoid–schizoid, a genius haunted by demons. Most recently, if perhaps least seriously, Mark Haddon (2003) contributes a new element by presenting the mathematical genius as a sufferer from Asperger's Syndrome, physically unable to construct human relationships, numbering his chapters by primes. The 'mathematician-problem' has become another object for consumption, a way of selling books, TV programmes, films, etc. In the old (1960s) situationist language of Guy Debord, the activity of mathematicians, especially at its most arcane and difficult level, has become part of the 'Spectacle', and so removed from historical thought.

With the destruction of history, contemporary events themselves retreat into a remote and fabulous realm of unverifiable stories, uncheckable statistics, unlikely explanations and untenable reasoning. (Debord 1990)

It will be an uphill struggle for this chapter to do any better, but it is our aim to try.

3 The Second World War

Pure mathematics (in the narrowest sense) has a meaning only as a means of education to a formal character-building that is consciously employed for service to the entire people.
   Mathematical character-building: that is, cultivation of the masculine principle in spiritual life. (E. A. Weiss, cited in Segal 2003, p. 193)

To assemble sufficient aircraft to implement the strategy required the diversion of several squadrons from Bomber Command to Coastal Command, a proposal that was fiercely resisted by Air Chief Marshal 'Bomber' Harris, who demanded of Churchill, 'Are we fighting this war with weapons or the slide rule?' Churchill puffed on his cigar and replied, 'That's a good idea. Let's try the slide rule.' The results of the strategy turned out almost exactly as Blackett and his colleagues had predicted. (Anecdote reported on www.orsoc.org.uk/conf/black.htm)

The involvement of mathematicians, and scientists in general, in the Second World War is in itself a vast field of study. In mathematics one naturally begins by thinking of the allied effort, the origins of operational research, the computer, the Enigma Code. All these are important, and were to lead to an increasing collaboration of mathematicians with the military *after* the war, which chapters in Nasar (1998) describe in some detail. However, the story begins some time earlier, with the rise of Nazism in Germany. Our two quotes set up a somewhat facile contrast, with the Nazis involved in cloudy rhetoric about the national spirit while the down-to-earth Allies see the point of using the slide-rule. Easy as this is, it is not a gross distortion.

Like many others, German mathematicians who were Jewish (Landau and Courant) or communist (Zorn) or in the case of Emmy Noether both, were surprised when they found that the Nazi régime actually proposed to dispense with their services, however distinguished they might be. For Noether it was easy—she did not have a titular post. For Landau, considerable manoeuvring was necessary, including a student 'boycott' of his classes; but he was eventually forcibly retired.

The combined expulsion of Jews from their university positions and exodus of many non-Jews (like Weyl and von Neumann) who found Nazi Germany uncongenial was catastrophic for mathematics in Germany. At the same time, in keeping with the Nazi programme, there was an attempt to define what German mathematics should be. This posed a challenge, given the nature of the subject. It was too easy to see *all* mathematics as part of the abstract intellectualism which the Nazis wished to overthrow, which of course would be unfortunate for any mathematicians however patriotic who wished to hold on to jobs in universities. Drawing on some of Brouwer's ideas (although Dutch, he was a strong German nationalist in the period), a group of mathematicians, some convinced Nazis or at least right-wing nationalists, some simply opportunists, presented an image of two opposing kinds of mathematics: crudely, Nordic/concrete/intuitive versus Jewish (or French)/abstract/cerebral. As the main spokesman Bieberbach put it:

[T]he whole dispute over the foundations of mathematics is a dispute of contrary psychological types, therefore in the first place, a dispute between races. (Quoted in Segal 2003, p. 365)

The turn towards intuitionism as a model for 'German mathematics' was not by any means universal, but for a short time it was influential, particularly under the ascendancy of Bieberbach, a time-server who succeeded in dominating the depleted German scene in the early 1930s. The (Jewish analyst) Landau's definition of π in his classic textbook as 'twice the smallest positive x for which $\cos(x) = 0$' (where $\cos(x)$ is defined as a series, $1 - (x^2/2!) + (x^4/4!) - \cdots$) was easy to characterize as such a retreat from the 'natural-intuitive'.

> There is no independent realm of mathematics, independent of intuition and life: the struggle over the foundations [of mathematics] that now rages is in reality a racial conflict: 'Political rootedness gives thinking its style!'
> Since German mathematics is rooted in blood and soil, the state ought to and must support and cultivate it ... ('New Mathematics', by 'P.S.', cited in Segal 2003, p. 267)

It would appear that the Nazi command—who had a high regard for rationalism when it came to organizing train time-tables—were too ideologically confused to make adequate use of the mathematicians whom they had left, except in special favoured fields like aeronautics[1] and, with a brief but deadly effect, rocketry. Reduced though they were, they could have participated more fully in the war effort, ranging as they did from dissidents to patriotic conservatives to committed Nazis; an effort was made to enlist their skills in the later part of the war, but by then it was too late.[2]

Germany's loss was to a quite outstanding extent America's gain, most particularly Princeton's. Overcoming fears, which were strong at Harvard, that they would end up with an excess of Jews, Princeton used money from Rockefeller to build up a research department. Even more fortunately, a gift from a New Jersey department store owning family, the Bambergers, endowed the Institute for Advanced Study to which Einstein went.

> Kurt Gödel, the Viennese wunderkind of logic, came in 1933, and Hermann Weyl, the reigning star of German mathematics, followed Einstein a year later ... Practically overnight, Princeton had become the new Göttingen. (Nasar 1998, p. 54)

For once, the high-flown language reflects the reality. By 1936, solid Cambridge men like G. H. Hardy and the young Alan Turing saw Princeton as a useful place to spend a year. [It has more or less retained its dominant position ever since, in the face of stiff competition from a dozen equally deserving American universities.] And, at Princeton and elsewhere in the United States (and indeed Britain), refugees from Hitler were willing to take part in the war effort; and (in contrast to the Nazis) the governments learned how to use them.

The most high-profile part of the story, the atom bomb project, is really a part of the history of physics rather than of mathematics (see later for the difference); but mathematicians were widely employed on it; often mentioned is von Neumann's 'conclusion that large bombs are better detonated at a considerable altitude than on the ground' (Macrae 1992, p. 209). Besides this limited and spectacular application, many of the most distinguished of them were involved in what could be seen as more routine work; planning resources, codebreaking, and ballistics. All of these, which might seem a distraction from 'real' theorem-proving mathematics, led to the development of new fields which were later promoted as the answer to the problems of peace (i.e. government and business) as well as war. Norbert Wiener, in the United States, contributed the field which he was to name 'cybernetics', the science of control, while his counterparts in Britain (Blackett and others) devised the equally ambitious science of operational research. Again, Allied commanders showed much less resistance to the use of such abstractions than did the Nazis.

> There had to be some pattern that the methodical Germans were using to plant their mines along the convoy routes to Britain. Johnny [von Neumann] was asked to work out mathematically what these patterns might be, and how best to counter them. (Macrae 1992, p. 207)

---

1. Here the practical Göring had some influence.
2. The best-known example, the brilliant Nazi mathematician Teichmüller, was conscripted and died on the Russian front in 1943. It seems unlikely that this would have happened in Britain or the United States.

Idealism—the conviction that the war was not mere national defence but the defence of civilization—powered the enlistment of many mathematicians, and other scientists, in the war effort. Conscientious objectors were rare, and although André Weil unusually decided not to enlist, he was prepared (as Oppenheimer was)[3] to cite the warlike Hindu scriptures in support of the war effort:

> The law is not 'Thou shalt not kill', a precept which Judaism and Christianity have inscribed—to what avail?—in their commandments. The *Gita* begins with Arjuna, 'filled with the deepest compassion', stopping his chariot between two armies, and ends with his acceptance of Krishna's injunction to go to combat unflinchingly ... Arjuna belongs to a caste of warriors, so his *dharma* is to go to combat. (Weil 1992, p. 124)

Indeed, from the Second World War onwards, mathematicians were to find that their *dharma* did not always involve simply sitting in libraries and proving theorems.

4 Abstraction and 'Bourbaki'

> On these foundations, I state that I can build up the whole of the mathematics of the present day; and if there is anything original in my procedure, it lies solely in the fact that, instead of being content with such a statement, I proceed to prove in the same way as Diogenes proved the existence of motion[4]; and my proof will become more and more complete as my treatise grows. (Bourbaki 1948)

> For Bourbaki the fields to encourage were few, and the fields to discourage were many. (Mandelbrot 1989)

The drive to a more abstract view of mathematics, which has been both admired and deplored as peculiar to the twentieth century, had its roots early on in the foundational enterprise. The schools of axiom-builders often claimed that it was not important what their axioms referred to: Hilbert was quoted as saying that one should be able to replace the words 'points, lines, planes' in the axioms of geometry with 'tables, chairs, beer-mugs'; and Russell characterized mathematics as the science 'in which we do not know what we are talking about, nor whether what we are saying is true'. However, if we think of the abstract viewpoint as one in which one aims systematically to lose sight of any actual real-world *objects* to which the discourse refers, and to concentrate on the *relations* and *structures* which connect those objects, then the high point of abstraction came in the 1940s and 1950s, and the leading spirits in carrying through a programme for making all mathematics more abstract were a strange revolutionary band of young French university teachers who formed a semi-secret society under the collective name 'N. Bourbaki'. Typically, the name derived from a juvenile prank—the invention of a mathematician whose name was borrowed from a Greek general under Napoleon III.

A high-spirited male clique from typically French élite-school backgrounds, the Bourbakists (Henri Cartan, Claude Chevalley, Jean Delsarte, Jean Dieudonné, André Weil, and a later 'second generation' after the war) did not set out to call the foundations of mathematics in question. Their aim was more straightforward, and more understandable: they felt that they had received atrocious and antique teaching. They had, as an alternative, learned of new ideas from Germany, particularly those of Hilbert and Emmy Noether as embodied in van der Waerden's ground-breaking new algebra textbook; and they decided to produce a series of textbooks which would form a complete new course of mathematics from the ground up. Did they aim for their books to be set texts?

---

3. Oppenheimer's famous description of the atomic bomb as 'brighter than a thousand suns' is from the *Bhagavad-Gita*.
4. That is, by walking.

Possibly. The folklore of the group is large, and growing, even though almost all of the founders are now dead; and there are interviews and more or less gossipy histories (e.g. Mashaal 2002) which show a series of photographs of young men meeting and presumably arguing about their project in the sunshine, in the French countryside. The contrast could not be more complete between the frozen impersonal texts of the *Éléments de mathématiques* and the apparently anarchic obscene atmosphere, part party, part student meeting, in which their content was hammered out.

> [W]e almost surprised ourselves when for the first time we approved a text as ready to go to press. This was the *Fascicule de Résultats* [volume of results] of set theory, adopted in its definitive form just before the war. A first text on this theory, prepared by Cartan, had been read at the 'Escorial Congress'; Cartan, who had been unable to attend, was informed by telegram of its rejection: '*Union intersection partie produit tu es démembré foutu Bourbaki*' [Union intersection subset product you are dismembered fucked Bourbaki]. (Weil 1992, p. 114)

An extract from the introduction to Bourbaki's *Algebra* is provided in Appendix A, to show with what severity he stated his aims and method. Before analysis could begin, the real numbers had to be defined; before the real numbers, the elements of topology; before that, the theory of sets. The student faced a long march before arriving (say) at the least upper bound theorem which we have seen causing such problems earlier on. The 'second-generation' Bourbakist Pierre Cartier provides a balanced evaluation of strengths and weaknesses of the project:

> Bourbaki knew where to go: his goal was to provide the foundation for mathematics. They had to submit all mathematics to the scheme of Hilbert; what van der Waerden had done for algebra would have to be done for the rest of mathematics. What should be included was more or less clear. The first six books of Bourbaki comprise the basic background knowledge of a modern graduate student.
>
> The misunderstanding was that many people thought that it should be taught the way it was written in the books. You can think of the first books of Bourbaki as an encyclopedia of mathematics, containing all the necessary information. That is a good description. If you consider it as a textbook, it's a disaster. (Senechal 1998)

This is slightly disingenuous; it is rare to come across an encyclopaedia which is equipped with a complete set of exercises. All the same, there were few who used the complete *Éléments* as their textbook. What was much more important about it was that its existence profoundly influenced the way in which a large number of mathematicians thought about their subject; and some of them used their thinking to write more readable textbooks of their own, in which the ideas of structures, the emphasis on the mapping rather than the object, and so on, became foregrounded. The writer of a textbook can, and usually does pick and choose from available material without necessarily overt plagiarism. In any case, neat Bourbaki was hard to plagiarize; but a watered down version became increasingly dominant outside France as well as inside.

The Bourbakists had no particular interest in axiom systems as a means of saving mathematics from contradiction.[5] They did see axiom systems as the basic tool in defining 'structures', which were to be central to the whole way in which the *Éléments* was presented. As has been pointed out, the idea of structure was one which, among so many other definitions, was never defined; but numerous individual structures (group, ring, topological space, uniform space, ...) permeated the text and were central to its particular way of thinking. It became almost a reflex in France, if one had fallen under the spell of the *Elements*, to speak not of defining a group but of 'providing (*munir de*) a set with a group-structure'.

---

5. Nor were they interested in a number of other things—probability theory, for example, and physics.

Let us look at an example. To define the sine and cosine functions in chapter VIII (general topology), Section 2 ('measure of angles'),[6] Bourbaki used the group isomorphism from the (multiplicative) unit circle U in the complex numbers $C = R^2$ to the quotient group R/Z. He then 'endowed' the set A of 'half-line angles' $(\widehat{\Delta_1, \Delta_2})$ (see Fig. 1) with a group structure, and showed it was isomorphic to U. Finally:

1. if θ is an angle in A, you define $\cos\theta$ to be the real part of the complex number in U corresponding to θ;
2. if x is a real number, you define $\cos x$ using some homomorphism from R to A.

At this point, Bourbaki points out the need to decide the least positive value of $x \in R$ which corresponds to $1 \in U$, and discusses the merits of 360, 400, and a number called 2π ('we will prove the existence of such a number later'). [This account is of course simplified; in Bourbaki it takes four pages, with everything proved.] As a reward, you finally get one of the author's rare pictures—the graphs of the trigonometric functions (Fig. 2).

This hardly qualified as a philosophy in the sense that the great systems of Russell, Brouwer, and Hilbert did, but it was certainly a practical ideology, and defined an orthodoxy about what one liked or disliked in mathematics. Much has been written, particularly by opponents, about the hegemony of Bourbakist ideas in France from 1945 on, and their insistence that their way

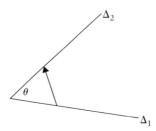

Fig. 1 The 'half-line angle' $(\widehat{\Delta_1, \Delta_2})$ is the angle of rotation from the first line to the second.

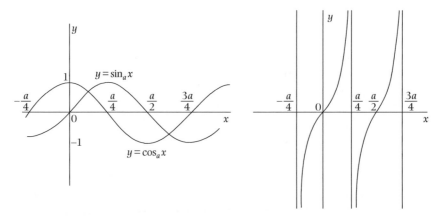

Fig. 2 The graphs of sin x, cos x, and tan x from Bourbaki's *Topologie Générale* VIII, p. 105.

---

6. Which, as we have mentioned, Landau did using series.

of teaching was the only right one. It never extended even over all of France, but where it was entrenched, it could be fairly intolerant. If installed as head of an unreformed mathematics department, a Bourbakist was capable of purging the library of most of its books and writing out orders for replacements—Bourbaki, naturally, but also the seminars of Henri Cartan, the works of van der Waerden, Eilenberg, MacLane, Steenrod, and subscriptions to the *Annals of Mathematics*, and the publications of the American Mathematical Society. More disastrous was the brief incursion of Bourbakism into the French high-school curriculum in the 1960s (paralleled by what, in the United States was called 'new math'); the idea of replacing times tables by theorems about $\mathbf{Z}$ caused confusion among teachers and students and was eventually withdrawn.

We have expressed doubts about the responsibility of mathematicians for 'modernism'. In the more minor case of the philosophical movement called structuralism, the case is clearer. In a text for the 'Que-sais-je' series explaining structuralism to the general public (1970), Jean Piaget cited Bourbaki as his first example before proceeding to the social sciences, adding that 'the structural models of Lévi-Strauss, the acknowledged master of present-day social and cultural anthropology, are a direct adaptation of general algebra' (p. 17). The influence appears directly in an anecdote of André Weil:

In New York, I had met the sociologist [sic] Lévi-Strauss, and we had hit it off quite well. I had solved for him a problem of combinatorics concerning marriage-rules in a tribe of Australian aborigines. (Weil 1992, p. 185)

Mathematics never consciously progressed to post-structuralism; nor did Lévi-Strauss repay the debt by teaching the Bourbakists some useful elements of anthropology.

Exercise 1. *(From Bourbaki, Algebra, chapter I, §1, §7.)*

(i) *Show that the only triplets (m, n, p) of natural numbers $\neq 0$, such that $(m^n)^p = m^{(n^p)}$ are: $(1, n, p)$, n and p being arbitrary; $(m, n, 1)$ where m, n are arbitrary; and $(m, 2, 2)$ where m is arbitrary.*

(ii) *If G is a finite group of order n, prove that the number of automorphisms of G is $\leq n^{\log n / \log 2}$ (show that there exists a system of generators $\{a_1, \ldots, a_m\}$ of G such that a_i does not belong to the subgroup generated by $a_1, a_2, \ldots, a_{i-1}$ for $2 \leq i \leq m$; deduce that $2^m \leq n$, and that the number of automorphisms of G is $\leq n^m$).*

5 The computer

The distinctive characteristic of the Analytical Engine, (from the earlier Difference Engine)... is the introduction into it of the principle which Jacquard devised for regulating, by means of punched cards, the most complicated patterns in the fabrication of brocaded stuffs... We may say most aptly that the Analytical Engine weaves algebraical patterns just as the Jacquard-loom weaves flowers and leaves. (Lovelace, Note A, on Menabrea's description of Babbage's engine (1843), in Fauvel and Gray 19.B.4, p. 392)

An *automatic computing system* is a (usually highly composite) device, which can carry out instructions to perform calculations of a considerable order of complexity—for example to solve a non-linear partial differential equation in 2 or 3 independent variables numerically. (von Neumann 1945, p. 7)

It may appear somewhat surprising that this can be done. How can one expect a machine to do all this multitudinous variety of things? The answer is that we should consider the machine as doing something quite simple, namely carrying out orders given to it in a standard form which it is able to understand. (Alan Turing, 'Intelligent Machinery', cited in Hodges 1985, p. 318)

Well, who did invent the computer? Once one examines the history carefully, there is obviously no single answer—unlike, say, the telephone. The claims of Charles Babbage and Ada Lovelace in the mid-nineteenth century are attractive, and they have the advantage of being British; but the machine which they designed was never built. Between their outlines and the actual machine-building stands one undoubted landmark, Alan Turing's 1937 paper 'On computable numbers'. This was Turing's first paper, and it was pure mathematics; it dealt with the one part of Hilbert's programme which Gödel had not demolished—did there exist a procedure for determining which formulae were provable (the *Entscheidungsproblem*)? The work was similar to Gödel's, but with an interesting difference of style. Turing's famous image for computing real numbers (the problem he aimed to prove undecidable) was via a machine, and the machine followed automatic instructions to read and write.

A section of his paper is reproduced as Appendix B. It is easy, to begin with, to point out that a Turing machine is not a computer, since it is infinite. ('In general the arrangement of the memory on an infinite tape is unsatisfactory in a practical machine', Turing was to observe 10 years later—ironically? (Hodges 1985, pp. 318–9)) It is also not a physical machine, merely a description of one. However, if one looks at its method of procedure, it is definitely 'computer-like'. The usual modern description—a slight variation on Turing's paper—uses the following specifications:

1. At a given time, the machine is in a *state* (one of a finite set) and in a given *position* on the 'tape'.
2. It now *reads* the symbol at its current place on the tape.
3. Having done this, it then can *change* the symbol in its current position, or *move* left or right, or a combination.
4. Which of the tasks specified in 3 is carried out is entirely determined by the machine's state 1, and by the symbol which it has read 2.

The tasks initially assigned seem distinctly boring (Turing starts by designing a machine which will compute the sequence '1010101...'), but since the 'tape' is infinite, it is easy to define the *universal* Turing machine which will compute any computable number.

Turing was at that stage as innocent of engineering as any Cambridge mathematician might be. 'Mrs Turing [Alan's mother] had a typewriter', says Hodges. Definitely less sophisticated as a machine than Ada Lovelace's Jacquard loom, the typewriter seems to have served well enough for what was at the time a theorem and a 'thought-experiment'.

The next stage is well, if not always accurately known and somewhat contested. The Americans are certain that they invented the computer (although a long-standing wrangle between von Neumann and his collaborators leaves the balance of credit uncertain), the British in their more modest way feel that they were responsible. Turing has an able advocate in Andrew Hodges; his tragic end, prosecuted under Britain's primitive homosexuality laws, victimized and finally driven to suicide in 1954 certainly inspires sympathy. The case is a romantic one:

And it was thus that in this remote station of the new Sigint[7] empire, working with one assistant in a small hut, and thinking in his spare time, an English homosexual atheist mathematician had conceived of the *computer*. (Hodges 1985, p. 295)

---

7. Signals intelligence.

John von Neumann, the other mathematician involved, a Hungarian refugee who turned Cold Warrior, has his supporters, particularly in the US military and business establishment; and a biography which is sympathetic to his role in both is available (Macrae 1992); naturally, he appears less the outsider than Turing.

> While all the other computer makers were generally heading in the same direction, von Neumann's genius clarified and developed the paths better than anyone else. (Shurkin, cited in Macrae 1992, p. 287)

Certainly large teams were involved on both sides of the Atlantic in the years following Second World War; the introduction of fast electronic components was crucial in making it possible to build machines which would perform the required tasks. John von Neumann was employed by the US Army from 1937 (the year of 'Computable Numbers') to advise on problems in ballistics. These were some of the difficult problems which he mentions in our second quote, but it was not until the end of the war that he learned of machines under construction which could help. These machines could, essentially, solve a single problem extremely fast; they belong to engineering history. They resembled the Turing machine which computes just one number.

There is no evidence that von Neumann had read 'Computable Numbers', although he was in Princeton and writing a letter of recommendation for Turing when it was published. In any case, the step which he and Turing took (and then had to get engineers to implement) was equivalent to replacing the one-task Turing machine by the universal one. While they were not abstract mathematicians in the post-1930 mould, they were both influenced by the logical 'revolution' of the early part of the century; in terms of this revolution, *instructions* like 'add' or 'move right' had the same status as signs on the paper, tape, or whatever as *numbers*. Hence we arrive at the idea of the 'programme', a stream of instructions and numbers which are *encoded* as numbers.

> Minor cycles fall into two classes: *Standard numbers* and *orders*... These two categories should be distinguished from each other by their respective first units ... i.e. by the value of i_0. We agree accordingly that $i_0 = 0$ is to designate a standard number and $i_0 = 1$ an order. (Von Neumann 1945, p. 45)

> The engineering work of producing various machines for various jobs is replaced by the office work of 'programming' the universal machine to do these jobs. (Turing in Hodges p. 293)

The two quotations express the same idea, arrived at by Turing and von Neumann almost simultaneously in the year 1945. Hidden from us, it happens all the time inside our machines, and it is probably second nature to us. To write (in old-fashioned programmers' language):

```
X = 1
FOR I = 1 TO 100
X = 1 + 1/X
NEXT I
PRINT X
```

is to write a sequence of instructions to the machine. In von Neumann's terms, '1' and '100' are numbers; while all the other signs ('=', 'for', 'next') are instructions, when suitably read. Here, with what Turing is already calling (if in quotes) 'programming', is the decisive *mathematical* input into the computer. The question of invention, for mathematicians, goes no further (fortunately). Development, under the relentless pressure of late capitalism's understanding of how much time could be saved, how many expensive workers replaced[8] has been much faster than any of the pioneers could have imagined. Mathematicians have, of course, continued to contribute directly and indirectly; and the fact that your machine not only can calculate partial differential equations—which you probably do rarely—but can respond when you type the letters 'Dear' by (a) storing an encoded form of the letters in memory, (b) displaying them on your screen in Times New Roman, and (c) asking you if you need help in writing a letter is, in itself, a form of debased practical mathematics. We have come a long way from Mrs Turing's typewriter.

Exercise 2. *(a) Devise a Turing machine which will change the natural number n (in binary digits) into $n + 1$ (e.g. 11 to 100, or 110 to 111). [Hint: Your machine will need to move to the left as it reads and changes the number, for obvious reasons.] (b) What is the programme fragment above intended to do?*

6 Chaos: the less you know, the more you get

Then [Lorenz] walked down the hall to get a cup of coffee. When he returned an hour later, he saw something unexpected, something that planted the seed of a new science. (Gleick 1987, p. 16)

It's the paradigm shift of paradigm shifts. (Ralph Abraham, cited Gleick 1987, p. 52)

The final nail in the coffin of a crude Marxist history of mathematics would seem to be provided by chaos theory. Surely, the argument would run, if mathematicians hope to gain something for their tedious profession their aim must be to persuade the Emperor that they can predict what is coming and so make more crops grow, defeat famines, warn against attack. Yet, it would seem, they can invent a theory whose main thrust is that, however precisely determined all these processes may be, they are unknowable. A butterfly in Brazil can cause a typhoon on the Isle of Wight. There is no point (or so some have said) in long-range weather forecasting—or, one might suppose, in long-range anything. The Emperor might as well sack the mathematicians and watch the butterflies.

And yet, if you search for 'chaos theory' + 'finance' + 'prediction' on Google, you come up with 4300 hits, among them the following (which sounds definitely optimistic):

With chaos, financial understanding grows exponentially, creating new software and hardware to understand and manage increasing risk. (Scholes 2002)

There seems to be a great deal of interest out there. Is this simply an application of the old deterministic mathematical principle that there's one born every minute? It is easy to be cynical, and chaos theory (as its name suggests) faces a continual risk of expanding beyond all reasonable bounds; and so becoming too formless for the historian to describe it in as centred and confident a way as (say) the Bourbakists. James Gleick's book (1987) reflects this, ranging over a great number of different bodies of work which call themselves chaos (or are so called by others), a great panorama with no central landmark on which the reader can fix attention. Even so, in that the main outlines of the subject were already fixed by the late 1980s, it is still a useful popular guide to the variety of ways in which it can be viewed.

Moreover, through the 1970s and 1980s, chaos theory not only became fashionable among mathematicians, and among the journalists like Gleick who try to find out what they are up to; it

---

8. The analogy with the Jacquard loom and its impact on the handloom weavers (a typical image of hardfaced early capitalism) is a striking one.

was found to be an easily teachable and attractive subject on university courses. This could not have been the case in the early days of Edward Lorenz's surprise 'discovery' referred to in our first quote; another case of the mathematician's inspiration through coffee. The folklore account, which seems reliable enough, makes clear that the use of computers was essential in the breakthrough. Lorenz was solving complicated systems of differential equations by using a primitive (by our standards) computer programme; and the discovery referred to was what is now called 'sensitive dependence on initial conditions'; or, popularly, 'the butterfly effect'. A tiny variation in the starting point $f(0)$ for the solution of an equation would lead to a large variation in $f(t)$ as t grew, and Lorenz created just such a tiny variation 'by mistake' (i.e. by leaving out the last three figures of the decimal for $f(0)$, which the computer had stored in memory but not displayed) (Fig. 3).

Lorenz's system was a particular one—a toy model for a weather system—and many classical systems (think of a pendulum, or the motion of a planet) do not have this behaviour, otherwise the edifice of mechanics becomes problematical. We only ever know the initial conditions approximately, and if a small error is going to increase beyond all bounds, then how is the model going to be of any use?

All the same, once sensitive dependence was discovered, it made sense to try to understand the phenomenon. Two things made the field much more easily accessible. The first was the idea, which came from the mathematicians around Steve Smale at Berkeley, that one can replace the hard study of differential equations by the much easier study of *maps* $f: X \to X$ when iterated. [Usually X is the line, or the plane, or a suitable subset.] One considers what happens when one repeats f indefinitely. getting a sequence of maps:

$$f_1 = f; \quad f_2 = f \circ f; \quad f_3 = f \circ f \circ f; \ldots$$

The second was the arrival in the 1980s of the desktop computer, vastly simpler, quicker and more graphically oriented than Lorenz's model, with the help of which a second-year university student (say) can study such equations and draw pictures. It is less fashionable now, but at the time it was a favourite occupation to write and run programmes in one's preferred language which would compute the iterations of some map and draw gaudily coloured pictures of its behaviour on

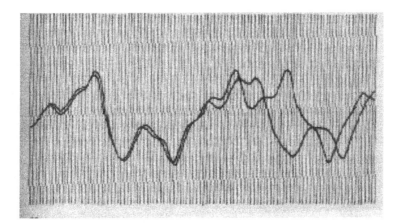

Fig. 3 The 'butterfly effect'. Two solutions of Lorenz's equation with very close starting points (at the left) eventually follow completely different paths (at the right).

the screen. The most popular of all was the *quadratic map* (just $ax^2 + bx + c$), in two forms: real (Feigenbaum) and complex (Mandelbrot).

The underlying mathematics, after these simplifications, is not particularly difficult, although difficult and interesting results have been proved about such systems. Also, naturally, not everything which was studied was chaotic; but quite simple maps could have both chaotic regions and other more stable ones. The theory was simple, and the results were often surprising. The presentation, particularly in Robert Devaney's classic textbook (1992), could be clear and (relatively) accessible, and one would have the benefit of the striking pictures, absent from the average differential equations course. What really seized the imagination of teachers, students, and popularizers alike was the possibility that the computer—necessarily a finite system—could provide an image of infinite complexity, the 'Mandelbrot set' being the universal icon which symbolized this. In one well-known example, 'Douady's rabbit' (Fig. 4), one is considering the behaviour of the *complex* function $f(z) = z^2 + c$ where $c = -0.12 + 0.75i$ under iteration. Within the grey area, $f^n(z) \to \infty$ as $n \to \infty$; while within the black area there is a periodic orbit of period 3:

$$z_0, z_1 = f(z_0), \qquad z_2 = f^2(z_0), \qquad f^3(z_0) = z_0$$

which is *attracting*; if z is in this area, $f^n(z)$ tends to cycle round the periodic orbit as $n \to \infty$. And the diminishing 'rabbit-ears' of the picture invite the viewer to visualize an infinite process, present in the idealized mathematics if only suggested on screen.

The inside and the outside are regions where f shows stable, non-chaotic behaviour, but the boundary which separates the two—the 'Julia set'—is, as might be expected, chaotic. This is true both in the obvious sense that an arbitrarily small deviation from the boundary will land you in one or other of the stable sets, and also in the sense that nearby points on the boundary behave completely differently under iteration. Considering such images one can ask, is chaos theory going to be used more as a guide for analysing systems or as a means of producing art?

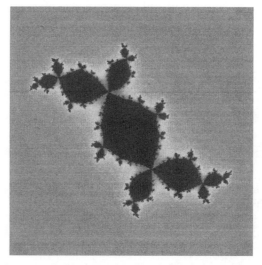

Fig. 4 'Douady's rabbit'. Let $f(z) = z^2 + c$ (a quadratic function), and $c = 0.12 + 0.75i$. Then the black area, a region in the complex plane, represents all z such that iterations of $f(z)$ do not tend to infinity.

Statistics, in the form of hypothesis testing developed in the early twentieth century, gave a useful way of dealing with the unpredictable; even apparently random processes are regular. We may not know when any particular carbon$_{14}$ nucleus will decay to nitrogen$_{14}$, but we know the statistics of the process as applied to a large number of nuclei, and so can use it within limits for reliable dating. Chaos theory can be seen as the mirror image of statistics, asserting that even some completely deterministic processes cannot be used for accurate prediction. How can these viewpoints be reconciled? It depends on one's point of view. No matter how much information one has about the weather on July 1, one's forecasts for July 15 will be limited by the variability which arises from sensitive dependence—this is chaos theory's input. On the other hand, observation of the weather over a long period makes possible some reasonable predictions about the mean July temperature and rainfall—this comes from statistics. Asked about the effect of the French Revolution in the 1950s, the Chinese Prime Minister Zhou Enlai replied 'It is too early to say'; and this is certainly a reasonable response to the problem of assessing the contribution of the very young 'paradigm shift', if there is one, associated with chaos theory.

Exercise 3. *(a) Consider the function $f(z) = z^2$ on the unit circle C, or $|z| = 1$ in the complex plane.*

Find all periodic points of period n, that is, points z such that $f^n(z) = z$. How many points of period 4 have prime period 4, that is, $f^n(z) \neq z$ for $0 < n < 4$?

Show that given z, and $\varepsilon > 0$, we can find z' and an integer n such that: (1) $d(z, z') < \varepsilon$; (2) $d(f^n(z), f^n(z')) > \frac{1}{2}$, where d denotes distance on the circle. (This is sensitive dependence, for a very simple map).

(b) Define $g(z) = z^2 + c$ where z is complex. Show that g has (in general) two fixed points z_0, z_1; and that the condition that one of them should satisfy $|g'(z)| = 1$ is that c lies on a curve (a cardioid) in the c-plane. (This curve bounds the largest area in pictures of the Mandelbrot set.)

7 From topology to categories

We cannot think any object except by means of the categories; we cannot know any object except by means of intuitions corresponding to these concepts. (Kant 1993, p. 117)

[M]athematics is about to go through a second revolution at this moment. This is the one which is in a way completing the work of the first revolution,[9] namely which is releasing mathematics from the far too narrow limits imposed by 'set'; it is the theory of *categories and functors* . . . (Dieudonné, 1961 lecture, cited in Corry 2004, p. 383)

'Algebraic topology', which we have seen in process of defining itself in the 1920s, had progressed by the 1950s to a massively successful and integrated subject. Indeed as the century progressed it was constantly growing and subdividing (like most other fields in mathematics) into further areas of specialization. Any kind of survey which took in Smale's horseshoe map (Fig. 5), Vaughan Jones's knot invariants, and the wide variety of contributions made by Michael Atiyah at the interface of topology, differential equations and even physics would become a whole chapter in itself. We should, though, find space to consider topology's illegitimate child category theory which succeeded in doing what, as we have seen, the Bourbakists failed to do: to provide a clear idea of what was meant by 'structure'.

---

9. Axiomatization and structure, briefly.

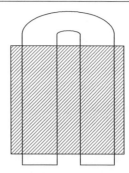

Fig. 5 The 'Smale horseshoe map' is often quoted as an example of chaotic behaviour. The map f takes the square, stretches it, and folds it over into the horseshoe shape. (Think of making puff pastry, if that helps). If the map f is repeated over and over again, most points in the square end up outside; those that remain inside (forever) wander around chaotically.

The reason why the idea came from topology is interesting in itself. As we have seen, Alexandrov, (with some help from Noether) came to understand that many basic topological 'invariants' associated to a space X a *group* $G(X)$. It was also realized—indeed it had been implicit for some time— that if you had a continuous map f from X to Y, then you would be given a homomorphism $G(f)$ from $G(X)$ to $G(Y)$. This is true[10] of the fundamental group $\pi_1(X)$, defined by looking at loops in X. f maps loops in X to loops in Y, and so—with appropriate care! (see a textbook)—defines a homomorphism of groups from $\pi_1(X)$ to $\pi_1(Y)$.

Even this observation was probably too trivial to deserve a formal language. However, let us give it one, following Eilenberg and MacLane in their path-breaking paper of 1942. We say that we have a *category* when we have a set of 'objects' (e.g. spaces or groups), and maps ('morphisms') between the objects (e.g. continuous maps or homomorphisms); and obvious rules for how the maps should behave:

1. given X, it has an 'identity morphism' 1_X;
2. given morphisms f from X to Y and g from Y to Z, the composition $g \circ f$ from X to Z is defined, and (of course);
3. $(h \circ g) \circ f = h \circ (g \circ f)$ if the compositions of morphisms are defined.

So, topological spaces (and continuous maps) form a category usually called **Top**; and groups (and homomorphisms of groups) form a second category, which we call **Gp**.[11] A *functor* from **Top** to **Gp** is a machine G which

(i) to any object X in **Top** assigns an object $G(X)$ in **Gp**;
(ii) to any **Top**-morphism f from X to Y assigns a **Gp**-morphism $G(f)$ from $G(X)$ to $G(Y)$;
(iii) such that identity maps go to identity maps, and compositions to compositions.

---

10. Strictly, it is not. The fundamental group needs a 'basepoint' to define it, and so is defined on the category of 'spaces with a basepoint'.
11. The quotation from Kant, which opens the chapter, and which places what he called 'categories' at the centre of thought, is probably irrelevant—there is no evidence that Eilenberg and MacLane had Kant in mind. And yet one wonders where the word came from.

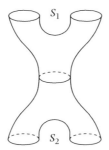

Fig. 6 This object (also called a 'string worldsheet') is (a) a surface, (b) the story of two strings which (reading time upwards) coalesce and then divide again, (c) a morphism in the category of strings from the two loops at the bottom to the two loops at the top. (See Baez, n.d).

You may (if you have done any of this kind of mathematics before) find that this is an amazingly enlightening idea; or you may sympathize with P. A. Smith, who allegedly said of Eilenberg and MacLane's work that he had never read a more trivial paper in his life (Corry 2004, p. 361). The ideas of category, functor and (still more important, but there is no space for it) natural transformation were in any case unexpectedly useful for the algebraic topologists of the 1940s and 1950s as they tried to pin down the ways in which one structure (group, or family of groups, etc.) carried information about another (space).

However, equally obviously, the way in which they are defined above has nothing to do with topology; and the ideas of category theory, besides becoming (naturally) a rapidly burgeoning field on their own account, were taken over in algebra, algebraic geometry, number theory, even analysis in the 1950s and 1960s. As with the 'old' abstract Bourbaki viewpoint, there were those who warned that the new categorical viewpoint was doing nothing but turn out trivialities ('general nonsense' was a term much in use among those who needed the theory but did not wish it to be thought that that was all they were doing). And, if we remarked with alarm in the last chapter that mathematics had moved from its traditional concerns to become centred on sets, category theory (as Dieudonné points out in his early celebration of the idea) is capable of finding sets too restrictive. Where next?

The attentive reader will notice that a category by no means escapes the set theoretic problems of the last chapter, and 'all groups' is much too large a set to be dealing with if one wants to avoid paradoxes. Sometimes this worries those who use categories; sometimes they assume that they, or the reader, is taking care in some specified way. Everything connects; and in the last 30 years, we have learned to see a category as defining a topological space; and to construct categories in which the sets of morphisms themselves have some extra structure. And (as a guard against the accusation of excessive abstraction) we could exhibit the category, crucial for string theory, whose *objects* (simplifying again) are sets of strings (circles); while the *morphisms* from one set of strings S_1 to another set S_2 are *surfaces* (see Fig. 6) whose upper boundary is S_1, and lower boundary S_2.

8 Physics

He now pushed away the paper, covered with formulae and symbols, on which the last thing he had written was an equation of state of water, as a physical example, in order to apply a new mathematical operation that he was describing. (Musil 1953 vol. 1, p. 128.)

All this was familiar to me from my research in high-energy physics, but until that moment I had only experienced it through graphs, diagrams, and mathematical theories. As I sat on that beach my former experiences came to life; I 'saw' cascades of energy coming down from outer space, in which particles were created and destroyed in rhythmic pulses; I 'saw' the atoms of the elements and those of my body participating in this cosmic dance of energy; I felt its rhythm and I 'heard' its sound, and at that moment I knew that this was the Dance of Shiva, the Lord of Dancers worshiped by the Hindus. (Capra 1983, p. 9)

One quite unexpected development, particularly of the mid to late twentieth century, has been that physics has become detached from mathematics. This does not mean that physicists in general see their subject in terms of the dance of Shiva; Capra is to be seen as a symptom, an index of the kind of statement which some people on the margins of physics now think they can get away with. The relation between physics and mathematics was close, almost incestuous until fairly recently. In the golden age of the eighteenth century, following Newton, physics or 'rational mechanics' could be seen as a particular kind of mathematics; the study of certain principles—first those laid down in the *Principia*, and later more abstract versions such as the Principle of Least Action. While physics necessarily did need experimentalists to advance (in new fields like electricity and magnetism, for example), it was possible to be a mathematician and study the diffusion of heat, or the vibrations of a drum, without leaving one's desk—as Kovalevskaya, in the late nineteenth century, solved the problem of the rotating top. Such, in the early twentieth century, was Musil's hero Ulrich, using formulae and symbols to describe an equation of state of water. We can see this happy situation as an outcome of the Scientific Revolution (Chapter 6), which saw the behaviour of the physical world as ordered by mathematical laws.

Indeed, which laws they were was not immediately important, and both the special and the general theories of relativity provided more employment for mathematicians, by replacing one mathematical description by another. However, by the 1920s it appeared that the physicists were becoming more impatient. Early formulations of the quantum theory were investigated and seen to work before their often ad hoc mathematical underpinnings were certified legal. Further, a combination of doubt about the continuum (Chapter 9) and uncertainty about measuring the extremely small raised some questions—which had been left in suspension—about whether traditional mathematics was as well-adapted to the universe as one had hoped. Nevertheless, the new quantum theory remained completely dependent on quite complex mathematical ideas. In his classic article 'The Unreasonable Effectiveness of Mathematics in the Natural Sciences', Eugene Wigner cites the example of the helium atom.

The miracle occurred only when matrix mechanics, or a mathematically equivalent theory, was applied to problems for which Heisenberg's calculating rules were meaningless. Heisenberg's rules presupposed that the classical equations of motion had solutions with certain periodicity properties; and the equations of motion of the two electrons of the helium atom, or of the even greater number of electrons of heavier atoms, simply do not have these properties, so that Heisenberg's rules cannot be applied to these cases. Nevertheless, the calculation of the lowest energy level of helium, as carried out a few months ago by Kinoshita at Cornell and by Bazley at the Bureau of Standards, agrees with the experimental data within the accuracy of the observations, which is one part in ten million. Surely in this case we 'got something out' of the equations that we did not put in. (Wigner, 1960)

Simplified, the helium atom is a three-body problem (nucleus and two orbiting electrons (Fig. 7)), and while experimenters could observe the energy levels, mathematical physicists—again working on paper—did calculations as Wigner describes to check whether theory and observation agree. They were, however, increasingly growing apart from the dominant trends in pure mathematics, in spite of a serious engagement in the field by, among others, Weyl, Noether, and van der Waerden,

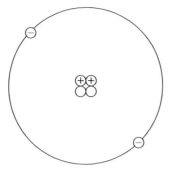

Fig. 7 The 'classical' model of the helium atom: two electrons orbiting a nucleus. The quantum model, which fails to distinguish between the electrons and smears them out over a wide radius, is too difficult to draw.

all 'modernists'. In the 1930s John von Neumann wrote a book (1996) one of whose aims was to show that the theory could be adequately developed without the bizarre functions which Dirac had introduced.[12] Twenty years later, Laurent Schwarz developed a respectable mathematical theory ('distributions') in which Dirac's functions made perfect sense. But physicists could not wait 20 years to be allowed to proceed.

The situation grew worse in the 1940s with the 'success' of quantum electrodynamics—the study of fields which were, at least potentially, infinite systems of particles. This is certainly a sign of the unreasonable effectiveness of something, but not of mathematics as most mathematicians would accept it. To quote an online encyclopedia's summary:

It was immediately noticed, however, that self-interactions of particles would give rise to infinities, much as in classical electromagnetism. At first attempts were made to avoid this by modifying the basic theory ... But by the mid-1940s detailed calculations were being done in which infinite parts were just being dropped—and the results were being found to agree rather precisely with experiments. In the late 1940s this procedure was then essentially justified by the idea of renormalization: that since in all possible QED processes only three different infinities can ever appear, these can in effect systematically be factored out from all predictions of the theory. (www.wolframscience.com/reference/notes/1056a)

Confused? A more sophisticated form of the theory, 'dimensional renormalization', involved writing the equations in $4 + \varepsilon$ dimensions (where they were not infinite), calculating the infinite term as $\varepsilon \to 0$ and removing it. It looks like nonsense to many mathematicians, but it gives accurate predictions.

And yet, theoretical physicists still behave, as they used to, like a kind of mathematician, writing down equations and manipulating them according to agreed rules to see if they work; the fact that most conventional mathematicians do not understand or believe the rules is immaterial. Moreover, they constantly find it useful to raid developing parts of 'pure' mathematics—Riemann surfaces, knots, complex three-manifolds, ...—for some idea which may be useful in a new model. Worse, there has been traffic the other way; important theorems in 'pure' mathematics have been proved by (in particular) Ed Witten by methods which many find suspect since they derive from the most high-flown of infinite physical procedures. Have we returned to the dark ages of Newton and Leibniz? An interaction of physics and mathematics is being preserved, but the power relations have shifted; and

---

12. Specifically, $\delta(x)$, infinite when $x = 0$ and zero otherwise; and, worse, the derivatives of $\delta(x)$.

for many mathematicians there is a strong temptation to see Fritiof Capra's intoxicated description of the Tao of modern physics as the literal truth.

9 Fermat's Last Theorem

I think I'll stop here. (Andrew Wiles, Cambridge, 1993)

Before beginning I would have to put in three years of intensive study, and I haven't that much time to squander on a probable failure. (Hilbert 1920, on being asked why he did not attempt to prove Fermat's Last Theorem)

And so to Andrew Wiles. One would like to say, after all the international excitement, that his proof was in some way peripheral to this story, an isolated result. This is obviously not so, although the reasons are quite complex. The fact that Wiles was stimulated in childhood by E. T. Bell's romantic personalized anecdotal book *Men of Mathematics* to nurse an ambition to solve the problem is in itself an index of the power which a certain view of the history of mathematics can exercise. The Last Theorem (FLT for short in what follows) states, if you have not seen it before, that the equation

$$x^n + y^n = z^n$$

has no solutions (x, y, z) in non-zero integers, for $n > 2$[13]; it was claimed by Fermat in the 1630s, but never proved, and has remained a challenge ever since. Andrew Wiles (of Princeton, however much the English like to claim him as their own) announced its proof in 1993; flaws were found in the proof, but a corrected version with help from Richard Taylor appeared in 1994 and is now accepted. His famous ending to his Cambridge lecture quoted above (translation: I have solved the most difficult problem there is, but I am too modest to say so) has become a favourite tag for mathematicians.

Without going into detail on the history of FLT, we should note its role in the development of ideals and factorization theory by Kummer and Dedekind in the mid-nineteenth century. There is an illuminating study on this by Catherine Goldstein (1995) which shows up the historicity of the problem; the difference between the seventeenth-century context of Fermat and the nineteenth-century one of Kummer:

At the end of the eighteenth century, number theory was still no more than a flower-filled country lane, disdainfully ignored by the great mathematical roads. Jean-Étienne Montucla, the first historian of mathematics, was still able to write: 'Geometry is still the general and only key to mathematics.' A woman, Sophie Germain, prevented by her sex from following a course of higher education, was still able successfully to solve certain cases of Fermat's problem by elementary methods and to maintain a real exchange with Gauss . . . Whatever the always keen interest Gauss always had in numbers, one is still very far from Kummer, who began his researches on this field as soon as he was appointed to a university post. (Goldstein 1995, p. 367)

The context of the late twentieth century is different again, of course. FLT, far from being *the* problem (if it had ever been so) had become marginal, and it plays little part in the work of such key number theorists as Hardy, Ramanujan, and André Weil. Interest in what one might have

---

13. There are of course plenty of solutions for $n = 2$; these are the 'Pythagorean triples'.

considered either a hopeless pursuit or a backwater was revived by Ken Ribet's work in linking it to a central concern of *modern* number theory—the modularity conjecture. Ribet, one could say, gave the theorem a little contemporary relevance; and Wiles produced not a 'classical' but a very contemporary theorem.

What of his work-habits? They also can be seen as belonging to the late twentieth-century setting. By the 1980s number theory had been a 'professional' study for 150 years; now, with diminishing funding and constant demands for publication to justify the researcher's existence, it had become (like the rest of mathematics) intensely competitive. Wiles's understanding that news of his work on the conjecture might stimulate others to enter the field may appear paranoid, but is perhaps not as unusual as Singh makes out with his images of loneliness, deviousness, silence, and withdrawal. And there is a contrast with the more social ethic of 20 years earlier in the chapters where he clarifies the crucial role of a number of others, most notably Taniyama, Shimura, and Weil (for the fundamental conjecture on elliptic curves), and Frey and Ribet (for the construction of a particular curve which links FLT to the Taniyama–Shimura–Weil conjecture). The fact that the latter conjecture seemed no easier than FLT itself[14] is also important if we are to have a balanced view of the history, rather than one centred on a 300-year-old problem. It is hard: already in 1993 t-shirts were on sale in Cambridge which read: 'Fermat's Last Theorem proved by Andrew Wiles at the Isaac Newton Institute'. Jaundiced mathematicians might well object that no single one of the statements was true, but history books will print the legend.[15]

What kind of mathematics are we talking about? This is late twentieth-century number theory of a very advanced kind. For all his 'seclusion', Wiles's paper is packed with references to work published during the previous five years. More generally, the key conjecture relates elliptic curves and modular forms, and without a great deal of theory about both of these abstruse and difficult mathematical objects developed in the previous century, there would certainly have been no result.

Singh makes a brave attempt to explain the various objects in terms of analogies (the idea that the *L*-series of a modular form is its 'DNA' is certainly striking, even if a number theorist would find it strange—do they replicate?). Elliptic curves are, in the crudest sense, easy: they are defined by an equation

$$y^2 = x^3 + ax + b$$

where $4a^3 + 27b^2 \neq 0$ (to ensure no double roots of the cubic, as Omar Khayyam knew). The picture (Fig. 8) shows one such; but to do number theory one must (a) think of a and b as rational numbers, and (simultaneously) (b) consider all solutions of the equation with (x, y) in C, a four-dimensional picture which we have difficulty drawing— although 'intrinsically', if we forget the space it is in, it looks like a torus (Fig. 9)

To give more details, to try to explain modular forms, etc., would take us outside the aim of this book, if not outside history; and other sources (e.g. van der Poorten 1996) can do it better than I can—you are encouraged to consult them.

---

14. In fact, Wiles's paper proves *enough* of the Taniyama conjecture to settle FLT, but not all of it; the full conjecture has since been settled by Breuil, Conrad, Diamond, and Taylor.

15. John Ford: *The Man Who Shot Liberty Valance*, 1962.

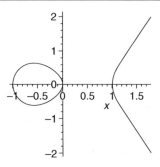

Fig. 8 Graph of $y^2 = x^3 - x$.

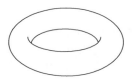

Fig. 9 A torus (again); this is the shape of a (complex) elliptic 'curve' in the complex projective plane; two real dimensions, one complex dimension.

Appendix A. From Bourbaki, 'Algebra', Introduction

HOW TO USE THIS TREATISE

1. The treatise takes mathematics at its beginning, and gives complete proofs. Consequently its reading presupposes, in principle, no mathematical knowledge, but only a certain habit of mathematical reasoning, and a certain ability to abstract.

    Nonetheless, the treatise is particularly aimed at readers who have at least a good knowledge of the subjects taught, in France, in courses of *'mathématiques générales'* (abroad, in the first or first two years of university), and, if possible, a knowledge of the essentials of the differential and integral calculus.

2. The first part of the treatise is devoted to the fundamental structures of analysis (on the meaning of the word 'structure' see book I, chapter 5); in each of the books into which this part is divided, we study one of these structures, or several structures which are closely related (book I, *Theory of Sets*; book II, *Algebra*; book III, *General Topology*; books to follow: *Integration, combinatorial topology, differentials and integrals of differentials*, etc.) . . .

The method of exposition followed in the first part is axiomatic and abstract; it proceeds on the whole from the general to the particular. The choice of this method was imposed by the principal aim of this first part, which was to provide solid foundations for all the rest of the treatise, and even for the whole of modern mathematics.

Appendix B. Turing on computable numbers

We have said that the computable numbers are those whose decimals are calculable by finite means. This requires rather more explicit definition. No real attempt will be made to justify the definitions given until we reach §9. For the present I shall only say that the justification lies in the fact that the human memory is necessarily limited.

We may compare a man in the process of computing a real number to a machine which is only capable of a finite number of conditions $q_1, q_2, \ldots, q_R$ which will be called 'm-configurations'. The machine is supplied with a 'tape', (the analogue of paper) running through it, and divided into sections (called 'squares') each capable of bearing a 'symbol'. At any moment there is just one square, say the rth, bearing the symbol $\mathfrak{S}(r)$ which is 'in the machine'. We may call this square the 'scanned square'. The symbol on the scanned square may be called the 'scanned symbol'. The 'scanned symbol' is the only one of which the machine is, so to speak, 'directly aware'. However, by altering its m-configuration the machine can effectively remember some of the symbols which it has 'seen' (scanned) previously. The possible behaviour of the machine at any moment is determined by the m-configuration q_n and the scanned symbol $\mathfrak{S}(r)$. This pair $q_n, \mathfrak{S}(r)$ will be called the 'configuration': thus the configuration determines the possible behaviour of the machine. In some of the configurations in which the scanned square is blank (i.e. bears no symbol) the machine writes down a new symbol on the scanned square: in other configurations it erases the scanned symbol. The machine may also change the square which is being scanned, but only by shifting it one place to right or left. In addition to any of these operations the m-configuration may be changed. Some of the symbols written down will form the sequence of figures which is the decimal of the real number which is being computed. The others are just rough notes to 'assist the memory'. It will only be these rough notes which will be liable to erasure.

It is my contention that these operations include all those which are used in the computation of a number. The defence of this contention will be easier when the theory of the machines is familiar to the reader. In the next section I therefore proceed with the development of the theory and assume that it is understood what is meant by 'machine', 'tape', 'scanned', etc.

2. *Definitions.*

Automatic machines.

If at each stage the motion of a machine (in the sense of §1) is *completely* determined by the configuration, we shall call the machine an 'automatic machine' (or a-machine). For some purposes we might use machines (choice machines or c-machines) whose motion is only partially determined by the configuration (hence the use of the word 'possible' in §1). When such a machine reaches one of these ambiguous configurations, it cannot go on until some arbitrary choice has been made by an external operator. This would be the case if we were using machines to deal with axiomatic systems. In this paper I deal only with automatic machines, and will therefore often omit the prefix a-.

Computing machines.

If an a-machine prints two kinds of symbols, of which the first kind (called figures) consists entirely of 0 and 1 (the others being called symbols of the second kind), then the machine will be called a computing machine. If the machine is supplied with a blank tape and set in motion, starting from the correct initial m-configuration, the subsequence of the symbols printed by it which are of the first kind will be called the *sequence computed by the machine*. The real number whose expression as a binary decimal is obtained by prefacing this sequence by a decimal point is called the *number computed by the machine*.

At any stage of the motion of the machine, the number of the scanned square, the complete sequence of all symbols on the tape, and the m-configuration will be said to describe the complete configuration at that stage. The changes of the machine and tape between successive complete configurations will be called the moves of the machine.

Solutions to exercises

These are perhaps unduly hard, but it is almost impossible to find easier ones; and of course, no exercises have been included on Fermat...

1. (a) If $(m^n)^p = m^{(n^p)}$, then $m^{np} = m^{(n^p)}$. This means that either $m = 1$ (since then all powers of m are the same) or $np = n^p$ and m is arbitrary; equivalently, $n^{p-1} = p$. This is possible (1) if $p = 1$, n arbitrary; or (2) if $n = p = 2$.

    (b) Finding a system of generators is easy; inductively, one chooses any a_1. If $a_1, \ldots, a_{i-1}$ have been chosen and do not generate the group, one chooses some a_i in G but not in the subgroup generated by $a_1, \ldots, a_{i-1}$. Since G is finite, eventually the process ends.

    From the characterization of the generators, one deduces that the products $a_{i_1}, \ldots a_{i_k}$ with $i_1 < i_2 < \cdots < i_k$ are all different. There are 2^m of these, so $2^m \le n$. And any automorphism f of G is determined by $f(a_1), \ldots, f(a_m)$. Since there are at most n possibilities for each of these, the number of automorphisms is $\le n^m$. Now using $m \log 2 \le \log n$, $m \le \log n / \log 2$, which proves it.

2. (a) The tape must of course not only contain the figures, but a symbol to tell the machine where the number stops; it must therefore read something like 'x1011', where 'x' means 'stop here' (otherwise the machine would have to move on in case of finding other figures!) At the start, the machine is at the rightmost position, and it reads and writes moving left. We can characterize the states as: S (start), N (not carrying), C (carrying), and F (finish). Rules are:
    1. If state is S and symbol read is 0, change state to N, change symbol to 1, move left. If symbol read is 1, change state to C, change symbol to 0, move left.
    2. If state is N and symbol read is 0 or 1, move left and do not change state or symbol. If symbol read is x, change state to F, do not change symbol, stop.
    3. If state is C and symbol read is 0, change state to N, change symbol to 1, move left. If symbol read is 1, change symbol to 0 and do not change state, move left.
    4. If state is C and symbol read is x, change symbol to 1, move left, write x, change state to F, stop.

    (b) The programme computes the sequence:

    $x_0 = 1$; $x_n = 1 + (1/x_{n-1})$. these are (the computer's approximations to) the successive steps in the continued fraction

    $$1 + \frac{1}{1 + (1/1 + \cdots)}$$

    This is the solution of $x = 1 + (1/x)$, or $(1 + \sqrt{5})/2$. How accurate the answer is depends on the number of places stored in the computer by the particular programme you are using. What is printed is the 100th approximation, which will probably be quite close. (In one version which I used, after about 25 steps the programme did not fix, but got into a two-cycle: $a, b, a, b, \ldots$ where a and b differed in the last decimal point. This illustrates the way in which the finiteness of computers affects the result.)

3. (a) $f^2(z) = z^4$, $f^3(z) = z^8$, and in general $f^n(z) = z^{(2^n)}$. Hence, $f^n(z) = z$ if $z^{(2^n)} = z$. Discount $z = 0$ (we are on the circle), and we are left with $z^{(2^n-1)} = 1$; z is a $(2^n - 1)$th root of 1, $\exp(2k\pi/(2^n - 1))$.

In the case $n = 4$, $2^n - 1 = 15$; $z = \exp(2k\pi/15)$. If $k = 0$, z is of period 1 (fixed), while if $k = 5, 10$, z is of period 2 (using $2^2 - 1 = 3$). The other twelve fifteenth roots of 1 give three 4-cycles.

Already with these facts, sensitive dependence looks likely. It is easy to *prove* if we write $z = \exp(i\theta)$; then what we have called $d(z, z')$ is just $|\theta - \theta'|$. Now f takes θ to 2θ. Given z and ε, let n be such that $(1/2^{n+1}) < \varepsilon$, and let $z' = \exp i(\theta + (1/2^{n+1}))$; then $f^n(z) = \exp i(2^n\theta)$ and $f^n(z') = \exp i(2^n\theta + \frac{1}{2})$.

(b) Fixed points are given by $z^2 + c = z$, or $z = \frac{1}{2}(1 + \sqrt{1-4c})$. Now $g'(z) = 2z$, and so at a fixed point $|g'(z)| = |1 + \sqrt{1-4c}|$. If $|g'(z)| = 1$, write $|2z| = 1, z = \frac{1}{2}\exp(i\theta)$. Then

$$c = \frac{1}{2}\exp(i\theta) - \frac{1}{4}\exp(2i\theta)$$

from the original equation. So c has coordinates $(\frac{1}{2}\cos\theta - \frac{1}{4}\cos 2\theta, \frac{1}{2}\sin\theta - \frac{1}{4}\sin 2\theta)$. This describes a cardioid as θ goes from 0 to 2π.

Conclusion

Yesterday all the past. The language of size
Spreading to china along the trade-routes; the diffusion
Of the counting-frame and the cromlech;
Yesterday the shadow-reckoning in the sunny climates...
Yesterday the classic lecture
On the origin of Mankind. But today the struggle.   (W.H.Auden, 'Spain', 1936.)

Only an alert and knowledgeable citizenry can compel the proper meshing of the huge industrial and military machinery of defense with our peaceful methods and goals, so that security and liberty may prosper together. Akin to, and largely responsible for the sweeping changes in our industrial-military posture, has been the technological revolution during recent decades. In this revolution, research has become central; it also becomes more formalized, complex, and costly.

A steadily increasing share is conducted for, by, or at the direction of, the Federal government. Today, the solitary inventor, tinkering in his shop, has been overshadowed by task forces of scientists in laboratories and testing fields. In the same fashion, the free university, historically the fountainhead of free ideas and scientific discovery, has experienced a revolution in the conduct of research. Partly because of the huge costs involved, a government contract becomes virtually a substitute for intellectual curiosity. For every old blackboard there are now hundreds of new electronic computers. The prospect of domination of the nation's scholars by Federal employment, project allocations, and the power of money is ever present and is gravely to be regarded.   (Speech by Dwight D. Eisenhower, 1961.)

At the end of this book, we look to the present and the future as the above quotations do. For the young Auden, the history of mathematics lay in the past, while the present belonged to agitation, rifles and comrades. For the old ex-president Eisenhower, technology was central, its nature was changing, and it was, as he was in a good position to know, controlled by what he termed the 'military-industrial complex'. Both of them belong to a time which seems remote, and mathematics continues to be produced — for what, and for whom? The military have certainly not reduced their demands, and they have money to spend.

Researchers at Duke, Georgia Institute of Technology, Stanford University and the University of Michigan will each take on different parts of developing the enabling mathematical underpinnings of this technology with $6 million in Defense Advanced Research Projects Agency (DARPA) funding, which will be administered through the U.S. Army.

The objective...is the development of 'detection and classification algorithms for multi-modal inverse problems.' That means developing mathematical rules – called algorithms – to 'train' and control multiple sensors that, with increasing precision, could detect invisible signals emanating from such targets, and trace those signals back to their sources – a technique called inversion.

'The targets could be land mines, targets under trees like tanks or troops, or targets in underground bunkers or caves,' said the overall administrator of the grant.   (Report in www.spacedaily.com, April 2002.)

Conclusion

Is the development of algorithms (the word comes by a circuitous route from the name al-Khwārizmī, chapter 5) to trace targets in underground bunkers[1] an indicator of how mathematics, and you, readers, as mathematicians, relate to today's 'struggle'?

This book is long enough already; and to raise serious questions about mathematics' present is really a question for another book. Still, in the fashionable language of course objectives, it could be useful to consider what you have learned. It's not usual today to see history as a source of 'lessons', and mathematicians rarely think that by understanding the past they might avoid repeating its mistakes. Still less do they appeal to the court of history, or, like politicians, claim that it will absolve them. We have seen mathematicians — Archimedes, Qin Jiushao, Galileo, Alan Turing — getting involved in 'history', making political choices which may have had little to do with their mathematical tastes. Do we judge them? We ourselves, having acquired their necessary skills, will inevitably go out into the world and (like anyone else) become immersed in history, if of a different sort. It may indeed be that we seek to be among the recipients of a large grant from the US Army.

Mathematics is traditionally seen as a peculiar kind of human activity. Hard, pure, exact and faultless it is divorced from the day-to-day concerns of ordinary people, and has little or nothing to say about their real-life problems. And yet it is a cliché by now that various applications of mathematics from computing to coding to accountancy pervade the world today. They are not pure, they are not always even reliable [2]. The student attracted by the traditional image of mathematics might wish rather to be studying one of the seven 'Millennium Problems' (www.claymath.org/millennium/). Although not all to be classed as pure research — they include problems in hydrodynamics and quantum field theory — they are all goals for the academically ambitious, lead to a prize of $1,000,000 each and could allow a mathematician to escape the job market and enjoy a life of leisure. A simple sum shows that few of this book's readers can hope to attain this goal.

For most of the world, a shrinking market and increased globalization constrain choices, though they may provide unexpected windfalls. The advice to train as a software developer, which the American Mathematical Society was tendering to graduates ten years ago, rings hollow as illustrated with human detail in *Business Week* in March 2004:

As Stephen and Deepa emerge this summer from graduate school – one in Pittsburgh, the other in Bombay – they'll find that their decisions of a half-decade ago placed their dreams on a collision course. The Internet links that were being pieced together at the turn of the century now provide broadband connections between multinational companies and brainy programmers the world over. For Deepa and tens of thousands of other Indian students, the globalization of technology offers the promise of power and riches in a blossoming local tech industry. But for Stephen and his classmates in the U.S., the sudden need to compete with workers across the world ushers in an era of uncertainty. Will good jobs be waiting for them when they graduate? 'I might have been better served getting an MBA,' Stephen says.

Worse, Deepa is likely in turn to find herself losing her job as the software companies find yet cheaper sources. Not only is mathematics, as we suggested above, less different from other human activities than its idealized, originally Greek model might lead us to suppose. Mathematicians as a caste are less sharply marked off by their unique abilities from others. The corporations which employ them may ask in addition for a range of skills from Powerpoint presentations to a

---

1. Having cited this as an application of mathematics to military uses, and so automatically undesirable, we should in fairness note that the UN in October 2004 recommended its application as preferable to the Israeli Army's wholesale demolition of houses in Gaza where there might (in their opinion) be a bunker.

2. The failure of a US Patriot missile in 1991 leading to the loss of 28 lives, which was due to accumulated 'rounding errors' is often cited.

nodding acquaintance with intellectual property law. As in ancient Babylon, the military-industrial complex is looking for trained personnel with the widest possible range; and mathematics as an isolated nerdy pursuit of those unable to communicate their insights is no longer the goal, if it ever was. This book would like to encourage readers by assuring them that the enriching culture of the history of mathematics will make them more marketable, but we have a legal and moral duty to be truthful. The knowledge that empires rise and fall is no use and not much comfort if you are out of work in Pittsburgh.

Teachers, who deserve special words of praise and encouragement, may escape the global market for longer — particularly if they choose to teach at a level or in a country where e-learning is too difficult or expensive to be widely applied. They will be subject to the usual pressures to produce results, and the mathematics to be taught will change constantly, particularly at the secondary or higher level. The varieties of mathematics available are evolving, as we saw in chapter 10, and computers, which do afford more scope for experiment and visualization than traditional mathematics, have often become cheap enough, even outside the developed world, to be considered essential learning aids. Teaching flourishes under conditions where it is freed from control and dogmatism; the understanding that mathematics has a history can help with this (as we saw Simone Weil claiming in the introduction). And here this book may be of use, in a modest way.

Is a different future possible? Nearly forty years ago, the universities which Eisenhower had seen as controlled by government were in revolt. The radical mathematician Steve Smale received the highest award (the Fields Medal) at the Moscow International Congress of Mathematicians, 1966; he chose the admittedly small and élite setting of the congress to launch an attack on the United States' war in Vietnam — and simultaneously (and with even less tact) against the Soviet invasion of Hungary eight years earlier. Attacked by his government, he was seen by his peers as the spokesman of a generation who wanted change. Remembered as an inspiring 'moment' by many mathematicians and others, the episode was also something of a one-off, an idiosyncratic statement which was not followed up by a widespread withdrawal of mathematicians from the world of army grants and corporate employment. In today's situation where mathematicians are increasingly involved (like all citizens) with the problems of war and global inequality, they need to consider their common destiny and how as participants in an international civil society they might take action to control it.

Bibliography

Abū Kāmil, Shujā ibn Aslam: *The Algebra of Abū Kāmil in a commentary by Mordecai Finzi*, tr. and ed. Martin Levey, Madison and London: Wisconsin University Press, 1966.
Abū-l-Wafā al-Buzjānī, *Kitāb fī mā yaḥtāju ilayhi al-ṣāni' min 'ilm al-handasah*, ed. and tr. S. A. Krasnova, in *Fiziko-matematicheskiye nauki v stranakh vostoka*, Vypusk I (IV), pp. 42–140, Moscow: 'Nauka', 1966.
Abū-l-Wafā al-Buzjānī, *Kitāb fī mā yaḥtāju ilayhi al-ṣāni' min 'ilm al-handasah*, ed. Ṣāliḥ Aḥmad al-'Alī, Baghdad: Baghdad University, 1979.
Adorno, Theodor and Max Horkheimer, *Dialectic of Enlightenment* (trans. John Cumming). London: Verso, 1979.
Aiton, E. J., *Leibniz: A Biography*. Bristol: Hilger, 1985.
Al-Bīrūnī (Abū al-Raiḥān Muḥammad ibn Aḥmad al-Bīrūnī), *Kitāb maqālid 'ilm al-hayā'. La trigonométrie sphérique chez les Arabes de l'Est à la fin du Xe siècle*, ed. and tr. Marie-Thérèse Debarnot, Damascus: Institut Français de Damas, 1985.
Al-Bīrūnī (Abū al-Raiḥān Muḥammad ibn Aḥmad al-Bīrūnī), *The Determination of the Coordinates of Position for the Correction of Distances between Cities*, tr. and ed. Jamil Ali, Beirut: American University of Beirut, 1967.
Al-Daffa, Ali Abdulla and J. J. Stroyls, *Studies in the exact sciences in medieval Islam*, New York: Wiley, 1984.
Al-Kāshī (Jamshīd Ghiyāt al-Dīn al-Kāshī), *Miftāḥ al-ḥisāb*, Cairo: Dār al-Kātib al-Arabī, 1967.
Al-Khwārizmī (Muḥammad ibn Mūsa al-Khwārizmī), *The Algebra of Mohammed Ben Musa*, tr. F. Rosen, London 1831, repr Hildesheim: Olms, 1986.
Al-Khwārizmī (Muḥammad ibn Mūsa al-Khwārizmī), *Le Calcul Indien (Algorismus)*, ed. and tr. from Latin texts by André Allard, Paris: Librairie Scienifique et Technique Albert Blanchard, 1992.
Alic, Margaret, *Hypatia's Heritage. A History of Women in Science from Antiquity through the Nineteenth Century*. Boston, MA: Beacon Press, 1986.
Althusser, Louis, *For Marx*. London: Verso, 1996.
Al-Uqlīdisī, *The Arithmetic of Al-Uqlīdisī* (trans. and ed. A. S. Saidan). Dordrecht: Reidel, 1978.
Archimedes, *Works* (tr. T. L. Heath) reprinted. New York: Dover, 2002.
Artmann, Benno, *Euclid: The Creation of Mathematics*. New York: Springer-Verlag, 1999.
Ascher, Marcia, *Ethnomathematics: A Multicultural View of Mathematical Ideas*. Pacific Grove, CA: Brooks/Cole Publishing Co., 1991.
Ashman, Keith and Philip Barringer, eds., *After the science wars : science and the study of science*, London: Routledge, 2000.
Bachelard, Gaston, *Le Nouvel esprit scientifique*, Paris: PUF, 1934.
Bachelard, Gaston, *The Formation of the Scientific Mind (Philosophy of Science)*. Manchester: Clinamen Press, 2003.
Baez, John, (at http://math.ucr.edu/home/baez/quantum/node1.html).
Barnes, Barry, David Bloor and John Henry, *Scientific knowledge : a sociological analysis*, London: Athlone, 1996.
Batterson, Steve, *Stephen Smale: The Mathematician who Broke the Dimension Barrier*. Providence, RI: AMS, 2000.
Beckett, Samuel, *Murphy*. London: Calder, 1963.
Berggren, J. L., review, *Isis* **94** (2003), 134–6.
Berggren, J. L., *Episodes in the Mathematics of Medieval Islam*. New York: Springer-Verlag, 1986.
Berkeley, George *The Analyst* (numerous editions; online at www.maths.tcd.ie/pub/HistMath/People/Berkeley/Analyst/Analyst.html), 1734.
Bernal, Martin, *Black Athena: The Afroasiatic Roots of Classical Civilization*. London: Free Association Books, 1987 and 1991; New Brunswick: Rutgers University Press, 1991.
Berry, Chuck, 'School Day', 1957.
Bolzano, Bernard, *Beytraege zu einer begruendeteren Darstellung der Mathematik. Erster Lieferung*. Prague, 1810.

Bonola, Roberto, *Non-Euclidean Geometry: with a supplement containing 'The Theory of Parallels' by Nicholas Lobachevski and 'The Science Absolute of Space' by John Bolyai*, New York: Dover, 1955.
Bos, Henk, *Lectures on the History of Mathematics*. Providence, RI: AMS and LMS, 1991.
Bourbaki, N., 'The Foundations of Mathematics for the Working Mathematician'. *Journal of Symbolic Logic* **14** (1948), 1–14.
Bourbaki, N. *Elements of Mathematics: Algebra 1–3*. Berlin and Heidelberg: Springer, 1998.
Boyer, Carl B., *The History of the Calculus and its Conceptual Development*. New York: Dover, 1949.
Boyer, Carl B., and Uta C. Merzbach, *A History of Mathematics*. New York: Wiley, 1989.
Brentjes, Sonja, 'Historiographie der Mathematik im islamischen Mittelalter'. *Archives Internationales d'Histoire des Sciences* **42** (1992), 27–63.
Burkert, Walter, *Lore and Science in Ancient Pythagoreanism* (trans. E. L. Minar, Jr.). Cambridge, MA, Harvard University Press, 1972.
Butterfield, Herbert, *The Origins of Modern Science, 1300–1800*. Toronto: Clarke, Irwin & Co, 1968.
Calinger, Ronald, *A Contextual History of Mathematics*. New York: Prentice-Hall, 1999.
Cameron, A., 'Isidore of Miletus and Hypatia: On the Editing of Mathematical Texts'. *Greek, Roman and Byzantine Studies*, **31** (1990), 103–27.
Cantor, Moritz, *Vorlesungen über Geschichte der Mathematik*, New York: Johnson Reprint Company, 1965.
Capra, Fritjof, *The Tao of Physics*. London: Fontana, 1976.
Carr, E.H., *What is History?* Basingstoke: Palgrave, 2001.
Chasles, Michel, *Aperçu historique sur l'origine et développement des méthodes en géométrie* (Mémoires couronnés par l'Académie Royale des Sciences et Belles-Lettres de Bruxelles). Brussels, Académie Royale, 1837.
Chemla, Karine, 'Similarities Between Chinese and Arabic Mathematical Writings: (I) Root Extraction'. *Arabic Sciences and Philosophy* **4** (1994), 207–67.
Chemla, Karine, 'What is at Stake in Mathematical Proofs from Third-Century China?' *Science in Context* **2** (1997), 227–51.
Cohen, I. B., *Introduction to Newton's Principia*, Cambridge: Cambridge University Press, 1971.
Cohen, Floris, H., *The Scientific Revolution: A Historiographical Inquiry*. Chicago, IL: University of Chicago Press, 1994.
Corry, Leo, *Modern Algebra and the Rise of Mathematical Structures*. Basel, New York, Boston, MA, Berlin: Birkhäuser, 2004.
Costa, Shelley, essay ' "Our" Notation from Their Quarrel: The Leibniz-Newton Controversy in Calculus Texts' (on www1.umn.edu/ships/9-1/calculus.htm, n.d).
Coxeter, H. S. M., *Regular Polytopes*. London and New York: Macmillan, 1963.
Crombie, A. C. (ed.), *Scientific Change: Historical Studies in the Intellectual, Social and Technical Conditions for Scientific Discovery and Technical Invention, From Antiquity to the Present*. New York: Basic Books/London: Heinemann, 1963.
Cullen, Christopher, *Astronomy and Mathematics in Ancient China: The Zhou bi suan jing* (Needham Research Institute Studies 1). Cambridge: Cambridge University Press, 1996.
Cuomo, Serafina, *Pappus of Alexandria and the Mathematics of Late Antiquity*. Cambridge: Cambridge University Press, 2000.
Cuomo, Serafina, *Ancient Mathematics*. London and New York: Routledge, 2001.
da Vinci, Leonardo, *The Notebooks of Leonardo da Vinci* (trans. R. C. Bell and E. J. Poynter and ed. Jean Paul Richter) online as Project Gutenberg release no. 5000 (various sites), 2004.
Dauben, Joseph W., *Georg Cantor*. Princeton, NJ: Princeton University Press, 1990.
de Roover, R., 'Aux origines d'une technique intellectuelle: la formation et l'expansion de la comptabilité à partie double', *Annales d'histoire économique et sociale* **9** (1937), S. 171–93; S. 270–97.
Debord, Guy, *Comments on the Society of the Spectacle* (trans. Malcolm Imrie). London: Verso, 1990.
Dedekind, Richard, *Essays in the Theory of Numbers*. New York: Dover, 1948.
Descartes, René, *Correspondance*, ed. Ch. Adam and G. Milhaud. Paris: Librairie Félix Alcan, 1939.
Descartes, René, *The Geometry of René Descartes* (with a Facsimile of the First Edition) (trans. D. E. Smith and Marcia L. Latham). New York: Dover, 1954.
Descartes, René, *Rules for the Direction of the Mind*, in *The Philosophical Works of Descartes* (trans. by E. S. Haldane and G. T. R. Ross). Cambridge: Cambridge University Press, 1968a.
Descartes, René, *Discourse on Method and the Meditations* (trans. and ed. F. E. Sutcliffe). Harmondsworth: Penguin, 1968b.
Devaney, R. L., *A First Course in Chaotic Dynamical Systems: Theory and Experiment*. New York: Addison Wesley, 1992.

Dick, Auguste (tr. H. I. Blocher), *Emmy Noether, 1882–1935*. Boston, MA, Basel, New York, Stüttgart: Birkhäuser, 1981.
Dieudonné, Jean, *Pour l'honneur de l'esprit humain*, Paris: Hachette, 1987.
Dieudonné, Jean, *A History of Algebraic and Differential Topology*. Basel, New York: Birkhäuser, 1989.
Dijksterhuis, E. J., *Simon Stevin*. The Hague: Martinus Nijhoff, 1970.
Dijksterhuis, E. J., *The Mechanization of the World Picture*. Princeton, NJ: Princeton University Press, 1986.
Dilke, O. A. W., *The Roman Land-Surveyors: An Introduction to the Agrimensors*. Newton Abbot: David and Charles, 1971.
Drake, Stillman, *Galileo Galilei et al., The Controversy of the Comets of 1618*. Philadelphia, PA: The University of Pennsylvania Press, 1960.
Duhem, Pierre, *The Origins of Statics* (trans. Grant F. Leneaux, Victor N. Vagliente, and Guy H. Wagener), *Boston Studies in the Philosophy of Science*, vol. 123. Dordrecht: Kluwer, 1991.
Dupont, Pascal and Clara Silvia Roero, *Leibniz 84: il decollo enigmatico del calcolo differenziale*. Cosenza: Mediterranean Press, 1991.
Dzielska, Maria, *Hypatia of Alexandria* (tr. F. Lyra). Cambridge, MA and London: Harvard University Press, 1995.
Englund, Robert K., 'Hard Work: Where Will It Get You? Labor Management in Ur III Mesopotamia'. *Journal of Near Eastern Studies* **50** (1991) 255–80.
Epple, Moritz, 'Knot Invariants in Vienna and Princeton during the 1920s'. *Science in Context* **17** (2004), 131–64.
Ernest, Paul, *Social Constructivism as a Philosophy of Mathematics*. Albany, NY: SUNY Press, 1998.
Euclid, *The Thirteen Books of Euclid's Elements* (ed. and intro. Sir Thomas Heath). New York: Dover, 1967 (the *Elements* can also be found on the Internet, particularly at aleph0.clarku.edu/ djoyce/java/elements).
Everdell, William R., *The First Moderns: Profiles in the Origins of Twentieth-Century Thought*. Chicago, IL: University of Chicago Press, 1997.
Fairbank, John King, *China: A New History*. Cambridge, MA: Belknap Press of Harvard University Press, 1992.
Fauvel, John and Gray, Jeremy, eds., *The History of Mathematics: A Reader*, Basingstoke: Macmillan/Open University, 1987.
de Fontenelle, Bernard le Bovier, *Éloges des académiciens*, Brussels: Culture et Civilisation, 1969.
Feyerabend, Paul, *Against Method: Outline of an Anarchistic Theory Of Knowledge*. London: New Left Books, 1975.
Field, J. V., *The Invention of Infinity: Mathematics and Art in the Renaissance*. Oxford: Oxford University Press, 1997.
Ford, Ford Madox, *Parade's End*. Harmondsworth: Penguin Modern Classics, 2002.
Forman, Paul, 'Weimar Culture, Causality and Quantum Theory 1918–1927'. *Historical Studies in the Physical Sciences* **3** (1971), 1–115.
Forman, Paul, 'Truth and Objectivity', in *Science*, vol 269 (1995), pp.565–568.
Foucault, Michel, *The Archaeology of Knowledge*. London: Routledge, 2002.
Fowler, David, *The Mathematics of Plato's Academy*, 2nd edn. Oxford: Oxford University Press, 1999.
Freudenthal, Hans, 'Y avait-il un crise de fondements des mathématiques dans l'antiquité?', *Bulletin of the Belgian Mathematical Society* **18** (1966), 43–55.
Galilei, Galileo, *Dialogues Concerning Two New Sciences* (tr. Henry Crew and Alfonso deSalvio). New York: Dover, 1954.
Galilei, Galileo, *Dialogue Concerning the Two Chief World Systems* (tr. Stillman Drake). Berkeley and Los Angeles, CA: University of California Press, 1967.
Gerhardt, C. I. (ed.), *G.W. Leibniz, Mathematische Schriften*, reprint. Hildesheim: Georg Olms Verlagsbuchhandlung, 1962.
Gibbon, Edward, *The Decline and Fall of the Roman Empire* (online at http://www.ccel.org/g/gibbon/decline/index.htm).
Gillies, Donald, (ed.), *Revolutions in Mathematics*. Oxford: Oxford University Press, 1992.
Gjertsen, Derek, *The Newton Handbook*. London and New York: Routledge and Kegan Paul, 1986.
Gleick, James, *Chaos: Making a New Science*. New York: Viking 1987.
Goldstein, Catherine, 'Working with Numbers', in Michel Serres (ed.), *A History of Scientific Thought*. Oxford: Blackwell, 1995, pp. 344–71.
Goody, Jack, *The Logic of Writing and the Organization of Society*. Cambridge: Cambridge University Press, 1986.
Grant, E. (ed.), *Physical Science in the Middle Ages*. Cambridge: Cambridge University Press, 1978.
Grattan-Guinness, I. (ed.), *From the Calculus to Set Theory: An Introductory History*. London: Duckworth, 1980.
Grattan-Guinness, Ivor, *The Fontana history of the mathematical sciences: the rainbow of mathematics*, London, Fontana, 1997.
Gray, Jeremy, *Ideas of Space: Euclidean, Non-euclidean and Relativistic*. Oxford: Oxford University Press, 1979.
Gray, Jeremy, 'Anxiety and Abstraction in Nineteenth-Century Mathematics', *Science in Context* **17** (2004), 23–47.

Greenberg, Marvin Jay, *Euclidean and Non-Euclidean Geometries: Development and History*. San Francisco, CA: W.H. Freeman and Company, 1974.

Greenblatt, Stephen J., *Renaissance Self-Fashioning from More to Shakespeare*. Chicago, IL: University of Chicago Press, 1980.

Guicciardini, Niccolò, *Reading the Principia: The Debate on Newton's Mathematical Methods for Natural Philosophy from 1687 to 1736*. Cambridge: Cambridge University Press, 1999.

de L'Hôpital, Guillaume F. A. *Analyse des infiniment petits, pour l'intelligence des lignes courbes*. Paris, Imprimerie Royale, 1696.

Hadden, Richard W., *On the Shoulders of Merchants: Exchange and the Mathematical Conception of Nature in Early Modern Europe*. Albany, NY: State University of New York Press, 1994.

Haddon, Mark, *The Curious Incident of the Dog in the Night Time*. London: Jonathan Cape, 2003.

Hall, Rupert A., *Philosophers at War: The Quarrel Between Newton and Leibniz*. Cambridge: Cambridge University Press, 1980.

Hardy, G. H., *A Mathematician's Apology*. Cambridge: Cambridge University Press, 1940.

Hartner, Willy and Mathias Schramm, 'Al-Bīrūnī and the Theory of the Solar Apogee (an example of originality in Arabic science)', in A. C. Crombie(ed.), *Scientific Change*, London: Heinemann, 1963, pp. 206–18.

Hasse, H. and Scholz, H., '*Die Grundlagenkrisis der griechischen Mathematik*'. Charlottenburg: Pan-Verlag Kust Metzner 1928.

Heath, T. L., *A History of Greek Mathematics*. Oxford: Oxford University Press, 1921; reprinted New York: Dover, 1981.

Heron of Alexandria, *The Pneumatics* (tr. Bennet Woodcroft). London: Taylor Walton and Maberly, 1851.

Hessen, Boris, 'The Social Origins of Newton's Principia', in Werskey (ed.) *Science at the Cross-Roads*. London: Frank Cass and Co., 1971.

Hilbert, David, *Mathematical Problems: Lecture Delivered before the International Congress of Mathematicians at Paris* (text can be found online at http://aleph0.clarku.edu/djoyce/hilbert/problems.html), 1900.

Hilbert, David and S. Cohn-Vossen. *Geometry and the Imagination*. New York: Chelsea, 1999.

Hodges, Andrew, *Alan Turing: The Enigma of Intelligence*. London: Unwin, 1985.

Hoe, Jock, ('John Hoe'), *Les systèmes d'équations polynômes dans le Siyuan yujian (1303)* (Mémoires de l'Institut des Hautes Études Chinoises, Collège de France, 6). Paris: L'Institut, 1977.

Hofmann, Joseph E., *Leibniz in Paris 1672–1676*. Cambridge: Cambridge University Press, 1974.

Hofstadter, Douglas R., *Gödel, Escher, Bach: An Eternal Golden Braid*. New York: Basic Books, 1979.

Hogben, Lancelot, *Mathematics for the Million: How to Master the Magic of Numbers*. London: George Allen and Unwin, 1936.

Høyrup, Jens, *In Measure, Number, and Weight: Studies in Mathematics and Culture*, Albany, NY: State University of New York Press, 1994.

Høyrup, Jens, 'Changing Trends in the Historiography of Mesopotamian Mathematics: An Insider's View'. *History of Science* 34 (1996), 1–32.

Husserl, Edmund, in Jacques Derrida (ed.) *Edmund Husserl's Origin of Geometry: An Introduction*. Lincoln and London: University of Nebraska Press, 1989.

Ibn Khaldūn, *The Muqaddimah* (tr. Franz Rosenthal), 3 vols. London: Routledge and Kegan Paul, 1958.

Iliffe, Robert, 'Is He Like Other Men?' in Gerald MacLean (ed.), *Culture and Society in the Stuart Restoration*, Cambridge: Cambridge University Press. 1995.

James I. M. (ed.), *History of Topology*. Amsterdam: Elsevier, 1999.

Jami, Catherine, *Les Méthodes rapides pour la trigonométrie et le rapport précis du cercle (1774). Tradition chinoise et apport occidental en mathématiques* (Mémoires de l'Institut des Hautes Études Chinoises, Collège de France, 32). Paris: L'Institut, 1990.

Joseph, George Gheverghese, *The Crest of the Peacock: Non-European roots of Mathematics*, Harmondsworth: Penguin, 1992.

Kanigel, Robert, *The Man who Knew Infinity: A Life of the Mathematical Genius Ramanujan*. London: Scribners, 1991.

Kant, Immanuel, *Critique of Pure Reason*. London: Everyman, 1993.

Katz, Victor J., *A History of Mathematics: an Introduction*, Reading MA: Addison-Wesley, 1998.

Kennedy, Edward S., *Studies in the Islamic Exact Sciences*. Beirut: American University of Beirut, 1983.

Kepler, Johannes, *Gesammelte Werke III (Astronomia Nova)*. Munich: C. H. Beck'sche Verlagsbuchhandlung, 1990.

Kepler, Johannes, *Gesammelte Werke IX* (Mathematical works). Munich: C. H. Beck'sche Verlagsbuchhandlung, 1999.

Khayyam, Omar ('Ūmar al-Khayyāmī), *The Algebra of Omar Khayyam* (trans. and ed. Daoud S. Kasir). New York: Teachers' College, Columbia University, 1931.
Klein, Jacob, *Greek Mathematical Thought and the Origin of Algebra* (tr. Eva Brann). Cambridge: M.I.T. Press, 1968.
Kline, Morris, *Mathematics: The Loss of Certainty*. New York: Oxford University Press, 1980.
Knorr, Wilbur, *The Origin of Euclid's Elements*. Dordrecht: Reidel, 1975.
Knorr, Wilbur, *The Ancient Tradition of Geometric Problems*. Boston, MA: Birkhäuser, 1986.
Knorr, Wilbur, *Textual Studies in Ancient and Medieval Geometry*. Boston: Birkhäuser, 1989.
Koestler, Arthur, *The Sleepwalkers*. Harmondsworth: Penguin, 1959.
Koyré, Alexandre, *Galilean Studies* (tr. John Mepham). Sussex: Harvester, 1978.
Kuhn, Thomas, *The Structure of scientific Revolutions*, Chicago: University of Chicago Press, 1970 ('1970a').
Kuhn, Thomas, 'Reflections on my Critics', in I. Lakatos and A. Musgrave ed., *Criticism and the Growth of Knowledge*, Cambridge: Cambridge University Press, 1970 ('1970b').
Lach, Donald F., *Asia in the Making of Europe*, several vols. Chicago, IL: University of Chicago Press, 1965–2000.
Lam Lay Yong, *A Critical Study of the Yang Hui Suan Fa*. Singapore: Singapore University Press, 1977.
Lam Lay Yong and Ang Tian Se, *Fleeting Footsteps. Tracing the Conception of Arithmetic and Algebra in Ancient China*, Singapore: World Scientific, 1992.
Libbrecht, Ulrich, *Chinese Mathematics in the Thirteenth Century: The Shu-shu chiu-chang of Ch'in Chiu-shao* (MIT East Asian Science Series, 1). Cambridge, MA: MIT Press, 1973.
Lloyd, G. E. R., *Magic, Reason and Experience*. Cambridge: Cambridge University Press, 1979.
Machamer, Peter (ed.), *The Cambridge Companion to Galileo*. Cambridge: Cambridge University Press, 1998.
Macrae, Norman, *John von Neumann: The Scientific Genius who Pioneered the Modern Computer, Game Theory, Nuclear Deterrence, and Much More*. New York: Pantheon Books, 1992.
Mancosu, Paolo, *From Brouwer to Hilbert: The Debate on the Foundations of Mathematics in the 1920s*. Oxford: Oxford University Press, 1998.
Mandelbrot, Benoit, 'Chaos, Bourbaki and Poincaré'. *Mathematical Intelligencer*, **11** (1989), 10–12.
Mao Zedong, *Where do Correct Ideas Come From?*, Beijing: Foreign Languages Press, 1963.
Martzloff, Jean-Claude, *Recherches sur l'Œuvre Mathématique de Mei Wending (1633–1721)* (Memoires de l'Institut des Hautes Études Chinoises, 16). Paris: L'Institut, 1981.
Martzloff, Jean-Claude, *A History of Chinese Mathematics* (tr. Stephen L. Wilson). Berlin: Springer-Verlag, 1995.
Marvell, Andrew, *The Complete Poems*. Harmondsworth: Penguin, 1972.
McAuliffe, Jane Dammen (ed.), *The Encyclopaedia of the Qur'an*. Leiden: Brill, 2001–.
Merton, Robert K., *Science, Technology and Society in Seventeenth-Century England*. New York, Evanston, and London: Harper, 1970.
Mashaal, Maurice, *Bourbaki-Une société secréte des mathématiciens*, Paris: Belin, 2002.
Monk, Ray, *Bertrand Russell: 1 (1872–1920), The Spirit of Solitude; 2 (1921–1970), The Ghost of Madness*. New York: Vintage, 1997, 2001.
Montucla, Étienne, *Histoire des mathématiques*, Paris, 1758.
Moritz, Robert E., *On Mathematics and Mathematicians*. New York: Dover, 1942.
Murdoch, John E., 'The Medieval Language of Proportions', in A. C. Crombie (ed.), *Scientific Change*. London: Heinemann, 1963, pp. 237–71.
Musil, Robert, *The Man Without Qualities* (trans. Eithne Wilkins and Ernst Kaiser). London: Secker and Warburg 1953.
Nasar, Sylvia, *A Beautiful Mind*. London: Faber and Faber, 1998.
Needham, Joseph, *Science and Civilisation in China, vol. 1. Introductory Orientations*. Cambridge: Cambridge University Press, 1954.
Needham, Joseph, *Mathematics and the sciences of the Heavens and the Earth*. Cambridge: Cambridge University Press, 1959.
Netz, Reviel, *The Shaping of Deduction in Greek Mathematics*. Cambridge: Cambridge University Press, 1999.
Neugebauer, Otto, *The Exact Sciences in Antiquity*. Princeton, NJ: Princeton University Press, 1952.
Neugebauer, Otto and Abraham J. Sachs, *Mathematical Cuneiform Texts*. American Oriental Society, American Schools of Oriental Research, New Haven CT, 1946.
Newton, Isaac, in H. W. Turnbull, J. F. Scott, and A. R. Hall (eds.), *The Correspondence of Isaac Newton*, 7 vols, Cambridge: Cambridge University Press, 1959–77.

Newton, Isaac, in D. T. Whiteside et al.(ed.), *The Mathematical Papers of Isaac Newton*, 8 vols, Cambridge: Cambridge University Press, 1967–81.
Nietzsche, Friedrich, *Untimely Meditations*, Cambridge: Cambridge University Press, 1983.
Nissen, Hans J., Peter Damerow, and Robert K. Englund, *Archaic Bookkeeping: Early Writing and Techniques of Economic Administration in the Ancient Near East*. Chicago, IL and London: University of Chicago Press, 1993.
Pascal, Blaise, *Pensées* (tr. A. J. Krailsheimer). Harmondsworth: Penguin, 1966.
Piaget, Jean, *Structuralism* (tr. C. Maschler). New York: Basic Books, 1970.
Popper, Karl, 'Replies to my Critics', in P. A. Schilpp, ed. *The Philosophy of Karl Popper*, vol. XIV of The Library of Living Philosophers, Illinois: Open Court, 1974, pp. 961–1197.
Proclus, *A Commentary on the First Book of Euclid's Elements* (trans. with introduction and notes, by Glen R. Morrow). Princeton, NJ: Princeton University Press, 1970.
Ptolemy (Claudius Ptolemaeus), *The Almagest* (trans. and annotated by G. J. Toomer). London: Duckworth, 1984.
Rashed, Roshdi, *The Development of Arabic Mathematics: Between Arithmetic and Algebra*. Dordrecht: Kluwer, 1994.
Rashed, Roshdi, interview 'The End Matters' (at http://cc.usu.edu/bekir/rashed/interview Islam Science.htm) 2003.
Reid, Constance, *Hilbert*. New York: Springer, 1970.
Richards, Joan, *Mathematical Visions: The Pursuit of Geometry in Victorian England*. New York: Academic Press, 1988.
Riemann, Bernhard, 'On the Hypotheses which lie at the Bases of Geometry', (trans. William Kingdon Clifford) *Nature*, vol. VIII, pp.14–17, 36, 37 (1873).
Ritter, James, 'Babylon 1800', in Serres (ed.), *A History of Scientific Thought*. Oxford: Blackwell, 1995.
Robson, Eleanor, *Mesopotamian Mathematics 2100–1600 BC: Technical Constants in Bureaucracy and Education* (Oxford Editions of Cuneiform Texts 14). Oxford: Clarendon Press, 1999.
Robson, Eleanor, 'Neither Sherlock Holmes nor Babylon: A Reassessment of Plimpton 322'. *Historia Mathematica* **28** (2001), 167–206.
Rodinson, Maxime, *Islam and Capitalism* (tr. Brian Pearce). London: Allen Lane, 1974.
Roero, Clara Silvia, 'The Passage from Descartes' Algebraic Geometry to Leibniz's Infinitesimal Calculus in the Writings of Jacob Bernoulli'. *Studia Leibnitiana*, Sonderheft 17, Wiesbaden: Steiner, 1989, pp. 140–50.
Rose, P. L., *The Italian Renaissance of Mathematics: Studies on Humanists and Mathematicians from Petrach to Galileo*. Geneva: Librarie Droz, 1975.
Rotman, Brian, *Signifying Nothing: The Semiotics of Zero*. London: Macmillan, 1987.
Russell, Bertrand, *The Principles of Mathematics*. Cambridge: Cambridge University Press, 1903.
Russell, Bertrand, *The Autobiography of Bertrand Russell*, vol. 1, 1872–1914. London: Allen and Unwin, 1967.
Said, Edward, *Orientalism*. New York and London: Vintage, 1978.
Schleiermacher, Friedrich D. E., *Hermeneutics: The Handwritten Manuscripts* (ed. Heinz Kimmerle and trans. James Duke and Jack Forstman). Oxford: Oxford University Press, 1978.
Scholes, Myron, 'Information Technology' (online at bizpark.asaban.com/psa/pdf/MyronScholes.pdf), 2002.
Seaford, Richard, *Money and the Early Greek Mind: Homer, Philosophy, Tragedy*. Cambridge: Cambridge University Press, 2004.
Segal, Sanford L., *Mathematicians under the Nazis*. Princeton, NJ: Princeton University Press, 2003.
Seifert, H. and Threlfall, W., *Textbook of Topology* (tr. W. Heil). New York: Academic Press, 1980.
Senechal, Marjorie, 'The Continuing Silence of Bourbaki—An Interview with Pierre Cartier, June 18, 1997'. *The Mathematical Intelligencer* **1** (1998), 22–8.
Serres, Michel, 'Gnomon', in Serres (ed.), *A History of Scientific Thought*. Oxford: Blackwell, 1995.
Sezgin, Fuat (ed.), *Islamic Mathematics and Astronomy*, vol. 22. Frankfurt: Institut für Geschichte der arabisch-islamischen Wissenschaften, 1999.
Sharaf al-Dīn al-Ṭūsī, (trans. and ed. R. Rashed) *Œuvres mathématiques. Algèbre et Géométrie au XIIe siècle*. Paris : Les Belles Lettres, 1986.
Shen Kangshen, John N. Crossley and Anthony W. -C. Lun, *The Nine Chapters on the Mathematical Art: Companion and Commentary*, Oxford and Beijing: OUP and Science Press, 1999.
Singh, Simon, *Fermat's Last Theorem*. London: Fourth Estate, 1997.
Sivin, Nathan, 2004 (at http://ccat.sas.upenn.edu/ nsivin/nakbib.html).
Smith, D. E., *A Source Book in Mathematics 1200–1800*. New York: Dover, 1959.
Sohn-Rethel, Alfred, *Intellectual and Manual Labour: Critique of Epistemology*. London: Macmillan, 1978.
Sokal, Alan and Bricmont, Jean, *Fashionable nonsense: postmodern intellectuals' abuse of science*, New York: Picador, 1998.

Spengler, Oswald, *Decline of the West*, London: George Allen and Unwin, 1934.
Steele and Imhausen (eds.), *Under One Sky: Astronomy and Mathematics in the Ancient Near East*. Münster: Ugarit-Verlag, 2002.
Stephenson, Bruce, *Kepler's Physical Astronomy*. New York and London: Springer, 1987.
Stevin, Simon, *The Principal Works of Simon Stevin. Vol. II: Mathematical works*, ed. D. J. Struik, tr. C. Dijkshoorn, Amsterdam: Swets and Zeitlinger, 1958.
Struik, Dirk, *A Concise History of Mathematics*, New York: Dover, 1986.
Struik, Dirk, ed. *A sourcebook in mathematics 1200–1800*, Cambridge MA: Harvard University Press, 1969.
Struik, Dirk, 'Concerning mathematics', *Science and Society* **1** (1936), 81–101.
Swetz, Frank J., *The Sea Island Mathematical Manual; Surveying and Mathematics in Ancient China*. Philadelphia, PA: Pennsylvania State University Press, 1992.
Tacchi Venturi, Pietro, *Opere storiche di Matteo Ricci*, 2 vols. Macerata, 1911–13.
Tannéry, Paul, *La geometrie grecque, comment son histoire nous est parvenue et ce que nous en savons*. Paris: Gauthier-Villars, 1887.
Tartaglia, Niccolo, *Quesiti et inventioni diverse* (facsimile reproduction) (ed. and intro. A. Masotti). Brescia, Ateneo di Brescia, 1959.
Taylor, Charles, review, film 'A Beautiful Mind' (at http://dir.salon.com/ent/movies/review/2001/12/21/beautiful mind/index.html), 2001.
Thomas, I., *Selections Illustrating the History of Greek Mathematics*, vol. II. London: Heinemann, 1939.
Thurston, William, in Silvio Levy, (ed.), *Three-Dimensional Geometry and Topology*, vol. 1. Princeton, NJ: Princeton University Press, 1997.
Torrelli, Roberto, *Philosophy of Geometry from Riemann to Poincaré*. Dordrecht: Reidel, 1978.
Trifkovic, Serge, 'The Golden Age of Islam is a Myth', extract from the book *The Sword of the Prophet: A Politically-Incorrect Guide to Islam*; reproduced in www.frontpagemag.com, November 15, 2002.
Vaiman, A. A., *Über die sumerisch-babylonische angewandte Mathematik*, (XXV Internationaler Orientalisten Kongress, Vorträge der Delegation des UdSSR). Moscow: Verlag für Orientalische Literatur, 1960.
Valéry, Paul, 'Le Cimetière Marin' (trans. C. Day Lewis) 'The Graveyard by the Sea' (online at http://homepages.wmich.edu/cooneys/poems/fr/valery.cimetiere.html), 1920.
van Dalen, Dirk, *Mystic, Geometer, and Intuitionist: The Life of L. E. J. Brouwer, vol. 1: The Dawning Revolution*. Oxford: Oxford University Press, 1999.
van der Poorten, Alf, *Notes on Fermat's Last Theorem*. New York: Wiley, 1996.
van der Waerden, B. L., *Science Awakening*. New York: Oxford University Press, 1961.
van der Waerden, B. L., *Geometry and Algebra in Ancient Civilizations*. Springer-Verlag, New York 1983.
van Egmond, Warren, *Practical Mathematics in the Italian Renaissance: A Catalog of Italian Abbacus Manuscripts and Printed Books to 1600*. Florence: Instituto e Museo di Storia delle Scienze, 1980.
van Heijenoort, Jean, *From Frege to Gödel*. Cambridge, MA: Harvard University Press, 1967.
Viète, François, *The Analytic Art: Nine Studies in Algebra, Geometry and Trigonometry from the Opus restitutae mathematicae analyseos, seu, algebra nova*, tr. T. Richard Witmer. Kent, Ohio: Kent State University Press, 1983.
von Helmholtz, Hermann, 'The Origins and Meaning of Geometric Axioms', in Russell Kahl (ed.), *Selected Writings of Hermann von Helmholtz*. Middletown: Wesleyan University Press, 1979, pp. 246–65.
von Neumann, John, *Mathematical Foundations of Quantum Mechanics*. Princeton, NJ: Princeton University Press, 1996.
von Neumann, John, *First Draft of a Report on the EDVAC* (1945 report, published online at a number of sites, e.g. www.virtualtravelog.net/entries/2003-08-TheFirstDraft.pdf).
Weil, André, *The Apprenticeship of a Mathematician*. Basel, NY, Boston, MA, Berlin: Birkhäuser, 1992.
Weil, Simone, *Simone Weil: an introduction*, ed. and intro. Siân Miles, London: Virago, 1986.
Wells, R. O., Jr., *The Mathematical Heritage of Hermann Weyl*. Providence, RI: American Mathematical Society, 1988.
Westfall, Richard S., *Never at Rest: A Biography of Isaac Newton*. Cambridge: Cambridge University Press, 1980.
Weyl, Hermann, 'Emmy Noether'. *Scripta Mathematica*, **3** (1935), 201–20.
Weyl, Hermann, *Philosophy of Mathematics and Natural Science*. Princeton, NJ: Princeton University Press, 1949.
Wigner, Eugene, 'The Unreasonable Effectiveness of Mathematics in the Natural Sciences', *Communications in Pure and Applied Mathematics*, **13** (1960), 1–14.
Youschkevitch, A. P., *Les Mathématiques arabes (VIIIe-XVe siècles)*. Paris: Vrin, 1976.

Index

Note: Page numbers in *italics* refer to figures.

π
  Archimedes' approximation 61–2, 157
  estimation in Keralan mathematics 167, 168
  Landau's definition 238
  transcendence of 221

abacus schools 141–3
Abbasid dynasty 109
Abū Kāmil 115–16
absolute measurement 200–1
abstraction of thought 43
  in Babylonian mathematics 18–19, 20–1, 24–6
  in Chinese mathematics 88
  twentieth century 240
acceleration, uniform 154
accounting methods 142
acute angle hypothesis (HAA) 190–1
Akkadians 15
al-Bāhir fi-l jabr (al-Samaw'al) 104, 117–20
Albert of Saxony 136
  discussion of squaring the circle 137–8
al-Bīrūnī 95, 124–5, 195
  ideas on velocity 152
Alexandrov, P. 224, 250
Alexander, J. W. 226, 227
al-Fārābī 113
algebra
  Analytic Art, The (Viète) 146, 147–8
  Descartes' notation 149
  Diophantus 66
  Islamic 103–4, 108
    abū Kāmil's work 115–16
    al-Khwārizmī's work 110–11, 125–6
    al-Samaw'al's work 117–20
    Omar Khayyam's work 116–17
  origins 110–14
  texts of 16th, 17th centuries 147
Algebra ('Bourbaki') 256
algebraic geometry, rings and ideals 230
algorithms
  Leibniz's use 174, 186
  military uses 260–1
al-Kāshī 104, *105*, 117, 120–3, 128–9
  Miftāh al-hisāb 104
  regular solids *128*
al-Khwārizmī (Muhammed ibn Mūsa) 125–6
  Hisab al-jabr wa al-muqābala 103, 110
Almagest (Ptolemy) 66–9, 75–6
al-Ma'mun 109

Al-Mas'ūdī 95
'alogoi' ratios 35
alphabetic writing, invention of 43
  al-Bāhir fi-l jabr 104
al-Uqlīdisī 104, 114
  arithmetic (*Kitāb al Fusūl fī al-Hisāb al-Hindī*) 106–7
  use of decimals 121
Ambassadors, The (Holbein) *142*
anagrams, use in scientific communications 165
Analyse des infiniment petits, pour l'intelligence des lignes courbes (de l'Hôpital) 179
analysis *see* calculus
Analyst, The (Berkeley, George) 179–80
Analytical Engine (Babbage) 243
Analytic Art, The (Viète) 146, 147–8
angle of parallelism (Π(p)) 202–3
angles of a triangle theorem, proof *210*–11
angle-sum, spherical triangles 200
anonymous publication 165
anxiety and modernism 218
Apollonius 40
  Conics 58
  planetary motion 67
apotomes 53
applications of mathematics 261
approximation procedures, in Chinese mathematics 96–7
Arabic mathematics *see* Islamic mathematics
arc, Keralan calculation *167*–8
Archimedes 40, 41, 60–2, 64–5
  approximation of π 157
  area of circle 157–8
  finite universe 194
  infinities 151, 153
  Measurement of the Circle 137, 138
  and Newton's *Principia* 177
  volume of cone 48–9
area
  Archimedes' method of measurement 61
  of cardioid *182–3*
  of circle *157*–8
    in Euclid's *Elements* 38
  of triangle, Heron's formula 64–5
area law, Kepler, Newton's version 177–8, 186
argument
  role in history 3
  scholastic methods 136–8
Aristotle, method of argument 137

arithmetic
  consistency of axioms 217
  Kitāb al Fusūl fī al-Hisāb al-Hindī
    (al-Uqlīdisī) 106–7
Arithmetic (Diophantus) 111
Arithmétique (Stevin) 149
Array (Fangcheng) Rule 89
'asamm' numbers 114
astrology 66
Astronomia Nova (Kepler) 151, 152–3
astronomy
  in Greek mathematics 66–9
  planetary motion 152
    Newton's ideas 176
  and ratios 50
  role of Islam 124
Atiyah, Michael 250
atom bomb project 239
Auden, W. H. 260
Axiom of Choice, set theory 218–19
Axiom of Comprehension 218, 231–2
axiomatization of geometry 206–7
axioms 222
  for arithmetic, consistency 217–18
axiom systems, Bourbakists ideas 241

Babbage, Charles 244
  Analytical Engine 243
Babylonian mathematics
  abstraction 18–19, 20–1, 24–6
  Fara period 27–8
  interpretation 7
  number system 22–4
  sources 17–20, 21–2
  units of measurement 20, 29
  Ur III period 28–30
  'uselessness' 26–7
Barrow, Isaac 49, 170
beginning of mathematics 14
Bieberbach 238
Beltrami 192
Berggren, J. L. 3, 103–4
  abū-l-Wafā 107
  al-Kāshī 122
  Greek historiography 35
Berkeley, George 162, 163
  The Analyst 179–80
Bernal, Martin 43
Bernoulli, Jakob and Johann 166, 178–9
  literature 163
  representation of curves 181
biography, St Andrews archive 4
Bishop Berkeley *see* Berkeley, George
Black Athena (Bernal, Martin) 43
Bolyai, Janos 189, 191, 192, 193
  construction of geometry 202
  isolation 203
Bolzano 218
  on application of geometry 199
Bombelli 146
Bonola, Roberto 193

Book of Changes (Yijing, I Ching,) 78
Bos, Henk 3, 163–4
  construction of curves 180–1
  independent variable 176
'Bourbaki' 240–3
  Algebra 256
Bradwardine, Thomas 135, 136
  ideas on infinity 197
Brouwer, L. E. J. 219–20
  intuitionism 231–2
Bruno, Giordano 197
bureaucracy as trigger for mathematics 16
'Burning of the Books', China 81
'butterfly effect' 246, *247*

calculating tradition, role in scientific revolution 141–3
Calculator's Key, (al-Kāshī) 104, 117, 120–3, 128–9
calculus 161–3
  Archimedes, possible use of 61
  Berkeley, George, *The Analyst* 179–80
  Bernoulli brothers' adaptation 178–9
  de l'Hôpital's contribution 179
  Keralan mathematics 167–9
  Leibniz, 1684 paper *185–6, 172–6*
  limits 215
  practical use 180–2
  Principia (Newton) 176–8
  priority dispute 165–6
  sources 163–4
  tangents, Newton's method 169–72, 183–5
  use of infinitesimals 182–3
calendar construction 50
  cooperation between Chinese and Near East 95
  Matteo Ricci, China 98
Cantor, Georg 215
  continuum hypothesis 217
Cantor, Moritz 1
Capra, Fritjof 252, 254
Cardano, Hieronimo 144, 145
cardioid, area of *182–3*
Carr, E. H. 3
Cartan, Henri 240
Cartier, Pierre 241
category theory 249–51
catenary 180–1
cell decomposition, topology 223–4
Ceyuan Haijing (Li Zhi) 90, *91*, 92
Ch'in Chiu-shao *see* Qin Jiushao
chaos theory 246–9
Chasles, Michel, descriptive geometry 198
Chemla, Karine, on Liu Hui 84
  Ruffini-Horner Procedure 97
Chevalley, Claude 240
China, early history 80–2
Chinese mathematics 78–80
  counting rods 85–8
  matrices 88–90
  Ming dynasty 98
  Nine Chapters on the Mathematical Art 82–4
  Qin Jiushao 90, 91
  Song dynasty 90–3

Index

sources 80
transfers of knowledge 95–8
Chinese Remainder Theorem 78, 91
chord of an angle (Crd θ) 68
Circle Limit III, (Escher) 192
circles
    Archimedes' work 61–2
    area of 157–8
        in Euclid's *Elements* 38
circular motion, heavenly bodies 66, 67
cissoid 184
city-states, Greek 43
coined money, introduction of 43
Commercium epistolicum 166
common measure (greatest common divisor) 54
common notions, Euclid 37, 38
completeness of a system 222
componendo rule 74
computable numbers (Turing) 220, 256–7
computers, invention of 243–6
computer science 236
conchoid, tangent to 184
cone, volume of, Archimedes 48–9
Confucianism 81
Conics (Apollonius) 58
conic sections
    Descartes 150
    in Greek mathematics 58
consistency of a system 222
constant of curvature (K) 203
construction of geometry 201–3
continuum, doubts 215–16, 252
continuum hypothesis, Cantor 217
conversion factors Babylonian 30
coordinate geometry, Descartes 149–51
Coordinates of Cities (al-Bīrūnī) 124–5
Copernicus
    influences 147
    theory of planetary motion 152
copying of manuscripts, editing problems 41
cosine *see* trigonometric ratios
cosine formula, in spherical and hyperbolic geometry 209
counting rod numbers 86
counting rods, Chinese 85–8
    use, description in *Nine Chapters* 83
counting symbols, invention of 16
Coxeter, H., S., M. 192
'crisis of foundations' 215
cube
    doubling of 6, 58–9
    multiplication of 60
    as Platonic solid 46
cubes, difference between (Viète) 148
cubic curve, graph 150, *151, 256*
cubic equations
    Omar Khayyam's work 116–17
    Tartaglia, Niccolò 144–5
cuneiform numbers 23
cuneiform script 15
Cuomo, Serafina, on Roman mathematics 69–70

curvature of space, constancy 204–5
curve-drawing machine, Descartes 150, *156*
curves
    description of 180
    generation by motion 169–70
cuts *215–16*
    definition 231
cybernetics 239
cycloid 170

D'Alembert 180
dal Ferro, Scipione 144
Dao De Jing (Lao-Zi) 81
Daoism (Taoism) 81
day length 50
de l'Hôpital, *Analyse des infiniment petits, pour l'intelligence des lignes courbes* 179
de Montmort, Pierre 1, 166
decimal fractions
    in Islamic mathematics 120–1
    Stevin's work 149
decimal place-value numbers, and Chinese counting rods 87–8
Dedekind, Richard 214–16, 230
Dedekind cut *215–16*, 231
deductive structure, Greeks 38, 44
Dehn & Wirtinger 226
Delsarte, Jean 240
democratization of mathematics 235, 237
Democritus 40, 48–9
Descartes 112, 133, 149–51, 162
    curve-drawing machine *156*
    finding of tangent to a curve 175
    on Greek mathematics 39
    ideas on infinity 197
    Newton's opinions 177
descriptive geometry *198*, 236
Devaney, Robert 248
diagrams, use in Greek mathematics 37, 38, 44
Dialogue on the Two Major World-Systems (Galileo) 153–4
Dialogues (Plato) 33
Dieudonné, Jean 10–11, 223, 240
differential geometry 183
differentiation
    relationship to integration 171–2
    see also calculus
dimensional renormalization 253
Diophantus
    algebraic notation 66
    Arithmetic 111
Dirac's functions 253
Dirichlet 229
Discourses on Two New Sciences (Galileo) 153–4
distributive law 5, 48
divergent series, Ramanujan's work 229
division of polynomials, al-Samaw'al *119*–20
division problems, Babylonian Fara period 27
documentation
    Babylonian mathematics 18–20, 21–2
    Chinese mathematics 79
dodecahedral space 225

dodecahedron 46
Douady's rabbit 248–9
double entry bookkeeping, invention 142
double false position method 83–4
doubling of cube 6, 58–9
doubling of square, Plato's *Meno* 34–5, 50, 51–2
Duhem, Pierre 135
dynasties, Chinese 80–2
Dzielska, Maria, on Hypatia 71–3

e, transcendence of 221
eccentric model, sun's movements 69, 75–6
ecliptic 67
Edinburgh school 11
Egypt, historical background 16–17
Egyptian mathematics 42
   solution of linear equations 21
Eilenberg 250–1
Einstein, Albert
   General Theory of Relativity 204
   move to Princeton 239
   Special Theory of Relativity 207
Eisenhower, Dwight D. 260
electrodynamics, quantum 253
elementary equivalence of knots 226–7
Elements (Euclid) 36–9
   comparison with *Nine Chapters on the Mathematical Art* 83
   proportion theory 47–8
Éléments de mathématiques, ('Bourbaki') 241–2
elliptic curves 255, 256
encryption 236
epicycle model, sun's (or planet's) movements 69
epistemological break 42
equal parallelograms, Euclid 37–8
equal ratios, Euclid 49
equant 152
equation of time 69
equations
   in Babylonian mathematics 18–19, 20–1, 25
   from Qin Jiushao's work 94
Eratosthenes 195
   doubling of cube 58–9
Escher, Moritz, *Circle Limit III* 192
ethnomathematics 14
Euclid 40, 45
   China, introduction of methods to 98
   'common measure' (greatest common divisor), method for finding 54
   Elements 4, 5, 36–9, 48
      comparison with *Nine Chapters on the Mathematical Art* 83
      Islamic interpretations 113, 114, 115, 127
      parallel postulate 194–6
         attempts at proof 190–1, 196
      proportion 139
      proposition I.16 207–8
      theory of ratios 3, 35, 45, 48, 49
      use of proof by contradiction 219
Euclidean geometry, continued validity 9–10

Eudoxus of Cnidus 3, 40, 48, 49
   theory of proportions 47
Euler 223
Euler's constant 229
Eupalinus, tunnel of 70
Eurocentrism 12–13, 17
   attitudes to Islamic mathematics 102
example, explanation by 111
excess and deficit rule 83–4
existence proof 137
exponential curve, tangent to (Leibniz) 175–6
external viewpoint 10–12
extreme and mean ratio 53
   see also golden ratio

'false position' solution of linear equations 21, 143
Fangcheng (Array) Rule 89
Fara period 15, 27–8
Fārs 108
Fauvel, John 2
Feigenbaum quadratic map 248
Fermat's Last Theorem 220, 254–5
   proof 235–6
first principles 36
Fixed Point Theorem, Brouwer 219–20
fluents 171, 172
fluxions 170, 171
formalists 222
Forman, Paul 220–1
Foucault, Michel 43
Fowler, David 3, 35–6
   on Eudoxus 48
   his reconstructions 49
   Meno 33
   on Hasse–Scholz thesis 47
fractions
   Archimedes, use of 62
   in Babylonian mathematics 23–4
free fall, Galileo's work 154
Frege, Gottlob 216
Frey Gerhard 255
functors 250–1
fundamental group, Poincaré 226

Galileo 133–4, 152, 153–4
   infinities 158–9
   influences 147
   motion of projectiles 150
Gauss 165, 191, 254
   caution 203
General Theory of Relativity (Einstein) 204
geocentric model of heavens 67
geodesic 209
geometric constructions, abū-l-Wafā al Buzjānī 107
geometric language, use in Greek mathematics 5, 45, 46–7
geometric proof, quadratic equations 112
geometric solutions, Plato's *Meno* 34–5
Géométrie (Descartes) 149–51
Géométrie Imaginaire (Lobachevsky) 204

geometry
  concept of infinity 194, 197
  construction of 201–3
  descriptive *198*, 236
  development of axiom systems 206–7
  projective *199*, 204
  status of 10, 189
  see also non-Euclidean geometry
Germain, Sophie 254
German mathematics, Second World
    War 238–9
Girard, Albert, angle-sum of spherical
    triangles 200
Gleick, James 246–7
globalization of technology 261
Gödel, Kurt 222
  move to Princeton 239
Gödel's Theorem 235
golden age
  Islamic mathematics 108–10
  Newton's belief in 177
  Song dynasty 90–3
golden ratio (section) 49, 53
Goldstein, Catherine 254
Göttingen mathematics 221–3
Gp category 250
gradients 169
gravitation, Newton's ideas 176–7
Gray, Jeremy 2, 193
greatest common divisor 54
Great Wall of China 82
Greek mathematics 40
  Archimedes 60–2
  astronomy 66–9
  Heron (Hero) 63–6
  Hypatia 71–3
  irrational numbers 46–7
  lack of 'hard' facts 3
  literature 35–6
  Newton's use of 176–7
  Plato 33–5
  proof by contradiction 219
  Ptolemy 66–9
  ratios 49–51
  second revolution 44–5
  sources, problems with 39–42
  spherical geometry 200
  theory & practice, interaction 57–60
Greek miracle (revolution), origin 42–4
  dating of 45
Gregory's series 167
group, fundamental (Poincaré) 226

HAA *see* hypothesis of the right
    (acute, obtuse) angle
'half-line angle' *242*
Halley, Edmond 176
Han dynasty 81–2
Hardy, G. H. 228, 229, 235
Hardy–Ramanujan asymptotic formula 229
harmonic series, Oresme, Nicholas 140

harvest yield record, Ur III period 28–9
Hasse-Scholz thesis 46–7
heavenly bodies, circular motion
    restriction 66, 67
helium atom 252–3
Helmholtz 191
  extract from 1876 paper 209–10
  publicization of hyperbolic geometry 205, 207
hermeneutics 7
Herodotus 42
Heron (Hero) 40, 63–6
  Metrics 73–4
  slot machine *64*
Heron's theorem 64–5, 73–4
hexagon, perimeter of 63
hexahedron *see* cube
Hilbert, David 214, 216, 217, 221–3, 232, 240
  axiom systems 206–7
  on Fermat's Last Theorem 254
Hipparchus 40
  planetary motion 67
Hippasos of Metapontum 46
Hippocrates of Chios 40, 60
  quadrature of lunes 56
Hisab al-jabr wa al-muqābala
    (al-Khwārizmī) 103, 110
historicism 6–7
historiography 4
Hogben, Lancelot, *Mathematics for the Million* 235
Holbein, *The Ambassadors* *142*
homomorphism 223, *224*, 250
horseshoe map (Smale) 249–*50*
Høyrup, Jens 3, 7
  Babylonians 19
  Islamic miracle 110, 124
  'subscientific' mathematics 65
Huygens 172
hydraulic project thesis, Egypt and Iraq 16
Hypatia 71–3
hyperbolic geometry 205
hyperbolic trigonometry 209
hypothesis of the right (acute, obtuse) angle (HRA, HAA,
    HOA) 190–*1*
  construction of a geometry 202–3
  Lambert's work 200–1

Iamblichus 46, 47
ibn al-Haytham 104
  parallel postulate 190, 196
I Ching (Yijing, Book of Changes) 78
icosahedron 46
idealism in wartime 240
ideals, Noether's work 230–1
incommensurability 8, 9
  impact on Greek mathematics 45–7
Indian numbers, possible derivation from
    counting rods 87–8
infinite series
  in Keralan mathematics 167–9
  Oresme, Nicholas 139–40, 155–6
infinite sets 219

infinitesimals 162
  Leibniz's use 173, 174
  Newton's use 170
  present day use 182–3
infinities 151–3
  use by Galileo 154, 158–9
infinity, as concept in geometry 194, 197
instantaneous velocity 152, 154, 170
integration
  relationship to differentiation 171–2
  see also calculus
internal viewpoint 10–12
Internet 4
  sources on OB mathematics 19
intrinsic equation 187–8
intuitionism 220, 221, 231–2, 238
  see also Brouwer, L. E. J.
inverse-square law 176
Iraq, historical background 14–16
  see also Babylonian mathematics
irrational numbers
  Dedekind cut 215–16, 231
  Greek knowledge of 35
  proof of 54
irrational ratios 46
Islamic mathematics 101–3
  algebra, origins 110–14
  al-Samaw'al 117–20
  golden age 108–10
  proportion 139
  role of religion 123–5
  'second generation' algebra 115–17
  sources of information 103–5
  spherical geometry 200
  translations 136
Islamic work on parallels 196
Islamic world
  general history 109
  transfers of knowledge with Chinese 95–8
isosceles triangles, Thales' statement 44

Jacquard loom 244, 246
James, I. M. 223
Japanese counting board 86
Jesuits, arrival in China 98
Jewish mathematicians, expulsion from Nazi Germany 238
Jia Xian 92
Jiuzhang suanshu see Nine Chapters on the Mathematical Art
job market 261–2
Jones, Vaughan 249
Joseph, George Gheverghese 12, 101, 167
  Eurocentrism 12–13
  Islam 101–2
  Keralan School 167
Julia set 248
Jyesthadeva 167, 168

Kant 193
Kepler
  area law, Newton's version 177–8, 186–7
  Astronomia Nova 151, 152–3
  influences 147
  Nova stereometria doliorum 157–8
Keralan mathematics 167–9
Khayyam, Omar 103, 116–17, 144
  algebra 103–4
  parallel postulate 190, 196
Kitāb al Fusūl fī al-Hisāb al-Hindī (al-Uqlīdisī) 106–7
Kitāb fī mā yahtāju ilayhī al-sani'min a'māl al-handasah (abū-l-Wafā al Buzjānī) 107
Klein, 192
Kline, Morris 217
knots 225–6
Knorr, Wilbur 3, 36
  circle squaring 137
  cube duplication 58–9
knotted torus 224
Koran (Qur'an) 124
Kovalevskaya, Sofia 231, 252
Kuhn, Thomas 8–9
Kummer 254

La Disme, Stevin 146, 149
Lambert, Johann Heinrich 200–1
Lambert's quadrilateral 201
Landau
  definition of π 238
  expulsion from Nazi Germany 238
Lao-Zi 81
Law of Contradiction 162
Law of the Excluded Middle, Brouwer's attack 219, 220
leap years 50
Legendre 165
Lehrbuch der Topologie (Seifert & Threlfall) 224, 225
Leibniz 161–3
  1684 publication 172–6, 185–6
  criticisms of work 179–80
  infinite series 167, 168
  priority dispute 165–6
Leibniz rule, first publication 173, 179
Leonardo of Pisa, *Liber abbaci* 141
L'Hôpital 163–4
Li Zhi 90
  'round town' 91, 92
Liber abbaci (Leonardo of Pisa) 141
limits 215
linear equations, solution by Egyptians 21
Listing J. B. 223
Liu Hui 82, 83
  commentaries 83–4
  use of negative numbers 88
Lobachevsky, Nikolai, I. 189, 191, 192, 193
  construction of geometry 202
  Géométrie Imaginaire 204
  isolation 203
Lobachevsky–Bolyai geometry 205
logarithmic curve, finding of tangent to (Leibniz) 175–6
Lorenz, Edward 246, 247
Lovelace, Ada 231, 244

Index

Ma Yize 95
machine construction, Heron 63–4, 65
machines, curve-drawing 150, *156*
MacLane 250–1
Madhava 168
Mandelbrot (complex) quadratic maps 248–9
manifolds 223
Mao Zedong 11
Martzloff, Jean-Claude 7, 80
   Chinese and Western equations 95
   Mei Wending 98
   'Pascal's triangle' 97
   rod numbers 85–6
Marxism 10–11, 125
Mathematical Cuneiform Texts (Neugebauer & Sachs, 1946) 19
Mathematics for the Million (Hogben, Lancelot) 235
mathematics, Hilbert's definition 222
   Russell and Weyl's definitions 214
matrices, in Chinese mathematics 88–90
mean-taker (mesolabe) 59
Measurement of a Circle (Archimedes) 61–2, 137, 138
Menaechmus, doubling of cube 58
Meno (Plato) 33–5, 36, 50, 51–2
mesolabe 59
metamathematics 222
Method, The (Archimedes) 61
method of double false position 83–4
Method of Fluxions and Infinite Series (Newton) 171–2, 183–5
Metrics (Heron) 64, 73–4
Miftāh al-hisāb (al-Kāshī) 104
'Millennium Problems' 261
military applications of mathematics 59, 60, 260, 262
Ming dynasty 98
Möbius 223
modernism and anxiety 218
Moerbeke, mathematical translations 136
Monge, Gaspard 236
   descriptive geometry *198*
months, length of 50
Montucla, Jean Étienne 1, 7, 254
mosques, alignment of 124–5, *131*, 195
motion, relationship to curves 169–70
Muhammad ibn Mūsā *see* al-Khwārizmī
multiplication, relationship to ratios 50
multiplication tables, Babylonian 23

Nash, John 237
Nasir al-Dīn al-Tūsī 190
natural numbers, definition 216
Nazi Germany 238–9
negative numbers
   as roots of quadratics 111
   use in Chinese mathematics 88
neoplatonism 72
Netz, Reviel 7, 36
   Greek mathematics, community 41
   origins 45
Newton, Isaac 161–2
   ideas on infinity 197

   literature 163, 164
   method for finding tangents 169–72, 183–5
   nature of space 207
   Principia 176–8, 186
   priority dispute 165–6
Nicholas of Cusa 153, 174
Nicomachus of Gerasa 57
Nine Chapters on the Mathematical Art 78–9, 80, 82–4
   matrices 88–90
Noether, Emmy 230–1
   expulsion from Nazi Germany 238
non-Euclidean geometries 189
   consequences 206–7
   construction of 201–3
   delays in development 203–5
   Escher, Moritz, *Circle Limit III* 192
   failure of Euclid's proposition I.16 207–8
   Helmholtz's 1876 paper 209–10
   Lobachevsky and Bolyai 191, 192, 193
   proof of consistency 191–2
   source material 193
   spherical geometry 199–201
   spherical and hyperbolic trigonometry 208–9
normal science 8, 9
notation, Leibniz's 173
Nova stereometria doliorum (Kepler) 157–8
number rings 230
numbers
   Babylonian 22–4
   Chinese 85
   Islamic 116, 120–1
   representation by counting rods 85, *86*
   Stevin's views 149
number theory, Ramanujan's work 228–9

objects in mathematics 216
octahedron 46
Old Babylonian (OB) period 15, 17–19
On computable Numbers (Turing) 244
On the Hypotheses which lie at the basis of Geometry (Riemann, Bernhard) 204
one, status as number 149
operational research 239
Oresme, Nicholas 136, 139–40
   influence on Descartes 150
   Quaestiones super Euclidem 155–6

Pappus 40
   on Heron 63–4
papyri, mathematical 17
paradigm 8
paradoxes of Zeno 139, 140
parallax of stars *205*
parallel angle ($\Pi(p)$) 202–3
parallel lines, Lobachevsky's definition *202*
parallelograms, equal (Euclid) 37–8
parallel postulate 194–6
   attempts at proof 190–1, *196*
partition number, Ramanujan's formula 228–9

Printed and bound by CPI Group (UK) Ltd, Croydon, CR0 4YY